HOW MAOISM WAS MADE

For over 100 years the *Proceedings of the British Academy* series has provided a unique record of British scholarship in the humanities and social sciences. Each themed volume drives scholarship forward and is a landmark in its field. For more information about the series, please visit www.thebritishacademy.ac.uk/proceedings.

PROCEEDINGS OF THE BRITISH ACADEMY • 267

HOW MAOISM WAS MADE

RECONSTRUCTING CHINA, 1949–1965

Edited by
JENNIFER ALTEHENGER AND
AARON WILLIAM MOORE

Published for THE BRITISH ACADEMY
by OXFORD UNIVERSITY PRESS

Oxford University Press, Great Clarendon Street, Oxford OX2 6DP

First edition published in 2025

British Library Cataloguing in Publication Data
Data available

Library of Congress Cataloguing in Publication Data
Data available

Typeset by Newgen Publishing Uk

ISBN 978-0-19-726781-3 (hardback)
ISBN 978-0-19-726806-3 (ebook)
ISBN 978-0-19-726807-0 (online)
ISSN 0068-1202

Contents

Figures and Tables

Figures

Table

Notes on Contributors

Jennifer Altehenger is Associate Professor of Chinese History and Jessica Rawson Fellow in Modern Asian History at the University of Oxford and Merton College. Her research focuses on the social and cultural history of modern and contemporary China. She is the author of *Legal Lessons: Popularizing Laws in the People's Republic of China, 1949–1989* (Harvard University Asia Center, 2018) and together with Denise Y. Ho of *Material Contradictions in Mao's China* (University of Washington Press, 2022). She is also the editor of the online resource 'The Mao Era in Objects' and is currently working on the history of furniture and mass production in socialist China.

Mary Augusta Brazelton is Professor in Global Studies of Science, Technology, and Medicine in the Department of History and Philosophy of Science at the University of Cambridge. Her research focuses on the history of science and medicine in twentieth-century China, with particular interests in Sino-French relations, China's role in international and global health, and transportation technologies. She is the author of *Mass Vaccination: Citizens' Bodies and State Power in Modern China* (Cornell University Press, 2019) and *China in Global Health: Past and Present* (Cambridge University Press, 2023). At Cambridge, she is also a fellow of Jesus College and a trustee of the Needham Research Institute.

Robert Culp is Professor of History and Asian Studies at Bard College, where he has taught since 1999. He is the author of *Articulating Citizenship: Civic Education and Student Politics in Southeastern China, 1912–1940* (Harvard University Asia Center, 2007) and *The Power of Print in Modern China: Intellectuals and Industrial Publishing from the End of Empire to Maoist State Socialism* (Columbia University Press, 2019). He is currently finishing a book-length project on the distribution of Chinese-language publications in overseas Chinese communities in Southeast Asia from the 1910s into the 1950s that is tentatively called 'Circuits of Meaning: Chinese Publishing Networks and the Sinophone Reader in Southeast Asia'. Another avenue of his work focuses on youth culture in twentieth-century China.

Juliane Fürst is head of the department 'Communism and Society' at the Centre of Contemporary History at Potsdam and Professor of Modern History at the Central European University in Vienna. She has published a monograph on the Soviet hippie movement titled *Flowers through Concrete: Explorations in the Soviet Hippieland* (Oxford University Press, 2021) and is the author of *Stalin's Last Generation: Soviet*

Post-War Youth and the Emergence of Late Socialism (Oxford, 2010) and, among other volumes, the editor of *Dropping Out of Socialism: Alternative Spheres in the Soviet Bloc* (Lexington Books, 2016) and *The Cambridge History of Communism Vol. III* (Cambridge, 2017). She is currently the Principal Investigator of the ERC-funded project 'Perestroika from Below: Participation, Biography and Emotional Communities 1980–1999'.

Jochen Hellbeck is Distinguished Professor of History at Rutgers University, and he specialises in modern Russian and European history. He has won fellowships from the Guggenheim Foundation, the New York Public Library Cullman Center, the Princeton Davis Center, the American Academy in Berlin, and the Freiburg Institute for Advanced Study. His first book *Revolution on My Mind* (Harvard University Press, 2006) explores personal diaries written in the Soviet Union under Stalin, and addresses the paradox of self-expression in an overtly repressive political system. His second book, *Stalingrad: The City that Defeated the Third Reich* (Public Affairs, 2015) is the first Western study to probe the meaning of the Battle of Stalingrad for the Soviet soldiers and civilians who defended the city in World War II (WWII). In preparation for this book, Hellbeck visited former soldiers who had fought at Stalingrad in their homes in Russia and Germany. His website, *Facing Stalingrad*, features their portraits and voices. Hellbeck's new monograph, *A War Like No Other*, will appear in 2025. It presents a new reading of WWII centring on the Soviet Union as both the prime target of Nazi aggression and the principal power to defeat Germany.

Christine I. Ho is Associate Professor of East Asian Art History at the University of Massachusetts Amherst. Her research focuses on modern and contemporary art, craft, and design in China. The author of *Drawing from Life: Socialist Painting and Socialist Realism in the People's Republic of China* (University of California Press, 2020), she has written on socialist visual culture, materiality, and artistic practice in book chapters and articles in *The Art Bulletin*, *Art History*, and *Archives of Asian Art*, among other publications. Her current research projects centre upon craft, decorative art, and muralism in modern China, as well as questions of method, evidence, and aesthetic categories in global modernism.

Wenyu Jing is a research associate in the Department of China Studies at the University of Cologne. She is currently working on her dissertation, 'Revolution and Tradition: The Village Cadres in China's People's Commune'.

Fabio Lanza is Professor of Modern Chinese History in the Departments of History and East Asian Studies at the University of Arizona. He is the author of *Behind the Gate: Inventing Students in Beijing* (Columbia University Press, 2010) and of *The End of Concern: Maoist China, Activism, and Asian Studies* (Duke University Press, 2017). He also co-edited (with Jadwiga Pieper-Mooney) *De-Centering Cold*

War History: Local and Global Change (Routledge, 2013). He is currently completing a manuscript on urban collectivisation in Beijing during the Great Leap Forward.

Toby Lincoln is Associate Professor of Chinese Urban History based in the Centre for Urban History at the University of Leicester. His publications include two books. *Urbanizing China in War and Peace, the case of Wuxi County* (University of Hawaii Press, 2015) explored how urbanisation was a total societal transformation that affected both city and countryside in the first half of the twentieth century. *China and Urban History* (Cambridge University Press, 2021) is the first book in English that describes the history of Chinese cities from their origins to the present for a general audience. His current research explores how Chinese cities were reconstructed after WWII and compares this with reconstruction in other countries around the world. He has also worked on aspects of heritage in contemporary China and has a longstanding interest in urban sustainability in the Anthropocene.

Sarah Mellors Rodriguez is Associate Professor of Chinese Humanities in the Department of Russian and East Asian Languages and Cultures at Emory University. Her work focuses on gender and sexuality and the history of medicine in modern China. In 2023, she published her first monograph, *Reproductive Realities in Modern China: Birth Control and Abortion, 1911–2021* (Cambridge University Press), which uses interviews and archival research to analyse how ordinary women and men navigated China's shifting fertility policies both before and during the One Child Policy era. *Reproductive Realities in Modern China* was awarded Honourable Mention in the Berkshire Conference of Women Historians Book Prize Competition. Dr Rodriguez's scholarship has also been published in *East Asian Science, Technology and Society*, *Social History of Medicine*, and *Nan Nü: Men, Women and Gender in China.*

Covell Meyskens is Associate Professor of National Security Affairs at the Naval Postgraduate School in Monterey, CA. His research examines national security and economic development in modern China. He is the author of *Mao's Third Front: The Militarization of Cold War China* (Cambridge University Press, 2020). He is currently working on his second book, *The Three Gorges Dam: Building a Developmental Engine for China and the World.* Meyskens has published book chapters in several edited volumes and refereed journal articles in *Cold War History*, *Twentieth Century China*, *positions: asia critique*, and the *Journal of Modern Chinese History.*

Aaron William Moore is the Handa Chair of Japanese–Chinese Relations at the University of Edinburgh. He is a scholar of comparative history and literature whose research languages include Chinese, Japanese, and Russian. He has published two books: *Writing War* (Harvard University Press, 2013), which examined hundreds of diaries by combat soldiers from Japan, China, and the US, and *Bombing the City* (Cambridge University Press, 2018), which described the air raid experiences of urban residents in Britain and Japan using diaries, letters, oral histories, and memoirs. In 2014 he was awarded the Leverhulme Prize for his work in transnational and comparative history. He is currently writing a monograph on the global history of youth in WWII, a series of articles on early twentieth-century science fiction and visions of the future in Asia, and, with Seth Jacobowitz, a translation volume entitled *The Hirabayashi Hatsunosuke Reader* (Bloomsbury, forthcoming). He has received research support from a variety of funders, including the AHRC, British Academy, the Chinese Academy of Social Sciences, the Leverhulme Trust, and the Taiwan Fellowship.

Aminda Smith is Associate Professor of history at Michigan State University. She is a historian specialising in modern Chinese history with a particular interest in the social and cultural history of Chinese Communism. She serves as co-director of the PRC History Group, an international scholarly organisation dedicated to increasing research and knowledge on the People's Republic of China. She has written widely on the global histories of the Chinese Communist Party and Maoism, including her first monograph, *Thought Reform and China's Dangerous Classes: Reeducation, Resistance, and the People* (Rowman & Littlefield, 2012).

Nicolai Volland is Associate Professor of Asian Studies and Comparative Literature. His research focuses on modern Chinese literature and culture in its transnational dimensions (including cosmopolitanism, transnationalism, translation, and transculturation), as well as reception and cultural consumption. He is also interested in film and visual culture, as well as print culture in modern China and Southeast Asia, and in oceanic/archipelagic studies. His book, *Socialist Cosmopolitanism: The Chinese Literary Universe, 1945–1965* (Columbia University Press, 2017), revisits Chinese literature of the 1950s and 1960s, situating Chinese cultural production within the broader context of the socialist world and showing how Chinese socialist literature engages with global aesthetic debates. Current projects investigate the Sino-French literary relationship across the long twentieth century, and rethinking Sinophone literature from an oceanic perspective. He is co-editor of *The Business of Culture: Cultural Entrepreneurs in China and Southeast Asia* (UBC Press, 2014) and *Comic Visions of Modern China* (MCLC special issue, Autumn 2008). He is a past president of the Association of Chinese and Comparative Literature (ACCL).

Xiaoxuan Wang is a social historian of modern China, whose research interests include religions in modern China and the history of the Chinese diaspora in the twentieth century. His publications on religion under Mao include *Maoism and Grassroots Religion: The Communist Revolution and the Reinvention of Religious Life in China* (Oxford University Press, 2020) and 'Solving the

"Religious Problem": The Great Leap Forward of "Religious Work" and Protestant Communities in Pingyang, Wenzhou in 1958' (*Twentieth-Century China*, 49.1, 2024), among others. His latest project explores the intricate relationship between European Chinese communities and the Chinese states during the Cold War era.

Benno Weiner is Associate Professor at the Department of History at Carnegie Mellon University. His research focuses on the ethnopolitics of twentieth-century state and nation-building along China's ethnocultural frontiers. He is the author of *The Chinese Revolution on the Tibetan Frontier* (Cornell University Press, 2020) and co-editor of *Conflicting Memories: Tibetan History under Mao Retold* (Brill, 2020).

Felix Wemheuer is Chair Professor for Modern China Studies at the University of Cologne. His publications include *Famine Politics in Maoist China and the Soviet Union* (Yale University Press, 2014) and *A Social History of Maoist China: Conflict and Change, 1949–1976* (Cambridge University Press, 2019). He hosts the YouTube channel 'Studying Maoist China'. From 2000 to 2002, Wemheuer studied 'History of the Communist Party of China' at the People's University in Beijing. Between 2008 and 2010, he was a visiting scholar at the Fairbank Center for Chinese Studies at Harvard University.

Emily Wilcox is Margaret Hamilton Professor of Modern Languages and Literatures (Chinese Studies) at William & Mary and a 2024 Guggenheim Fellow in Dance Studies. Wilcox is the author of *Revolutionary Bodies: Chinese Dance and the Socialist Legacy* (University of California Press, 2018, winner of the 2019 de la Torre Bueno Prize from the Dance Studies Association), which was released in Chinese by Fudan University Press in 2023. Wilcox is co-editor of *Corporeal Politics: Dancing East Asia* (University of Michigan Press, 2020), *Inter-Asia in Motion: Dance as Method* (Routledge, 2023), and *Teaching Film from the People's Republic of China* (MLA, 2024). Wilcox directed the American Dance Festival's 'Planting Seeds: ADF and Modern Dance in China' oral history project and co-created the University of Michigan Library's Chinese Dance Collection. She was a Fulbright Scholar at the Beijing Dance Academy from 2008 to 2009 and postdoctoral researcher at the Shanghai Theatre Academy from 2011 to 2013. Her current project examines dance exchanges in China during the Cold War.

Shellen Xiao Wu is Professor and L.H. Gipson Chair in Transnational History at Lehigh University. Her new book, *Birth of the Geopolitical Age: Global Frontiers and the Making of Modern China* (Stanford University Press, 2023) traces the global history of the frontier in the twentieth century, with an emphasis on China. Her first book, *Empires of Coal: Fueling China's Entry into the Modern World Order, 1860–1920* (Stanford University Press, Studies of the Weatherhead East Asian Institute, 2015), examines how the rise of science and industrialisation destabilised global systems and caused widespread unrest and the toppling of ruling regimes in China and around the world. Wu has received fellowships from the Institute of Advanced Studies in Princeton, the National Humanities Center, the Luce/ACLS Program in China Studies, Fulbright, and the Mellon Foundation. She has published articles in *Nature*, *The American Historical Review*, *International History Review*, and other leading journals in history, history of science, and Asian Studies.

Acknowledgements

A project as large as this required multiple hands in its making. As editors, we wish to thank, first and foremost, the British Academy for the financial support it provided in 2018. This enabled us to assemble a 'dream team' of early and mid-career historians of modern China from Europe and North America, all participating in an extraordinary conference on the early People's Republic of China (PRC) held at the British Academy in London. All of the participants, discussants, chairs, and subsequent contributors to this volume significantly shaped our understanding of the period. The Department of Asian Studies at the University of Edinburgh helped bring some of the contributors to Scotland in order to further refine our ideas. Portia Taylor at the British Academy was very patient as the volume kept experiencing delays throughout the pandemic period. We were fortunate to have many anonymous reviewers, as well as the review process organised by the British Academy, giving constructive critique to individual chapters and the volume as a whole. The editors wish to mention Michael Schoenhals and Stephen Smith in particular for providing critical feedback throughout our process of organising this volume.

Introduction: The Early PRC Between Theory and Practice

JENNIFER ALTEHENGER AND AARON WILLIAM MOORE

REVOLUTIONS NEED TO be explained. After the People's Republic of China (PRC) was established in October 1949, many different publications tried to render the concepts of the day, such as 'socialism' and 'communism', accessible to a wide domestic audience. It was a rich world of printed works and propaganda that included dictionaries, a genre well placed to provide handy explanations of terms and their meanings. In 1953, the editors of the *Dictionary of New Knowledge*, a popular resource published by Chunming Bookstore, used a colourful two-page illustration in the dictionary's endpaper to help readers make sense of what socialism might mean. Titled 'Build Our Great Country', the illustration combined drawings, statistics, and captions to recount the Chinese Communist Party's (CCP) master narrative of the 'Great achievements of the last three years (1949–1952)' and the challenges that had been successfully overcome along the way. Readers could learn that, since 1949, the country had forged close ties with the Soviet Union and worked to secure peace and that the People's Volunteer Army had aided Korean brethren in fighting off the 'US Imperialist Aggressors'. Meanwhile, looking at domestic developments, the editors illustrated that the CCP had successfully moved against counterrevolutionaries, carried out Land Reform, and protected its citizens in hygiene campaigns against local illnesses (but also the bacteriological warfare carried out by the US enemy). Newly strengthened and secured, readers were told that China had embarked on an extraordinary path of 'constructing industry', a path illustrated through colourful statistics and charts: total output across key sectors – steel, oil, coal, electricity, yarn, and paper – had increased significantly and the image celebrated 'soaring developments' in rural industrial construction, especially in cotton output. China, in other words, was ready for its first Five-Year Plan and for the next step of 'building a new society'.[1]

[1] Chunming chubanshe (ed.), *Xin mingci cidian*. Shanghai: Chunming chubanshe, 1953, endpaper.

Proceedings of the British Academy, **267**, 1–18, © The British Academy 2025.

Making China into a 'great country' has been an image that is central to the CCP's vision of its past and present, which was born through the tribulations of the twentieth century. In 1949, the CCP inherited a country that was in considerable disorder; the Chinese people had known almost constant war from 1937 to 1949, which inflicted inestimable damage to both life and livelihood.[2] Right from the start, the party proclaimed, repeating the view of Josef Stalin, that China's social and economic base were not developed enough to transition to socialism immediately. Instead, China would first enter a period of 'New Democracy' during which private and public ownership would continue to co-exist – a preliminary state of partial socialism prior to true socialism. The argosy of crises faced by the party leaders after 1949 was truly daunting: millions of casualties, collapsed agricultural productivity, runaway inflation, encirclement by hostile powers, insecure borders, rampant banditry, and a state apparatus both smashed by invaders and compromised by people whom the party considered 'collaborators' and 'counterrevolutionaries'.[3] To remake a devastated country with a skeletal party apparatus organised under the exigencies of total war was an endeavour that seemed doomed to failure. Nevertheless, despite the difficulties that plagued Chinese society for much of the first half of the twentieth century and after, by the middle of the 1950s, the CCP had secured the borders, stabilised the currency, embarked on large-scale industrial and infrastructural projects, and brought some degree of stability to the political system.

In March 1953, only a few months after the *Dictionary of New Knowledge* was published and just as the editors finished the foreword to this edition, Josef Stalin died. Within a few months, party leaders suddenly announced that the 'New Democracy', a state that they had earlier said would last for many years, was over. The country, leaders argued, was now in the 'transition to socialism', which was to be a period of rapid collectivisation and nationalisation of industry and business. The second half of the 1950s was characterised by new political purges as the party began to turn on people suddenly labelled 'rightist' in 1957. Many of those who had spent years working to support the party were stripped of their livelihoods and sent to the countryside. Meanwhile, in 1958, the government initiated the Great Leap Forward (1958–61), which was supposed to advance China in industry and

[2] Zhang Xianwen, *Zhongguo kang-Ri zhanzhengshi, 1931–1945*. Nanjing: Nanjing daxue chubanshe, 2001; Odd Arne Westad, *Decisive Encounters: The Chinese Civil War, 1949–1950*. Stanford: Stanford University Press, 2003. Diana Lary, *The Chinese People at War: Human Suffering and Social Transformation, 1937–1945*. Cambridge University Press, 2010. Rana Mitter, *China's War with Japan, 1937–1945: The Struggle for Survival*. London: Allen Lane, 2013.

[3] Frederick C. Tiewes, *Politics and Purges in China*, 2nd edition. Armonk: ME Sharpe, 1993; Margherita Zanasi, 'Globalizing Hanjian: The Suzhou Trials and the Post-World War II Discourse on Collaboration', *The American Historical Review* 113:3 (2008): 731–51; Li Lifeng, 'Tugaizhong de suku: yizhong minzhong dongyuan jishu de weiguan fenxi', in *1940 nendai de zhongguo*, eds. Zhonguo shekeyuan and Sichuan shifan daxue, v. 1. Beijing: Shehui kexue wenxian chubanshe, 2009, pp. 371–89; Harold Miles Tanner, *The Battle for Manchuria and the Fate of China: Siping, 1946*. Bloomington: Indiana University Press, 2013.

agriculture, make it largely self-sufficient, and complete the transition to socialism. It was a high point of revolutionary optimism when, for a brief period, many thought socialism had been achieved. It was soon followed by entrenchment as well as a realisation and admission that socialism would take several more years after all. Whatever achievements people had made during the Leap came at a high price for substantial parts of the Chinese population, especially people living in the countryside, as many areas suffered from a devastating famine.[4] By the early 1960s, for a few years, the country wavered between a more gradual approach to building towards socialism and new mass campaigns in the form of the Socialist Education Movement, which began in 1963.

Subsequent historians have reflected on the first years of the PRC and provided much insight into every topic covered in 'Build Our Great Country'. New research has made it possible to evaluate which information was more aspiration than reality and what complex histories were unfolding when Hu Jitao and Wang Feipeng designed the 1953 illustration.[5] Studies have revealed how Land Reform and the 'Campaign to Suppress Counterrevolutionaries' proceeded in different localities and how the Korean War affected domestic politics and society.[6] They have traced how Sino-Soviet friendship was forged and broke apart a decade later.[7] We know much more about how state authorities industrialised the country and how and why statistics such as those in the illustration were produced.[8]

[4] Felix Wemheuer, *Famine Politics in Maoist China and the Soviet Union*. New Haven: Yale University Press, 2014.

[5] See especially the edited volumes by Jeremy Brown and Paul Pickowicz (eds.), *Dilemmas of Victory: the Early Years of the People's Republic of China*. Cambridge, MA: Harvard University Press, 2007 and Julia Strauss (ed.), *The History of the PRC (1949–1976)*, The China Quarterly Special Issues New Series, No. 7. Cambridge: Cambridge University Press, 2007.

[6] See, for example, Julia Strauss, 'Paternalist Terror: The Campaign to Suppress Counterrevolutionaries and Regime Consolidation in the People's Republic of China', *Comparative Studies in Society and History* 44 (2002) 1: 80–105; Luo Pinghan, *Tudi gaige yundongshi*. Fuzhou: Fujian renmin chubanshe, 2005; Brian DeMare, *Land Wars: The Story of China's Agrarian Revolution*. Stanford: Stanford University Press, 2019; Aminda Smith, *Thought Reform and China's Dangerous Classes: Reeducation, Resistance, and the People*. Lanham: Rowman and Littlefield, 2013. On the Korean War, see Shu Guang Zhang, *Mao's Military Romanticism: China and the Korean War, 1950–1953*. Lawrence: Kansas University Press, 1995; David Cheng Chang, *The Hijacked War: The Story of Chinese POWs in the Korean War*. Stanford: Stanford University Press, 2020.

[7] Odd Arne Westad, *Brothers in Arms: The Rise and Fall of the Sino-Soviet Alliance, 1945–63*. Washington, DC: Woodrow Wilson Center Press, 1998; Zhihua Shen and Yafeng Xiao, *Mao and the Sino-Soviet Partnership, 1945–1959: A New History*. Lanham: Lexington Books, 2015; Zhihua Shen and Yafeng Xiao, *Mao and the Sino-Soviet Split: 1959–73: A New History*. Lanham: Lexington Books, 2018; Lorenz M. Luethi, *The Sino-Soviet Split: Cold War in the Communist World*. Princeton: Princeton University Press, 2008; Austin Jersild, *The Sino-Soviet Alliance: An International History*. Chapel Hill: The University of North Carolina Press, 2014.

[8] See, for example, the chapters in 'Part II: 1950–the present' in Debin Ma and Richard von Glahn (eds.), *The Cambridge Economic History of China Vol. 2*. Cambridge: Cambridge University Press, 2022, 531–828; Shu Guang Zhang, *Beijing's Economic Statecraft during the Cold War, 1949–1991*. Baltimore: Johns Hopkins University Press, 2014. On statistics, see Arunabh Ghosh, *Making it Count: Statistics and Statecraft in the Early People's Republic of China*. Princeton: Princeton University Press, 2020.

Careful work in archives, with oral histories and other materials has shed light on how campaigns did or did not affect people's everyday lives and how life varied between city and countryside.[9] These studies, and many more, have demonstrated how the party moved from New Democracy to the Great Leap Forward, the subsequent famine, and into the 1960s, a decade as rich in developments as the 1950s yet often overshadowed by a focus on the 'Great Proletarian Cultural Revolution'. Thanks in part to the opening of the archives in China, the extraordinary efforts of China scholars over the last few decades have illuminated a variety of historical developments in the history of the early PRC with unprecedented depth and detail. This volume turns to the concepts that tied these historical developments together, in the dictionary's illustration and in Chinese history itself: building and making. What did it mean to different people in different places and at different times to build a 'new society' and system? What role did state authority, ideology, and 'grassroots' mass mobilisation play? What can we learn by tying together, as this volume does, seemingly disparate case studies of 'building socialism' from diverse fields such as urban planning, water management, ethnic policy, thought reform, law, dance, publishing, and art?

Building a New China

In the years after 'national liberation' in 1949, the CCP engaged in a nationwide process of 'building'. In Chinese this was known as *shehui zhuyi jianshe*, a term that has often been translated as 'socialist construction', though it could also be translated more freely as 'building socialism'. The term is capacious and imprecise; in some respects, its meaning resembles what the United Nations has long termed 'capacity-building'. According to the 1992 United Nations Conference on Environment and Development, capacity-building …

> … encompasses a country's human, scientific, technological, organizational, institutional and resource capabilities. A fundamental goal of capacity-building is to enhance the ability to evaluate and address the crucial questions related to policy choices and modes of implementation among development options, based on an understanding

[9] On everyday life, campaigns and the rural-urban divide, see Kuisong Yang, *Eight Outcasts: Social and Political Marginalisation in China under Mao*. Oakland: University of California Press, 2019; Jeremy Brown, *City vs. Countryside: Negotiating the Divide*. Cambridge: Cambridge University Press, 2012; Gail Hershatter, *The Gender of Memory: Rural Women and China's Collective Past*. Oakland: University of California Press, 2011; Covell Meyskens, *Mao's Third Front: The Militarization of Cold War China*. Cambridge: Cambridge University Press, 2020; Kimberley Ens Manning and Felix Wemheuer, *Eating Bitterness: New Perspectives on China's Great Leap Forward and Famine*. Vancouver: UBC Press, 2011; Wemheuer, *Famine Politics in Maoist China and the Soviet Union*; Neil J. Diamant, *Revolutionizing the Family: Politics, Love, and Divorce in Urban and Rural China*. Berkeley: University of California Press, 2000; Jacob Eyferth, *Eating Rice from Bamboo Roots: The Social History of a Community of Handicraft Papermakers in Rural Sichuan, 1920–2000*. Cambridge, MA: Harvard University Asia Center, 2009.

> of environmental potentials and limits and of needs as perceived by the people of the country concerned.[10]

There were many parallels between the CCP's agenda of 'building socialism' and the 'modernisation' projects that characterised much of post-war international policy, but there were, and continue to be, important differences. By building infrastructure, training citizens, and developing bureaucratic processes, the CCP aimed not only to rebuild devastated cities, factories, and transport infrastructure, but also to create an entirely new kind of society. The years of the first Five-Year Plan (1953–57) reflected many of the experiences of the New Economic Policy (NEP, 1921–28) in the Soviet Union where 'the fundamental elements of a new Soviet social order, culture, and national identity' was 'in the making'.[11] Referencing Lenin's 1918 comments regarding the absence of 'socialist [social] relations', Milovan Djilas pointed out how revolutionary societies felt compelled to 'build' society:

> I quote Lenin, but I could quote any leader of the Communist revolution and numerous other authors, as confirmation of the fact that settled relationships did not exist for the new society, but that someone, in this case the 'soviet power', must therefore build them. If the new 'socialist' relationships had been developed to the fullest in the country in which Communist revolution was able to emerge victorious, there would have been no need for so many assurances, dissertations, and efforts embracing the 'building of socialism'.[12]

Chinese socialism in fact emerged out of the pre-revolutionary 'nation-building' (*jianguo*) projects of the Kuomintang (KMT) government, but the CCP significantly reimagined that process as a revolutionary transformation of the social relationships with which the 'nation' was constituted. Capitalist 'modernisation' projects also adopted a theory of society, which defined gender and reproductive roles within the centrality of the heteronormative and patriarchal family, defended the primacy of ownership and the exploitation of labour, and in most cases maintained racial and ethnic hierarchies. Yet the CCP's project went further. Chinese socialism attempted to build an entirely different system through transformations in and at 'work' (*gongzuo*), where building and making took place. Revolutionising 'work' was not only aimed at 'building capacity', or even the transfer of techniques

[10] UN Conference on Environment and Development, 'Agenda 21: The UN Sustainable Development Plan from Rio de Janeiro 1992' (New York: Cosimo Reports, 1992). The term 'capacity-building' has now been replaced by the term 'capacity development' and there have been debates about what the term means. For a very helpful analysis of understandings of 'revolution' and 'development' in post-war China and in an international context, see Sigrid Schmalzer, *Red Revolution, Green Revolution: Scientific Farming in Socialist China*. Chicago: Chicago University Press, 2016.

[11] William G. Rosenberg, 'NEP Russia as a "Transitional" Society', in *Russia in the Era of the NEP: Explorations in Soviet Society and Culture*, eds. Sheila Fitzpatrick, Alexander Rabinowitch, and Richard Stites. Bloomington: University of Indiana Press, 1991, p. 4. Rosenberg is paraphrasing Moshe Lewin in his *The Making of the Soviet System: Essays in the Social History of Interwar Russia*. Warrington: Methuen, 1985.

[12] Milovan Djilas, *The New Class: An Analysis of the Communist System*. London: Harcourt Brace Jovanovich, 1957, p. 20.

and technology.[13] It was to transform the social relationships that made production possible in the first place. Such a grand project required attention to activities as diverse as industrial work, sexual reproduction, farming and handicraft, visual art, and scientific research.

In Mao's writings, party documents, and media, the project of building a new society was often accompanied by references to 'making revolution' or 'doing revolution' (*gan geming*). It was a vague and ideological phrase, seldom used to describe people's everyday activities, but it underscored that the revolution was more of a process than a fixed idea. Revolutionary change would not simply happen on its own. If 'making revolution' was the general road to socialism, actual socialism would come about through work and labour (*laodong*). By participating in all kinds of work (*canjia gongzuo*) everywhere, people were supposed to increase 'production' (*shengchan*), which was essential to the building project. The concrete dimension of revolution through work and production was so important to Mao that he wrote a seminal essay in 1937 entitled 'On Practice' in which he outlined the connection between knowing and doing, between theory and practice.[14] After the founding of the PRC in October 1949, the CCP made this duality of theory and practice one of the country's central ideological tenets. Many in the party believed that the constant process of moving between theory/study and practice/work would be one of the methods to, first, attain socialism and then, eventually, communism, known in Chinese as the 'Great Harmony', and the classless society. Emphasising the importance of practice, Mao explained in 1958 that 'the revolution must continually advance. The Hunanese often say, "Straw sandals have no pattern – they shape themselves in the making"'.[15]

China's transformation was beset with challenges, however. If party leaders presented the final goal of communism as inevitable, dictated by the Marxist doctrine of historical development, the means to get there – revolution, socialist construction, and daily work – had to be adaptable. Observing the struggles of ordinary Chinese people to democratise the workplace or overcome a rapidly growing bureaucracy, Mao theorised the fluctuating situation in his writings on 'contradictions' (*maodun*): for him, the process of overcoming constantly emerging contradictions was the nature of socialism itself. As he famously stated in March 1949, a few months before the PRC was established: 'To win countrywide victory is only

[13] On the topic of 'work', see Jacob Eyferth (ed.), *How China Works: Perspectives on the Twentieth-Century Industrial Workplace*. Abingdon: Routledge, 2006.

[14] Mao Zedong, 'On Practice: On the Relation Between Knowledge and Practice, Between Knowing and Doing', (July 1937), www.marxists.org/reference/archive/mao/selected-works/volume-1/mswv1_16.htm.

[15] Mao Zedong, 'Speech at the Supreme State Conference', 28 January 1958, cited in Elizabeth Perry, 'Debating Maoism in Contemporary China: Reflections on Benjamin I. Schwartz', *Chinese Communism and the Rise of Mao, Harvard-Yenching Institute Working Paper Series*, 2020, 8, and Perry in turn cites from John Bryan Starr, *Continuing the Revolution: The Political Thought of Mao*. Princeton: Princeton University Press, 1979, ix. See also www.marxists.org/reference/archive/mao/selected-works/volume-8/mswv8_03.htm, accessed 20 November 2022.

the first step in a long march of ten thousand *li* [...] The Chinese revolution is great, but the road after the revolution will be longer, the work greater and more arduous'.[16] Socialism was supposed to transform material conditions and the social and economic relations of everyday life and, in this process, each new stage presented new challenges and contradictions. These had to be surmounted through study and action, or 'work and study' (*gongzuo yu xuexi*), which in turn produced new contradictions that again had to be addressed by study and action.

Our volume thus uses the term 'Maoism' as an entry point into studying how different people constantly tried to combine 'knowing and doing', theory and practice, during the historical period between 1949 and the mid-1960s. The term 'Maoism' has a complex trajectory, and this volume does not attempt to fix what it was or was not. Nor do contributors suggest that it was a naturalised term that people used during the time under analysis – indeed, Maoism is a foreign construct with no direct equivalent in Chinese. In 1951, Benjamin Schwartz, a scholar of Chinese communism at Harvard University, famously used it to draw attention to the duality of theory and practice in the ideological writings of the CCP. For Schwartz, 'Maoism' described a 'pragmatic strategy of revolution', closely related to but nonetheless distinct from Marxism-Leninism and Stalinism. Schwartz was not the first to use the term, as Aminda Smith demonstrates, but his work popularised it far more widely than before and brought it into the mainstream of academic discussions about China.[17] This conception of Maoism, as Elizabeth Perry explained, 'derived from an effort to understand the lived experience of Mao and his comrades as they groped in fits and starts toward a workable strategy of revolution.'[18] Knowledge about these 'practical lessons drawn from the experience of concrete political struggle' could be found in Mao's writings, which formed part of 'Mao Zedong Thought' (*Mao Zedong sixiang*).[19] Mao Zedong Thought was composed of a carefully curated and often heavily edited collection of speeches and writings that Mao and other party leaders started to build in the revolutionary base region of Yan'an during the early 1940s. It was canonised with the publication of the first volume of the *Selected Works of Mao Zedong* in 1951, a series of five volumes the last of which was published in 1977 after Mao's death.[20] Throughout the period of the PRC under Mao's rule (1949–76, sometimes called the 'Mao era' or the 'Mao period'), this canon was a core component of what was known in China as 'Marxism-Leninism-Mao Zedong Thought'.

[16] Mao Zedong, 'Report to the Second Plenary Session of the Seventh Central Committee of the Communist Party of China', 5 March 1949, www.marxists.org/reference/archive/mao/selected-works/volume-4/mswv4_58.htm.

[17] Aminda Smith, 'A Note on the Origins of Maoism', *The PRC History Review* 8:1 (June 2023): 1–4; Aminda Smith, *Maoism Is…* (unpublished manuscript, cited with author's permission); Fabio Lanza, *The End of Concern: Maoist China, Activism, and Asian Studies*. Durham: Duke University Press, 2017.

[18] Perry, 'Debating Maoism in Contemporary China', 6.

[19] Perry, 1.

[20] Daniel Leese, 'A Single Spark: Origins and Spread of the Little Red Book in China', in *Mao's Little Red Book: A Global History*, ed. Alexander C. Cook. Cambridge University Press, 2014, pp. 23–42.

Mao Zedong Thought was a crucial part of what one might call Maoism, formulating its theoretical base, but it was never the whole. For many years, when scholars studied the developments after 1949, they viewed these works to be the primary means of understanding the period and 'Maoism'. Even Mao himself, however, would have critiqued this approach as too limited and neglectful of the crucial agency of 'the people' (*renmin*). Inspired by Aminda Smith's approach to 'Maoism' as concept and practice, this volume focuses on the people who worked to make Chinese socialism a reality through a series of social, cultural, and economic practices and how, in doing so, they also helped create the theoretical understanding of this process itself.[21] They were the workers and farmers who formed the core constituents of the 'New Democracy'. The educated classes, scientists, and cultural actors had to go among the working people, labour with them, and learn from them; only by achieving a shared consciousness through a democratised and collective form of social, cultural, and economic production, could the 'mass line' be achieved. The ideology of the party both emerged from and guided this process, which was supposed to be ongoing until the ultimate achievement of communism.

The chapters in this volume examine how the efforts to try and find what Elizabeth Perry called 'a workable strategy' continued beyond the realm of policy-making after 1949 as people had to search for strategies daily. We investigate how thousands of people, who were not part of party leadership and mostly not even party members, pursued these efforts across many different areas of life and work. Understood in this way, Maoism does not encompass everything anyone in China did during this period. In many aspects of their lives, people were not concerned with building a new society, with the question of how to put theory into practice, or with socialism at all, as Jeremy Brown, Matthew Johnson, and many others have demonstrated.[22] It encompasses those moments when people across the country pursued, often in small steps and actions (and with great enthusiasm), what they believed were contributions to working towards socialism as outlined by the party's proclamations. It also encompasses moments when people were involuntarily caught up in such projects, were forced to participate and build, where they resisted participation, or became subject to the routine violence that characterised the life of many in the Mao period.[23] In bringing together a broad range of case studies, we seek to show how different people and groups tried to create a polity and society that they understood as Chinese and socialist (or which they thought might eventually become socialist), how they used what they knew and the legacies of the years 'pre-liberation', and how they tried to contend with the constant stream of theoretical and practical problems that they encountered in their work.

[21] Smith, 'A Note on the Origins of Maoism'.

[22] Brown and Pickowicz, *Dilemmas of Victory*; Jeremy Brown and Matthew Johnson, *Maoism at the Grassroots: Everyday Life in China's Era of High Socialism*. Cambridge, MA: Harvard University Press, 2015.

[23] Jeremy Brown and Matthew Johnson mentions this in their introduction, Brown and Johnson, *Maoism at the Grassroots*, 2, referring also to Joseph W. Esherick, Paul G. Pickowicz, and Andrew G. Walder (eds.), *The Chinese Cultural Revolution as History*. Stanford: Stanford University Press, 2006, 12–13.

How foreign scholarship made Maoism

Scholars have often tried to define 'Maoism', and these efforts have often fallen short because the inadequacy of such terms to describe complex historical processes. Nevertheless, 'Maoism' has long been a launchpad for attempts to understand Chinese socialism among scholars internationally, even if the resources to do so were initially very limited. Attempts to describe the CCP's revolutionary project on the mainland initially developed from first-hand observers who were often economists and social scientists, to later historians who 'knew' socialism through the critical examination of material objects like government documents and popular art. As Angela Romano and Valeria Zanier pointed out, the view that China 'opened' with Nixon's visit in 1972 and the death of Mao in 1976 ignores ongoing exchanges that did not involve American Sinologists, which mirrors older misguided assumptions that before the Opium War, China was 'closed'.[24] Nevertheless, while the scholars in this volume had extensive access to documentary evidence when considering the early PRC, the past is as foreign country as China was for the Cold War era scholars who tried to understand it from afar (or during curated visits). Our historical understanding of how 'Maoism' was made is a palimpsest of the Cold War and post-Cold War scholarly theory, variations in international relations, and ever-changing academic research practice. It is therefore necessary to reflect on the scholarly past, as the image of Maoism is made and remade by academic work.

The first attempts to describe socialism in China were not works of history or written by trained historians, because they were composed while the events were unfolding. In addition to access to growing numbers of documents in university, government, and private collections outside of China, observers who published in the 1950s often had personal experiences in China from which to draw. Foreigners had worked for years in Republican China as journalists, educators, missionaries, and medical or diplomatic personnel, while others had served in treaty-port colonial administrations or in wartime military intelligence, and some had spent their childhood and youth in China.[25] Personal experience was particularly important for authors who were reporters, activists, or government agents in China before 1949, like Agnes Smedley, Kaji Wataru, and Franco Calamandrei. For most of the foreign

[24] Angela Romano and Valeria Zanier, 'Circumventing the Cold War: The Parallel Diplomacy of Economic and Cultural Exchanges between Western Europe and Socialist China in the 1950s and 1960s: An Introduction', *Modern Asian Studies*, 51:1 (2017): 1–16. Also see Pär Cassel, 'Solitary Swedish Sinologists: Three Hundred and Fifty Years of Swedish China Studies', *Journal of Chinese Studies*, forthcoming; Nicolai Volland, *Socialist Cosmopolitanism: The Chinese Literary Universe, 1945–1965*. New York: Columbia University Press, 2017; Laura De Giorgi, 'Italians in Beijing (1953–1962)', in *Contact Zones in China: Multidisciplinary Perspectives*, eds. De Giorgi *et al.* De Gruyter Oldenburg, 2020, pp. 81–96. The Japanese government did not give permission for its citizens to visit the PRC, but the Chinese government allowed many Japanese to visit via neighbouring countries.

[25] Graham Peck, *Two Kinds of Time*. Boston: Houghton Mifflin, 1950; Jacques Guillermaz, 'La politique agraire du parti communiste chinois', in *Revue Militaire d'Information* 318 (July 1960): 4–25, *Histoire du Parti communiste chinois: 1921–1949*. Paris: Payot, 1968, and *Le Parti Communiste Chinois au pouvoir (1er Octobre 1949–1er Mars 1972)*. Paris: Payot, 1972.

military and political officialdom, the revolution was a catastrophe, and they were wont to criticise Chinese socialism; others were either members of their national communist parties at home or had in any case developed good relations with the CCP before 1949.[26] Throughout the early PRC period, foreign academics, reporters, and servicemen either framed the revolution as a peasant movement led by Mao Zedong, or a totalitarian terror state modelled after the Soviet Union.[27]

Despite the divisive politics of the Cold War, foreign evaluations of socialist transformation were not entirely without empirical basis; in fact, knowledge about Chinese socialism was generated both from econometrics and Marxist analysis, both of which focused on many of the same topics despite their methodological incommensurability.[28] Chinese authors, often economists by training, who left the mainland and settled in Taiwanese, Hong Kong, and non-Sinophone universities, were crucial in shaping early understandings of the PRC system.[29] Chinese writers based in the PRC sometimes had their work on the socialist system translated, or co-written, into foreign languages, pursuant to proving their revolutionary bona fides.[30] Working from data produced by the PRC government, newsprint, and occasionally personal survey experience, scholars outside of China attempted to understand how socialism functioned in economic terms.[31] Basic information provided

[26] Even for European communists, movements within China were controlled. De Giorgi, 'Italians in Beijing', p. 85. On Italian student exchanges, see Sofia Graziani, 'The Case of Youth Exchanges and Interactions between the PRC and Italy in the 1950s' in *Modern Asian Studies*, v. 51, n. 1 (2017); other articles in this special issue explore student exchanges as well. American visits were comparatively rare, but sometimes resulted in longitudinal studies. William Hinton, *Fanshen: A Documentary of Revolution in a Chinese Village*. New York: Monthly Review Press, 1966, and *Shenfan: The Continuing Revolution in a Chinese Village*. New York: Vintage, 1984.

[27] Richard L. Walker, *China under Communism: The First Five Years*. New Haven: Yale University Press, 1955.

[28] Ghosh, *Making It Count*, pp. 1–21.

[29] C.K. Yang, *The Chinese Family in the Communist Revolution*. Cambridge, MA: M.I.T. Press, 1954, and *A Chinese Village in Early Communist Transition*. Cambridge, MA: M.I.T. Press, 1959; Li Choh-ming, *Economic Development of Communist China*. 1959; Jerome Ch'en, *Mao and the Chinese Revolution*. London: 1965. Li Ta-chung and Yeh Kung-chia, *Economy of the Chinese Mainland: National Income and Economic Development, 1933–1959*. Princeton: Princeton University Press, 1965.

[30] Tchao Chou-li (Zhao Shuli) and Alain Roux, *Les révolutions chinoises-le matin des villageois*. Paris: Éditions du burin, 1973. Zhao's most famous work on China's transition to socialism was *Sanliwan*. Beijing: Tongsu duwu chubanshe, 1955; Chao Shu-li, *Sanliwan Village*. Beijing: Foreign Languages Press, 1957; Tchao Chou-li, *Le Village de Sanliwan*, trans. Marthe Hou. Beijing: Foreign Language Press, 1960. For more on Zhao's status as a model writer for rural socialism, see Yi-tsi Mei Feuerwerker, 'Zhao Shuli: The "Making" of a Model Peasant Writer', in *Ideology, Power, Text: Self-Representation and the Peasant 'Other' in Modern Chinese Literature*. Stanford: Stanford University Press, 1998, pp. 100–45.

[31] Christopher B. Howe, 'The Chinese Economy and 'China Economists' as Seen through the Pages of "The China Quarterly."' *The China Quarterly*, 200 (2009): 923. Also see Howe and Kenneth R. Walker, *The Foundations of the Chinese Planned Economy: A Documentary Survey, 1953–65*. London: The MacMillan Press, 1989. Howe also showed usable data for 1971 and 1972, however, in *Wage Patterns and Wage Policy in Modern China, 1919–1972*. Cambridge: Cambridge University Press, 1973. John Wong, *Land Reform in the People's Republic of China: Institutional Transformation in Agriculture*. New York: Praeger Publishers, 1973.

by government publications, combined with curated visits and interviews with emigres, provided clarity on issues such as industrial policy, wage differentials, social reform campaigns, and the link between Chinese socialism and pre-war developmentalism under the KMT.[32] Economic historians were not always critical of Chinese state socialism, sometimes praising Communist efforts as part of a longer history of development and industrialisation.[33] Rene Dumont, Christopher Howe, John Wong, and Liu Ta-Chung all approached the problem of socialist transformation from different methodological viewpoints while sharing a penchant for data collection and a focus on the domestic economy.

Early research on the PRC based in area studies, meanwhile, had a different methodology and focus, including political ideology, theory, and social relations, working largely with Chinese documents collected in non-Sinophone areas, including the Harvard-Yenching Library, the Hoover Institution, and the Tōyō Bunko in Tokyo.[34] Although the American institutions would enjoy the most wealth and support due to US power and strategic interests in the region, European and Japanese institutions were successfully training experts throughout the Maoist period. In the UK, for example, investments in training staff and building collections were triggered by both the Scarborough Report in 1947 and the Hayter Report in 1961 – although the main body of books at London's School of Oriental and African Studies (SOAS) was organised in the 1970s. The study of China had a long tradition in Germany, but in West Germany 'Modern Sinology' developed swiftly from the 1970s once it was possible for scholars and students to travel to the PRC.[35] Across the German–German border, socialist countries such as the German Democratic Republic had privileged access to information about China and to the country itself until the Sino-Soviet split.[36] Meanwhile, in Japan, politicians, academics, and even

[32] Howe, *Wage Patterns and Wage Policy in Modern China*; Martyn King Whyte, 'Inequality and Stratification in China', *The China Quarterly, 64* (1975): 684–711. Franz Schurmann, *Ideology and Organization in Communist China*. Berkeley: University of California Press, 1966.

[33] Audrey Donnithorne, *China's Economic System*. New York: Praeger, 1967; John Gurley, *China's Economy and the Maoist Strategy*. New York: Monthly Review Press, 1976; Carl Riskin, *China's Political Economy: The Quest for Development since 1949*. Oxford University Press, 1987.

[34] Mary Wright helped to build Hoover's collection in the late 1940s, eventually producing the edited volume *China in Revolution: The First Phase, 1900–1913*. New Haven: Yale University Press, 1968; the collection she curated influenced many later studies, including Benjamin Schwartz, *Chinese Communism and the Rise of Mao*, Cambridge, MA: Harvard University Press, 1951. European scholars also relied on university collections, including but not limited to those in the US, to document the revolutionary longue durée in China: René Dumont, *Révolution dans les Campagnes Chinoises*. Paris: 1956; Lucien Bianco, *Les Origines de la Révolution Chinoise, 1915–1949*. Paris: Editions Gallimard, 1967; Lucien Bianco, 'Les paysans et la révolution: Chine 1919–1949', in *Politique Étrangère* 33:2:3 (1968), pp. 117–41; Jean Chesneaux, *Peasant Revolts in China*. London: Thames and Hudson, 1973.

[35] Susanne Weigelin-Schwiedrzik, 'In Search of Modern China: the Development of Sinology in East and West Germany during the Cold War', in *Sinology during the Cold War*, ed. Antonina Luszczykiewicz and Michael C. Brose. London: Routledge, 2022, 43–71.

[36] Thomas Kampen, 'Ostasienwissenschaften in der DDR und in den neuen Bundesländern', in *Wissenschaft und Wiedervereinigung. Asien- und Afrikawissenschaften im Umbruch*, ed. Hagen Krauth and Ralf Wolz. Berlin: Akademie Verlag, 1998, 269–306.

former war reporters like Hino Ashihei, made the journey to the mainland to witness the putative successes of CCP policy, which they described in both popular and academic works.[37] The Tōyō bunko continued to collect Chinese materials, including Japanese celebratory assessments of the 'Great Leap Forward' (GLF); during this period, Japanese intellectuals (and not just on the left) were fascinated by the transformations occurring in 'New China' (*Shin-Chūgoku*), wondering whether Japan should take a similar course.[38] Foreign Sinologists also relied on Chinese-speaking institutions outside of the PRC, including the Bureau of Investigation Archives in Taiwan and the Hong Kong Universities Service Center, to conduct foundational research during the Cold War.[39] Although the political economy was also important to area studies experts, they were more willing to understand the PRC system on its own terms, including the CCP's descriptions of social and cultural revolution.

As access to the PRC became difficult around the time of the GLF, scholarship necessarily turned more towards topics such as party ideology, focusing on the writings of PRC leaders whose work was widely available outside of China.[40] In some cases, the study of Chinese socialism as an ideology coincided with self-described 'Maoist' movements abroad, in South Asia, Latin America, Western Europe, and North America.[41] Still, the focus on intellectual and political leaders frequently froze the study of socialism within the genre of political biography. As Alain Roux commented, the debates over the histories of paramount leaders like Mao Zedong had become 'useless controversies' (*controverse inutile*), trapped within the dialectical tradition of CCP hagiography and anti-CCP 'secret history' (*histoire indiscrète*, 秘史).[42] Other scholars extrapolated from the writings of CCP leadership the

[37] Hino Ashihei, *Akaikuni no tabibito*. Asahi Shinbunsha, 1955. For a scholarly treatment of Hino's account of the early PRC, see Stefano Romagnoli, 'Il "paese rosso" di Hino Ashihei: la Nuova Cina tra l'ombra del passato e la critica del presente', *Riflessioni sul Giappone antico e moderno* (2014), pp. 175–94.

[38] For example, a group of Japanese academic visitors in 1959 declared that reports of a famine were all 'lies'. Nakamura Motoya, Morikawa Hiroki, Seki Tomohide, Ienaga Masayuki, *Chūkaken no sengoshi*. Tokyo daigaku shuppankai, 2022, pp. 44–5.

[39] Kenneth R. Walker, *Planning in Chinese Agriculture: Socialisation and the Private Sector, 1956–1962*. London: Frank Cass and Co. Ltd., 1965; John Israel, *Student Nationalism in Revolutionary China, 1927–1937*. Stanford University Press, 1966; Lyman van Slyke, *Enemies and Friends: United Front in Chinese Communist History*. Stanford University Press, 1968; Mark Selden, *The Yenan Way in Revolutionary China*. Cambridge, MA: Harvard University Press, 1971; Kenneth G. Lieberthal, *Revolution and Tradition in Tientsin, 1949–1952*. Stanford: Stanford University Press, 1980; Elizabeth J. Perry, *Rebels and Revolutionaries in North China, 1845–1945*. Stanford: Stanford University Press, 1980.

[40] Enrica Collotti Pischel, *Le origini ideologiche della rivoluzione Cinese* [ideological Origins of the Chinese Revolution]. Torino: Einaudi, 1959, and *La rivoluzione ininterrotta. Sviluppi interni e prospettive internazionali* [Continuous Revolution: Internal Developments and International Perspectives]. Torino: Einaudi, 1962; Ch'en, *Mao and the Chinese Revolution*; Stuart R. Schram, *The Political Thought of Mao Tse-tung*. Harmondsworth, 1969; Maurice Meisner, *Li Ta-chao and the Origins of Chinese Marxism*. Cambridge, MA: Harvard University Press, 1970; Frederic Wakeman, *History and Will: Philosophical Perspectives of Mao Tse-Tung's Thought*. 1973.

[41] Camille Robcis; '"China in Our Heads": Althusser, Maoism, and Structuralism', *Social Text* 1 March 2012; 30 (1 (110)): 51–69.

[42] Alain Roux, 'Mao, Objet Historique', *Vingtième Siècle: Revue d'histoire* 101 (Jan 2009): 95–108.

theoretical and social experience of Chinese socialism – although without access to the mainland, the authors sometimes had to revise these works after the Reform era.[43] Roughly coinciding with the Vietnam War, the intellectual study of Maoism met with leftist critiques of twentieth-century capitalism, which was an ephemeral opportunity to illuminate both.[44]

After the death of Mao Zedong in 1976, fieldwork in mainland China began en masse again. Visitors to the mainland in the 1970s included scholars such as Paul Pickowicz, Guido Samarani, Anthony Saich, Michael Schoenhals, and many others who would play a major role in writing about the early PRC using historical methodology. Archival work in mainland China was still difficult, even with a formal affiliation inside a local work-unit, but scholars had increasing access to library collections in major cities such as Shanghai and Beijing. Use of published and curated material (vetted by the Party), such as *wenshi ziliao* and broadsheet commentaries on politics and society, informed new approaches to the history of Chinese socialism. In addition to the 'cultural turn' and its focus on art, music, and literature, 1980s' studies of early socialist China included considerations of ethnicity, gender, and science and technology; scholars also revisited official documents and leading intellectuals to highlight the importance of language in the revolution.[45] Reengagement with the mainland, perhaps unsurprisingly, encouraged scholars to ask different kinds of questions, and make a new field: PRC history.

By the 1990s, the impact of China's new openness was increasingly obvious, and a kind of 'golden age' of Chinese research began. The Academies of Social Science (*Shekeyuan*) in Beijing and its other cities facilitated scholarly access to new materials. Historians were now able to conduct projects in prefectural (*xian*), provincial (*sheng*), and municipal archives, including topics such as the history of medicine, reportage, war, gender, sexuality, economy, and art. Working in both Republican and PRC archives, historians quickly noted trans-revolutionary structures, including the pre-1949 roots of institutions that previous scholars assumed were Communist innovations, like self-criticism in life-writing, the state welfare

[43] Mark Selden, *The Political Economy of Chinese Development*. New York: M.E. Sharpe, 1993. For the earlier historiography's more positive evaluation of CCP policies, see Chalmers A. Johnson, *Peasant Nationalism and Communist Power: The Emergence of a Revolutionary China*. Stanford: Stanford University Press, 1962; Selden, *The Yenan Way in Revolutionary China*, and *China in Revolution: The Yenan Way Revisited*.

[44] Lanza, *The End of Concern*, p. 5.

[45] Arif Dirlik, 'Revolutionary Hegemony and the Language of the Revolution: Chinese Socialism between Present and Future', in *Marxism and the Chinese Experience*, eds. Dirlik and Maurice Meisner. New York: ME Sharpe, 1989, p. 38. The focus on language continued into the early 1990s, including David E. Apter and Tony Saich, *Revolutionary Discourse in Mao's Republic*. Cambridge, MA: Harvard University Press, 1994. PRC era ethnicity became a major topic in the 1990s, including work by both American and European scholars, for example: Dru C. Gladney, *Muslim Chinese: Ethnic Nationalism in the People's Republic*. Cambridge, MA: Council on East Asian Studies, Harvard University Press, 1991; Joakim Enwall, *A Myth Becomes Reality: History and Development of the Miao Written Language*, 2 vols. Stockholm: Stockholms Universitiet, 1994.

system, and even the work-unit.[46] At the same time, local Chinese archives began disposing of old files, many of which ended up in paper recycling facilities, online retailers, and antique or junk shops, giving rise to the study of discarded documents, which Michael Schoenhals dubbed 'garbology' (*lajixue*).[47] The expanded access to archival materials, whether in organised state institutions or in markets, mirrored the rapid growth of Chinese-language programmes across Europe, Japan, and North America.

Before the COVID-19 pandemic shut down all access to China, the Chinese government had already begun tightening access to historical collections, including online used booksellers such as Kongfuzi, ostensibly due to issues such as GDPR and fraud. The contributors to this volume were the direct beneficiaries of the new transnational system of historical and Sinological training that exponentially grew from the 1990s. We all fear that restrictions will tighten further in China, repeating the experience of Soviet historians in earlier years. This volume is, in part, a showcase of research done during those years when scholars of modern China had an unprecedented opportunity to study how the PRC was made by artists, scientists, farmers, local cadres, and many more.

Making the early PRC work

This volume explores the notion of 'making' in the early PRC not simply in the context of political economy and infrastructure, but also in the development of new identities, production and dissemination of knowledge, and support for and transformation of arts, design, and religion. Our chapters conclude in the mid-1960s, before the Cultural Revolution launched a widespread struggle over what kind of communism China was supposed to work towards; while it is crucial to see the continuation of revolution (and debates about it) throughout the period from 1949 to 1976, inclusively, it is unfortunately beyond the scope of our project. We keep our focus on the early PRC period, up to the mid-1960s, during which time making, as 'building', 'study', and 'work', were central to how socialist transformation was discussed. In four sections, chapters examine different forms of making, from politics, economy, and society to the self, knowledge, and the material world.

The first section starts by illuminating some of the most visible forms of making, which included the reconstruction of cities, ethnic boundaries, political economy, and agriculture directly before and after the establishment of the PRC in 1949. As

[46] Jan Kiely, *The Compelling Ideal: Thought Reform and the Prison in China, 1901–1956*. New Haven: Yale University Press, 2014; Mark W. Frazier, *The Making of the Chinese Industrial Workplace: State, Revolution, and Labor Management*. Cambridge: Cambridge University Press, 2002; Morris L. Bian, *The Making of the State Enterprise System in Modern China: The Dynamics of Institutional Change*. Cambridge, MA: Harvard University Press, 2005.

[47] Michael Schoenhals, *Spying for the People: Mao's Secret Agents, 1949–1967*. New York: Cambridge University Press, 2013, p. 12.

Toby Lincoln shows, in many respects early Chinese socialism was a continuation of efforts by the KMT government, which also claimed to be led by a revolutionary party, as Chinese cities were rebuilt after World War II (WWII) and the Civil War. Benno Weiner shows how socialist construction went well beyond infrastructure, to the definition of ethnicity. The transition to new agricultural practices faced failures not only in the pastoral borderlands, however, but also in the heartland of China itself, which Jing Wenyu and Felix Wemheuer argue was a product of how the revolutionary legal system criminalised peasant resistance strategies. Still, the political economy of the PRC was not simply dictated from above, as Fabio Lanza reveals in his examination of the urban commune system – it created opportunities, however fleeting, for the creation of new forms of social and economic organisation. 'National construction' (*jianguo*) was a keyword of Chinese modern history, but *How Maoism Was Made* shows how the revolution extended 'capacity-building' to systemic transformation, from material and legal infrastructure to ethnic identity and quotidian economic practices. As others also show in this volume, while the pre-1949 KMT era is essential to understanding the origins of these terms and concepts, the 1950s substantially redefined them in order to serve the process of socialist revolution, sometimes with devastating consequences.

The government after 1949 was not merely content to guide visible behaviours in areas of making; as Liu Shaoqi explained in his pamphlet on 'Self-Cultivation' (*xiuyang*), socialism required revolutionary action, but also changes in thought, linking together theory and practice.[48] As Aaron William Moore shows, PRC citizens worked hard to contribute to the remaking of the self, which is revealed in the diaries of educated people from former KMT strongholds. Rooting out the old ways of thinking, including the oppression of women, the exploitation of farmers, and the pre-eminence of the ownership class, required a massive, nationwide effort; this was a project so revolutionary that it was immediately attacked abroad as 'brainwashing', which Aminda Smith argues was not far from the truth while also desirable to many. Sarah Mellors Rodriguez's chapter reveals how even the most private matters, like sexuality, had to be rethought – albeit with mixed results. The CCP project of reinventing the self only exposes how universal such 'brainwashing' is in the modern era, whether in socialist or capitalist systems – but it also shows how perilous the focus on self-cultivation was, as Liu Shaoqi's pamphlet became a later target of activists during the Cultural Revolution.[49]

While some readers will be familiar with the PRC government's interventions in economy and social discipline, this volume shows how the revolution changed cultural practices, including design, art, and religion. As elsewhere in the capitalist

[48] Liu Shaoqi, 'Lun Gongchandangyuan de xiuyang', July 1939, accessed 25 May 2023, www.marxists.org/chinese/liushaoqi/1967/035.htm. Liu's views on self-cultivation were reprinted as pamphlets after 1949.

[49] 'Chedi pipan daducao *Xiuyang*', Shanghai: Shanghai renmin chubanshe, April 1967, accessed December 2023, https://chineseposters.net/posters/pc-1967-008; Hongqi zazhi bianjibu, '*Xiuyang* de yaohai shi beipan wuchan jieji zhuanzheng', Kunming: Renmin chubanshe, May 1967.

world, Chinese people had been accustomed to thinking of art and design as an elite practice, but the new government placed ordinary farmers and workers as central to the production of everything from public murals to the chairs people sat on in their own homes. Jennifer Altehenger describes how state officials, architects, and woodworkers all attempted to make material environments by standardising measurements for the furniture objects that surrounded people in their daily lives. Christine I. Ho, meanwhile, shows through the example of mass muralism during the GLF how public space became a showcase for art based on the subjectivity and artistic practices as well as content choices of the working masses. Xiaoxuan Wang goes further, to say that the remaking of socioeconomic production created a new platform for the re-emergence of religious expression – while Maoism aimed for the destruction of 'superstition', it paradoxically remade faith in a new form. Cadres did not simply aim to unify cultural expression across China, as our authors show, but to disrupt the systems of exploitation that persisted in cultural spheres from religious authorities to family enterprises in design and the arts; what was (re-)made in the wake of socialist revolution could not be foreseen even by party leaders.

In order to 'seek truth from facts', however, the very basis of capitalist knowledge had to be challenged, in science, engineering, and the humanities. Rather than a facile narrative of epistemological failure, the authors here show how the PRC's educated classes endeavoured to expose the politics of science, technology, and medicine. As Shellen Xiao Wu shows on the example of geographers, this was remade using the expertise of the past – a continuation of China's growing technocratic powers that is attracting more attention from modern historians.[50] Robert Culp then describes how Chinese scholars challenged the hierarchies of expertise as part of the process of making knowledge serve socialist transformation. While the knowledge of the previous government was crucial, and remained part of research in socialist China, new intellectual networks asked, whom does science and engineering serve? Both Wu and Covell Meyskens argue that PRC knowledge was increasingly focused on meeting the changing requirements of socialist economic transformation, especially in the area of energy production. A change in priorities for science, technology, and medicine also meant a change in the way research and applied fields were conducted, which consequently produced new forms of knowledge – sometimes with perilous results, but also in ways that benefitted working people more than ever before. In so doing, knowledge production reified the notion of science for the people, which provided essential legitimacy for the government.

All these transformations in political economy, culture, and knowledge are hardly understandable without analysing the PRC's international and transnational relationships; Chinese socialism defined itself against governments and

[50] Matsumoto Toshiro, *'Manshūkoku kara shin-Chūgoku e: Anshan tekkōgyō kara mita Chūgoku tōhoku no saihen katei*. Nagoya: Nagoya daigaku shuppansha, 2000. Victor Seow, *Carbon Technocracy: Energy Regimes in Modern East Asia*. Chicago: University of Chicago Press, 2021.

systems outside of its borders. Mary Augusta Brazelton continues the examination of Chinese scientific knowledge in the 1950s, showing how bacteriological research was inextricably bound up with international struggles of war crimes and new notions of expertise, scientific truth, and the origins from which such scientific truth might be derived. Science, as Brazelton writes, 'was incorporated into the making of a new socialist order'. Nicolai Volland captures the longer pre-1949 history of transnational socialist publishing. He demonstrates how integral information about the Soviet Union was to domestic Chinese understandings of revolution after liberation. Yet such information did not simply travel from the Soviet Union to China; it involved publishers, editors, translators, authors, and many others, and it involved decisions about which bits of knowledge about and from the Soviet Union were beneficial to the Chinese socialist project. Finally, Emily Wilcox reminds us that, even though Chinese leadership was often focused on looking to the Eastern European and Soviet world, the immediate regional context, in this case Korea and Japan, remained vital to the everyday process of making socialist China. She takes as her example Japanese and Korean dance teachers, illustrating how, for a brief but crucial moment in the early 1950s in the crucible of a single dance academy, they helped to lay the groundwork for what would become Maoist China's performance and arts culture.

The chapters in this volume are an example of our current moment in scholarship on the early decades of the PRC, produced during a period of relative archival openness in China. Historians of the Soviet Union experienced similar vicissitudes. Two experts on Soviet history, Jochen Hellbeck and Juliane Fürst, reflect on whether we can now speak of revolutionary 'life cycles' in the socialist world, comparing the 'making' efforts in both Russia and China. From the early personal accounts of the Chinese revolution to the current, archivally driven histories featured in this volume, the study of socialist transformation has gone from a preoccupation with political economy and ideological formation, to a focus on grassroots 'making' and the interplay of the grassroots with the 'making' of statecraft. *How Maoism Was Made* offers a particular way of writing the history of this period, having benefitted from the period of archival openness, and it will be up to the reader to decide whether we have produced a new form of understanding Chinese socialism.

Part I

Society and Political Economy

1

Making a 'New Changsha': Reconstructing China's Devastated City, 1945–1959

TOBY LINCOLN

IN 1959, CAO Ying, the secretary of Changsha's municipal committee, wrote the introduction to a history of the city commemorating the 10 years since its takeover.[1] On entering the city Communist cadres found, 'a scene of devastation, factories and shops had closed down … prices were rising sharply, many people were out of work, criminals were running amok, people could not earn a living, and many workers were on the breadline'. Cao argued that this was the result of 'imperialism, feudalism, bureaucratic capitalism … over 20 years of bloody control by the Kuomintang (KMT) reactionary clique, and continuous destruction from the ravages of war'.[2] Writing in a similar vein, the new mayor of Changsha, Yan Zixiang, declared in 1949, 'Speaking of the problem of construction, we should first note past urban construction under the reactionaries. They really did not care about the fate of the nation, and completely sold out to the imperialists'. Yet, Yan acknowledged that the KMT, 'repaired some government buildings and public housing'.[3] Although Cao and Yan recognised that the KMT overcame obstacles to reconstruct Changsha, they also expressed widely held negative views about the Nationalist Government.

[1] This research was partly funded by an Arts and Humanities Research Council leadership fellowship entitled 'Post-war Urban Reconstruction in China, 1937–1958'. I would like to thank Aaron Moore and Jennifer Altehenger for their careful reading of earlier drafts of this manuscript and assistance throughout the process of putting this volume together. Anonymous reviewers and other contributors to the volume also made valuable comments. I would also like to thank Helena Lopes for organizing a talk at the University of Bristol at which I first presented this research. I bear sole responsibility for any mistakes that remain.

[2] Cao Ying, 'Zai shengli de daolu shang jixu qianjing - qingzhu Changsha jiefang shizhounian', p. 4, in *Changsha jiefang shinian 1949–1959*, (1959).

[3] 'Yan Zixiang shizhang shizheng gongzuo baogao', *Changsha shizheng*, 1 (December 1949), 7.

From such inauspicious beginnings, Changsha recovered quickly. By July 1950, new businesses were opening and many unemployed were given aid in the form of work relief, which was food for labour on municipal construction projects. The following year, drinking water was available on tap for some residents for the first time in the city's history, the shore of the Xiang River that had previously been a slum was now a wide road, and the new municipal authorities had constructed parks. By 1952, the economic recovery was declared complete with total agricultural and industrial production nearly four times as high as in 1949, and like others across Maoist China the city was being transformed into a socialist production city – a 'New Changsha'.[4]

In this chapter, I explore how the Nationalist and Communist governments made a 'New Changsha', after the end of World War II (WWII). The language Chinese Communist Party (CCP) cadres used to describe the city in the first months after its takeover helped to consolidate the party's grip on power, and as time went on the reconstruction of the urban built environment came to represent the process of creating revolutionary socialism. However, this language also dismissed KMT efforts at reconstruction. Scholars have discussed how the CCP denigrated KMT governance and saw Nationalist spies, real and imagined, everywhere, even within the party itself. They, along with criminals, beggars, prostitutes, and drug addicts were either parts of the old pre-liberation society that was responsible for all the ills of China, or their plight was created by the evils of that old society.[5] Denise Ho describes how Fangua Lane in Shanghai, a street that was a slum before 1949, was transformed into a model district in the early 1960s. It was presented as an example of *fanshen*, standing up, in which Chinese people had transformed the old society. Indeed, it was a microcosm of New China, a promise of the socialist urban future that the party wanted to create for city dwellers across the country.[6] Changsha was not a model city, but the narrative CCP cadres wove around urban reconstruction in the months after 1949 drew a similar distinction between the old pre-revolutionary society and the new one the party was trying to create. One of the effects of this discourse about the KMT's failure to solve the problems in Changsha and other cities in China is that the impact of long years of war with the Japanese has remained largely unexamined. In short, the KMT has been unfairly maligned precisely because its reconstruction work has been ignored.

Changsha is a particularly good place to study post-war reconstruction in China. The site of one major fire, which the Nationalist Government started in 1938 as part of its scorched earth policy to try to delay the Japanese advance, four battles, and

[4] *Changsha jiefang shinian*, pp. 3–4, 9–11.

[5] Michael Dutton, *Policing Chinese Politics A History*. Durham: Duke University Press, 2005, pp. 13, 140–1; Aminda Smith, *Thought Reform and China's Dangerous Classes: Re-education, Resistance, and the People*. Lanham: Rowman and Littlefield, 2012, pp. 56–9.

[6] Denise Ho, *Curating Revolution Politics on Display in Mao's China*. Cambridge: Cambridge University Press, 2018, pp. 2, 91, 93.

a period of Japanese occupation lasting more than one year, Changsha may well have been China's most devastated city. George Sayles, a correspondent for the China Press, an American newspaper based in Shanghai, described what he saw in October 1945.

> The ruins stretch for block after block. Here and there are clusters of wooden houses which have been hastily constructed to provide shelter for the people gradually working their way back into the city. In some few streets, the original stone and brick buildings are comparatively undamaged, and in these streets you see the normal bustle of Chinese life. But in general, Changsha is a broken city, and one that will take years to rebuild.[7]

At first, the task of reconstructing this broken city fell to KMT officials and the people of Changsha. Historians now take a more positive view of Nationalist state-making and have investigated how in some ways the KMT state was a forerunner of Maoist China. However, even much of this later work does not discuss KMT reconstruction of Chinese cities, choosing instead to focus on poor governance, corruption, and of course the CCP victory in the Civil War.[8] Historians of the rural communist revolution in China have pointed to regional differences in base areas across the country, some of which were behind Japanese lines. They have studied how nationalism, socioeconomic policies, effective propaganda, and brute force were all part of the revolutionary process.[9] In doing so, they have drawn attention to the importance of locality and contingency in the Chinese Communist revolution and the later construction of the Maoist state. As Joseph Esherick argues in his recent study of the history of Yenan, 'the agency and political choices of individual and group actors were critical in determining the particular events whose cumulative effect was the revolutionary process'.[10]

By contrast, historians have paid less attention to how the urban Chinese Communist revolution was also a process of contingent events that includes a thorough assessment of the period after the end of the war with the Japanese. There are

[7] George Sayles, 'Correspondent Reports on 900 Mile Trip Across China', *The China Press*, 13 October 1945.

[8] Jeremy Brown and Paul G. Pickowicz, 'The Early Years of the People's Republic of China: An Introduction', in *Dilemmas of Victory the Early years of the People's Republic of China*, eds. Jeremy Brown and Paul G. Pickowicz. Cambridge, MA: Harvard University Press, 2007, pp. 1–18. See especially chapters two, four, seven, and fifteen in this volume for studies that argue for continuities across the 1949 divide. On CCP underground activity and the Civil War see: Joseph K.S. Yick, *Making Revolution in China the CCP-GMD Struggle for Beiping-Tianjin 1945–1949*. Armonk: M.E. Sharpe, 1995, pp. 48–51, 61–5, 71, 73; Odd Arne Westad, *Decisive Encounters the Chinese Civil War, 1946–1950*. Stanford: Stanford University Press, 2003, pp. 8–9, 10–13, 70–9, 86–9.

[9] Suzanne Pepper, 'The Political Odyssey of an Intellectual Construct: Peasant Nationalism and the Study of China's Revolutionary History: A Review Essay', *Journal of Asian Studies* Vol. 63 (1) (February 2004), 105–25; Joseph W. Esherick, *Accidental Holy Land The Communist Revolution in Northwest China*. Berkeley: University of California Press, 2022, pp. xii–xiv.

[10] Esherick, *Accidental Holy Land*, p. xxiv.

some exceptions, particularly in the study of labour history and social welfare.[11] Turning to cities, Mark Baker describes how Zhengzhou's importance as a railway hub underlay its commercial revival after 1945. It was only later that problems dealing with refugees, the militarisation of the city, and most importantly the taxes imposed on the local population and the protests that followed reveal a city struggling to cope with the twin demands of rebuilding itself after one war, while at the same time fighting another. Meanwhile, in northeastern Anshan, although the CCP called for a complete redesign of this city, which they argued had been built primarily for the Japanese, Victory Square, a physical manifestation of Communist power, was an expansion of work undertaken during the colonial era. Meanwhile, a pre-war elite housing district known as Taiding was occupied by high-level cadres and managers of the Anshan Steel Works, just as it had been used by Japanese managers of those same steel works before 1949.[12] All of these studies reveal different post-war histories of Chinese cities, demonstrating how localised stories of urban revolution in China challenge the simplistic and erroneous narrative of post-revolutionary socialist transformation that ignores reconstruction before 1949.

In Changsha, KMT officials and urban residents talked about wartime destruction all the time, and the latter used it to elicit sympathy for any difficulties they encountered. In fact, both the KMT and the CCP wanted to make a 'New Changsha' – one that was rebuilt along modern urban design principles and incorporated the latest developments in urban infrastructure. However, in seeking to create a socialist new Changsha, the CCP largely blamed the city's ills on the Nationalists, and in doing so their predecessors' part in urban reconstruction was forgotten. Later, during the transition to socialism under the first Five-Year Plan, cadres criticised the chaotic reconstruction of Changsha in the immediate aftermath of liberation. Their reframing of the city's immediate past continued as the party continually remade Maoist cities.

Reconstruction under the KMT

In February 1946, United Nations Relief and Rehabilitation Administration (UNRRA) personnel in Changsha found a 'city in ruins', with a population estimated to be around 400,000, and refugees from the surrounding countryside

[11] Robert Cliver, Nara Dillon, and Mark Frazier among others have all drawn attention to how developments in the late 1940s laid the foundation for welfare provision in socialist work-units after 1949. Nara Dillon, *Radical Inequalities China's Revolutionary Welfare State in Comparative Perspective*. Cambridge, MA: Harvard University Asia Center, 2015, pp. 78–154; Mark Frazier, *The Making of the Chinese Industrial Workplace: State, Revolution, and Labor Management*. Cambridge: Cambridge University Press, 2002, pp. 67–85; Robert Cliver, 'Second-class Workers: Gender Industry, and Locality in Workers' Welfare Provision in Revolutionary China', in *The Habitable City in China Urban History in the Twentieth Century*, eds. Toby Lincoln and Xu Tao. New York: Palgrave Macmillan, 2017, pp. 113–32.

[12] Mark Baker, 'Civil War on the Central Plains: Mobilization, Militarization, and the End of Nationalist Rule in Zhengzhou, 1947–1948', *Twentieth-Century China*, Vol. 45 (3) (October 2020), 266–84; Christian Hess, 'Sino-Soviet City: Dalian between Socialist Worlds, 1945–1955', *Journal of Urban History*, Vol. 44 (1) (2018), 9–25; Koji Hirata, 'Steel Metropolis: Industrial Manchuria and the Making of Chinese Socialism, 1916–1964', Ph.D. thesis, pp. 255–65.

desperate for work. Rivers and wells were contaminated, there was no sewage disposal, and human waste was simply dumped at the edge of the city.[13] A KMT survey found that 36,460 people out of a pre-war population of around 700,000 died during the war, with a further 56,536 injured. 97,283 buildings were destroyed and of those left standing nearly all were damaged. In China, only Guilin, Hengyang, and Yichang had suffered anything like as much destruction.[14]

Rubble lay everywhere in Changsha. Officials announced that every inhabitant should provide a day's labour to help clean up or pay for someone else to take their place. Fifteen trucks and other tools were provided, work began on 1 June 1946, and three weeks later the municipal government estimated that 88,199 cubic metres of rubble had been cleared.[15] Work continued, but in 1948 the police estimated that there was a staggering three million tonnes of rubble still in the city with 10 tonnes being added per day. A team of 166 people was able to clear 5 tonnes per day, and so 30 trucks were provided to help them try to keep pace with the growing mountain of rubbish.[16] In this way, the ruins of the city were at least partially cleared through a mass mobilisation campaign, well before the CCP took control of the city.

Cleanup was a precursor to reconstruction, and much of this was described in *One Year of Municipal Government in Changsha*, published in 1947. Like hundreds of similar compilations published throughout Republican China, this document was intended for government officials and other Changsha residents involved in the city's governance, and was divided into sections including administration, construction, education, finance, and health and hygiene. In the introduction, KMT officials emphasised the impact of the war, noting that:

> the price of goods fluctuated, those with evil intentions seized their opportunities to incite unrest, labour disputes and the like ... the people of the city suddenly faced disorder, cold, malaria, and other diseases. Disbanded soldiers, refugees, and children filled the streets, and everyone in the city could hear their cries of hunger.[17]

Faced with such a difficult situation, municipal officials had their work cut out. Much of their actions are perhaps better described as urban management, and the distinction between that and reconstruction is important to note. Chinese scholars of cities in the early Maoist period have for the most part focused on urban management and the construction of the new political regime, the latter denoted by the term

[13] 'Hunan Regional Office Monthly Report No. 1 February 13–28, 1946', United Nations Archives (hereafter UNA), S 1121 0000 0221, Regional Office Hunan (Changsha).

[14] 'Gengniu qijuyifu ji qita sunshi gujia xi an touxiang shizhi', Academia Historica, 126 1464–1; 'Zhanhou ge chengshi yiban gaikuang diaochabiao', *Gonggong gongcheng zhuangkuang* 2 (1947), 79–81.

[15] 'Hunan weekly report, 1 June 1946'; 'Hunan weekly report, 22 June 1946', UNA S 1121 0000 0221 Regional Office Hunan (Changsha).

[16] 'Changsha', *Jianshe Pinglun*, Vol. 1, no. 7 (1948), 39.

[17] 'Xuyan', in Wang Jietou, *Yinian lai zhi Changsha shizheng* (June 1947), Changsha Municipal Archives (hereafter CMA) 3-1-63.

jianzheng, which refers to the creation of grassroots party and security structures.[18] The repair and reconstruction of the built environment of Changsha is my focus here. Still, the KMT did far more than simply work on urban infrastructure, and this supports the thesis that we should take their efforts in governing China after the war seriously. Local governance was reorganised and political training provided, although municipal officials complained that this was inadequate and would not prevent corruption. Income from taxation increased by four times between May 1946 and May 1947, and although rising prices affected government expenditure, money spent on education also increased by four times over the same period. As the economy began to recover, new businesses opened up, and by the summer of 1947, the city had registered 188 enterprises.[19] Urban transport improved, and the newly established Kaimin Bus Company found itself in dispute with the Rickshaw Pullers Union after it sought to add a second route in cooperation with some rickshaw pullers. In response pullers damaged the buses, and the dispute was only resolved after the provincial government took away the licence for the second route.[20]

Changsha's infrastructure was in a terrible state. Between May 1946 and June 1947, 22,000 people worked to clear out over 30km of drains, which helped to reduce flooding. Over the same period, several kilometres of roads were repaired. The Chinese National Relief and Rehabilitation Administration (CNRRA), the sister organisation to the UNRRA, provided 150 tonnes of flour and 50 tonnes of canned food for the workers, which points to the extent of the work relief programme the government was using to feed some of the city's population. Officials also gave some thought to Changsha's green spaces. A Park Management Committee was established and, by the end of the summer in 1947, 3,750 seedlings had been planted in Tianxin Park, the city's only pre-war park, and over 500 square metres of grass laid. Zhongshan Park on the site of the old army headquarters was larger, with nearly 6,000 square metres of grass.[21] As in other cities across China, and indeed many of those abroad, wartime destruction provided planners an opportunity to rethink the use of urban space.

The most controversial infrastructure project in Changsha was the widening of the road along the Xiang River that ran north to south along the western edge of the city. It demonstrates how the KMT and CCP had similar aims in urban reconstruction, but somewhat different governmental capacities and styles. The KMT was less effective but there were legal mechanisms people could use to register their dissatisfaction with officials and their policies. The CCP was more effective, but

[18] Li Guofeng, *Chujin da chengshi – Zhonggong zai Shijiazhuang jianzheng yu guanli de changshi (1947–1949)*. Beijing: Shehui kexue wenxian chubanshe, 2008, p. 6. See also James Z. Gao, *The Communist Takeover of Hangzhou the Transformation of City and Cadre 1949–1954*. Honolulu: University of Hawai'i Press, 2004, pp. 74–96, 137–40.

[19] Wang, *Yinian lai zhi Changsha*, pp. 1–7, 11, 18.

[20] 'Hunan weekly report 6 July 1946', UNA S 1121 0000 0221 Regional Office Hunan (Changsha); Wang, *Yinian lai zhi Changsha*, p. 17.

[21] Wang, *Yinian lai zhi Changsha*, pp. 15–16.

Changsha urbanites had less room for protest. In the summer of 1946, the KMT municipal government published its plan for land requisition to allow for demolition of straw shacks and other buildings. After the Great Fire of Changsha in 1938, the New Changsha Urban Area Construction Plan had been submitted to the Executive Yuan for approval, and now that the war was over, this could be implemented. The land required along the riverbank stretched 4km from Beida Road in the north to Xihu Road in the south and was 60m wide. Fewer than half of the buildings had survived the war intact and both former inhabitants of Changsha and refugees from the countryside now lived in ramshackle houses and straw shacks. Their demolition would make way for new docks, warehouses, and company offices on the riverbank, a park running down the middle of the highway, and tree-lined pavements along the sides of the road. Officials argued that both national and provincial regulations on urban reconstruction gave them power to requisition land, and they planned for compensation to be paid to landowners. Wartime disruption meant it was difficult for officials to find many of the landowners, who were able to apply for compensation for their demolished homes and money to help with the move. Those who were not landowners, but were affected by the scheme, were not entitled to compensation so the government offered to help them to apply for a reconstruction loan from the Agricultural Bank of China.[22]

The announcement of the plan to requisition the land provoked a chorus of protest from residents and business owners living along the river, who either wanted compensation or the work abandoned altogether. One of the affected companies was Butterfield and Swire, which had apparently not been consulted about the plans or offered compensation for the proposed destruction of two godowns.[23] Meanwhile, Chinese complaints explicitly referenced Changsha's wartime destruction. A letter to the mayor, Wang Hao, from a group of business owners and workers is representative. The writers described how they and other residents had fled Changsha when the Japanese military invaded and had returned to find their home a desolate wasteland. They built shacks out of bamboo, which were easily destroyed by fire or flood. Then, it was announced that on 25 September they would have to move again, which only added to the misery already caused by the war. They lacked funds to rebuild a second time and asked the government for help.[24] Zhou Yunsen represented another group of businessmen along the river. He argued that Changsha was not as prosperous as Beijing, Shanghai, Tianjin, or Wuhan, where roads were only 40m wide, and so questioned the need for a 60m wide road. There were 100,000 people living along the road, many of whom had suffered during the war, and with winter coming demolition of their homes now would only add to their misery. Finally, Zhou requested that people who had been forced to flee their homes

[22] 'Changsha shizhengfu xiuzhu yanjiang dadao ji yingjian dongjiao xinqu zhengshou tudi jihua shu', CMA, 6-1-92.

[23] 'British Embassy, Nanking. 9 July, 1946', CMA, 6-1-90.

[24] 'Baogao, sanshiwunian jiuyue ershiri yu Chaozongmen wai shang he bian', (20 September 1946), CMA, 6-1-90.

during the war should receive relocation money and compensation according to the value of their land. In a second letter, Zhou took a similar line, but added that the Land Law stipulated the government had to provide 30 days' notice of any land requisition and set up a Land Price Appraisal Committee to work out how much compensation should be given to landowners.[25] The appeals to the government to halt the road widening scheme fell on deaf ears, and work began on 25 September.

Funding came from the Hunan Provincial Government, which initially provided 100 million yuan, with a further 200 million loan in February 1947. This was insufficient as the cost rose from an original estimate of 5.5 billion to 6.4 billion yuan, an indication of how hyperinflation hampered reconstruction efforts. By summer 1947, only 30 per cent of the work had been completed.[26] There is little information on what happened to those whom municipal authorities evicted, but a few families were lucky enough to find new homes in social housing built with the assistance of the CNRRA. The 54-unit 'village' of two, three, and four roomed buildings housed 133 families, who paid a reduced rent of 3,000 yuan a month to the municipal government.[27] Complaints continued after the demolition and relocation work had started. In November 1946, residents and business owners wrote to Chiang Kai-shek, the heads of the Ministry of Interior and the Executive Yuan, and the head of the Hunan Provincial Government among others. They repeated previous complaints about the width of the road and the pace of demolition work, and concluded that, with 150,000 homeless in the city, Changsha needed all the housing it could get, even if it was substandard. Fifty-eight Hunan delegates to the National Assembly made similar arguments, an indication of how the issue was receiving national attention.[28] In December, businessman Yin Jiesheng wrote to the Executive Yuan claiming that officials illegally used force to demolish buildings.[29]

By now, Changsha's local conflicts between municipal officials, businessmen, and local residents had come to the attention of the highest levels of the KMT state apparatus. The Executive Yuan responded to complaints with phone calls to the Hunan Provincial Government, and Ha Xiongwen, the head of the Construction Office, which was the government department responsible for post-war reconstruction, was sent to Changsha to investigate.[30] As a result, in March 1947 the

[25] 'Changsha yanhejie gongshang jumin daibiao Zhou Yunsen deng', (26 September 1946); 'Changsha yanhejie gongshang jumin daibiao Zhou Yunsen deng', (26 September 1946), CMA, 6-1-90.

[26] Wang, *Yinian lai zhi Changsha*, p. 14.

[27] 'Changshi diyi shanchiu xincun ye yi jianzhu wanjun', *Shanchiu yuekan*, 28 (1947), 14; 'Hejian zhuzhai zujin fangzu pingmin', *Shanchiu yuekan*, 28 (1947), 14.

[28] 'Changsha shi yanjiang gongshangjie quanti jumin daidian', (26 November 1947), Academia Historica, 126 1464–2; 'Changsha shi yanjiang gongshangjie quanti jumin shiyuwan ren xiang dangzheng dangju ji shehui renshi qingyuan zhi gongtong yijian', (26 November 1947), Academia Historica, 126 1464–2; 'Zhaochao Xiangji guoda daibiao lintong Xiangshengshi zhengfu yuan tong quanwen', (December 1946), Academia Historica, 126 1464–1.

[29] 'Juchengren Changsha shi yanjiang gongshang jumin qingyuan daibiao Yin Jiesheng', (December 1946), Academia Historica, 126 1464–1.

[30] 'Xingzheng yuan mishu chu gonghan', (23 January 1947), Academia Historica, 126 1464–1.

Executive Yuan ordered the Changsha Municipal Government to limit the width of the road to 40m. In May, licences were issued to those who wanted to build on the land that was now vacant between 40 and 60m and, by August, houses, shops, and warehouses were already under construction. Then, the city government announced that while the width of the southern section of the road would be limited to 40m, to the north the road foundations had already been laid to a width of 60m. Officials revoked construction licences and demolished some of the buildings.[31] However, in December, lack of funds forced the municipality to come up with a new plan for the road, which would be 40m wide in total, with a 20m highway of crushed stone, and tree-lined pavements alongside.[32] In the meantime, representatives from Changsha had travelled to Nanjing to ask for help in suing the municipal government.[33]

There is no information on whether the lawsuit ever made it to trial, and it is not clear whether the chaos surrounding land requisition, demolition, and relocation was simply bad management or corruption. There was certainly plenty of the latter in Changsha and Nationalist-run cities across the country in the immediate post-war years. Early in 1946, Horatio Hawkins, the acting regional director of the UNRRA in Hunan, reported that there 'seems to be widespread impressions of the wealth and gullibility of UNRRA-CNRRA, and speculators and promoters are attempting to take advantage of this, through projects for sales of land, contracts for buildings, and sales, storage, and loans of relief supplies'.[34] While it was more likely a combination of poor management and bad actors, the people of Changsha made it very clear in their complaints that they felt corruption was at the root of the problem.

The KMT post-war record in Changsha is certainly mixed. Governance was sorely lacking while the Civil War created problems such as inflation, which hampered reconstruction. However, some infrastructure was rebuilt, and officials presided over a partial recovery of the city. Moreover, as letters about the demolition of the road along the river reveal, urban residents recognised the problems that wartime destruction created, and were prepared to work with local officials on reconstruction. This meant that in Changsha at least, while reconstruction was not finished in 1949, CCP cadres could build on the foundation the Nationalists had made.

Reconstruction under the Communist Government

The People's Liberation Army (PLA) occupied Changsha peacefully on 6 August 1949, because Nationalist forces surrendered to the Communists. This peaceful takeover was like that of Hangzhou, but unlike cities such as Changchun and

[31] 'Juchengren Changsha yanjiang gongshangjie qingyuan daibiao Yin Jiesheng', (2 August 1947), Academia Historica, 126 1464–2.
[32] 'Changshashi zhengfu xiujian yanjiang dadao jihua gaiyao', (26 December 1947), Academia Historica, 126 1464–2.
[33] 'Changsha shi yanjiang gongshangjie quanti jumin daidian'.
[34] 'Horatio Hawkins to Col. Ralph W. Olmsted, Director of Operations, UNRRA – China Office, Shanghai, 28 March 1946', UNA S 1121 0000 0221 Regional office Hunan (Changsha).

Xuzhou, which saw significant fighting during the Civil War. Takeover was the responsibility of the army, which quickly set up military control committees in cities across China. In Hangzhou, for example, the military control committee established 16 bureaus to manage the city, presided over the takeover of municipal government and factories, organised study sessions for managers, and assisted with relief for refugees and the poor.[35] Its work in Changsha was made easier as the city remained relatively peaceful. A report to Mao Zedong at the end of August claimed that, in the two weeks since 'liberation', there had only been four cases of small-arms fire in the city and some theft from local factories. The railway was already repaired, and the power station was up and running. The military control committee had established a bank and a trading company, and was collecting taxes in Renminbi (RMB), which had already largely replaced the currency used under the Nationalists.[36] However, as the leaves fell from the trees that autumn, spies and enemy agents were apparently still at large, KMT soldiers stole from factories and businesses, and the CCP-appointed police chief was assassinated.[37]

By the end of 1949, the CCP had made more progress taking over the city. Of the 2,041 officials working in the old provincial government, half stayed in Changsha, with many remaining in post. Many businesses, utilities such as the water management company, and private banks were all now in Communist hands, and together they employed 2,972 managers and 7,501 workers. Banks were providing loans, and the trading company set up by the military control committee was assisting with sales of textiles. Transport organisations employed 15,000 people, and the post and telegraph offices were operating once more. The police registered some 200 'KMT agents', but more apparently remained in the city.[38] Across China, the police estimated there were over one million KMT spies and diehard supporters, two million bandits, and millions more vagrants, beggars, and other people whom the new regime felt were undesirable.[39] They, like the poverty-stricken slum dwellers of Fangua Lane in Shanghai, were all lumped together as remnants of the old pre-liberation society, and were either deemed a threat to the new regime or in need of political education. However, Beijingers remembered beggars as street artists rather than a threat to social order, a further reminder that post-1949 generalisations about the old society were rarely so simple.[40]

Neither the work that cadres undertook in Changsha nor the discourse of urban revolutionary transformation that they weaved around it was peculiar to the city.

[35] Gao, *Hangzhou*, pp. 80–92.

[36] 'Wang Shoudao gei Mao zhuxi, Binglin, Deng guanyu jinru Changsha hou guancha qingkuang de baogao, 21 August 1949', in *Hunan heping jiefang jieguan jianzheng shiliao*, eds. Hunan sheng danganguan. Changsha: Hunan renmin chubanshe, 2009, Vol. 1, p. 382.

[37] 'Zhonggong Hunan shengwei guanyu Changsha ji qi fujin shu xian heping jieguan jingyan jiben zongjie, 27 October 1947', in *Hunan heping jiefang jieguan jianzheng shiliao*, eds. Hunan sheng danganguan, Vol. 2, pp. 631–3.

[38] 'Wang Shoudao gei Mao zhuxi, Binglin, Deng guanyu jinru Changsha'.

[39] Dutton, *Policing Chinese Politics*, pp. 149–51.

[40] Smith, *Thought Reform*, p. 63.

As elsewhere, Changsha cadres established grassroots party organisations, began the task of transforming labour-management relations, improved security, implemented new curricula in schools and universities, and asserted control over the press.[41] Moreover, repair of urban infrastructure, whether the damage was the result of Japanese invasion, fighting during the Civil War, or simply neglect, was happening elsewhere. By 1952, across the country over 1,000km of drainage pipes were repaired and some 20 million tonnes of rubble were cleared. In Nanjing, 20,000 workers cleaned drains, improved flood defences, and constructed housing. In Chengdu, 90,000 people worked to dredge 50km of rivers and canals, clean 80km of drains, and remove 20,000 tonnes of rubbish from the streets.[42]

However, few other cities in China had suffered from as much destruction as Changsha. Rubble still lay in piles across the city, and in September 1949 2,400 tonnes was removed from the streets.[43] Work continued in 1950. Writing in the journal *Changsha Municipal Government*, cadres in the Construction and Health departments castigated reactionary KMT officials for their poor urban management, but admitted that four battles and one year of Japanese occupation meant that 'everywhere was dilapidated, decayed, and dirty ... From the perspective of construction, there were few modern buildings, 80 per cent of houses were wooden, the streets were winding, the road surfaces bumpy, and there were no streetlights or equipment for running water and sewage'.[44] Floods and disease since liberation added to Changsha's woes but had not stopped infrastructure repairs. Municipal officials issued 285 construction licences to registered building companies. Workers, many probably employed through work relief schemes, cleared nearly 20km of drains, removing 58,606 cubic metres of waste, which consequently reduced flooding in the city. They put up 1,285 streetlights, planted over 2,500 trees alongside roads and nearly 30,000 in parks and other scenic areas, cleaned some of the wells supplying drinking water from the river, and repaired some of the 198 public toilets in the city.[45]

Changsha Municipal Government was published monthly throughout the early 1950s, and consisted of government reports, speeches, records of meetings, and other information on city affairs. It was classified for internal party use only, and as such its audience was limited to government cadres. Information was factually correct, and as with all party publications however large their intended audience, writers knew it was important to hold to whatever the party line was at the time. Like the archival records below that I use to describe the continuing work to widen the road along the Xiang River, the party line in the months after takeover reinforced

[41] 'Yan Zixiang shizhang shizheng gongzuo baogao', pp. 5–6; For a description of similar work in Shanghai see: Wakeman, *Cleanup*, pp. 42–52.

[42] Cao Hongtao, Chu Chuanheng (eds.), *Dangdai Zhongguo de chengshi jianshe*. Beijing: Zhongguo shehui kexue chubanshe, 1990, pp. 26–8.

[43] 'Yan Zixiang shizhang shizheng gongzuo baogao', p. 7.

[44] Jianshe ju, Weisheng ju, 'Yinian lai de shizheng jianshe', *Changsha shizheng*, 3 (April 1950), 39.

[45] Jianshe ju, Weisheng ju, 'Yinian lai de shizheng jianshe', pp. 39–41.

distinctions between the new and old society.[46] Officials certainly emphasised Communist achievements in a ceremony that took place on 2 October 1951 to celebrate five major urban construction projects. These included the provision of running water and the widening of the road along the Xiang River. In his speech, the deputy mayor of Changsha said that the achievements of municipal construction were, 'an initial victory that the whole of the city's population have supported'.[47] Wang Shoudao, the provincial head, warned that projects should not be too ambitious, but noted that they would be of great benefit to the people, who would work hard to overcome any problems. Cadres wanted everyone in Changsha to feel a sense of ownership of reconstruction, and thereby a stake in the new urban society that they were building.

The provision of running water was a project the CCP inherited from the KMT. In September 1946, the Running Water Company Planning Committee was formed. Two machines for drawing water were ordered from the CNRRA in February 1947, and were finally delivered in June.[48] However, rising raw material costs and the Civil War put a stop to the work. In 1950, the Running Water Project Plan was published. Water was to be drawn from the Xiang River south of Changsha, treated, and then pumped through pipes into the city. By January 1950, a 21m high water tower had been constructed with two concrete pipes out into the river through which two pumps drew in the water. Outside the tower, factory housing was complete, and a reservoir was under construction. There were as yet no steel pipes to bring water into the city, the motors purchased under the KMT were not powerful enough to pump the required amount of water, and many raw materials were still hard to come by.[49] The report by the Changsha Running Water Company published for the ceremony celebrating five municipal construction projects picked up the story nearly two years later. Obfuscating the eight destructive years of total war with the Japanese, its authors stated that under the reactionary KMT there had been loud calls for running water for 20 years, but that it had not been possible to undertake the project. In March 1950, engineers concluded that the work conducted before 1949 was of poor quality and so a new plan was drawn up. Cost issues continued to hamper the project, and by November 1951, although enough pipes had been laid to bring water to the city, which helped with firefighting, it was not yet suitable for drinking. Cadres claimed that the city's unemployed had unbridled enthusiasm for the provision of running water and had apparently worked non-stop for two days and two nights to lay the pipes.[50]

[46] For discussions of these types of sources see Michael Schoenhals, 'Elite Information in China', *Problems of Communism*, 34 (September–October 1985), 67.

[47] 'Wuxiang shizheng jianshe zaocheng shifu juxing qinghuhui', *Changsha shizheng*, 10 (January 1951), 39.

[48] Wang, Yinian lai zhi Changsha, pp. 8, 16.

[49] 'Changshashi zilaishui linshi gongcheng chu', *Changsha shizheng*, 2 (January 1950), 41–5.

[50] Changsha zilaishui gongsi, 'Changsha zilaishui diyiqi diyi jieduan gongcheng de jiandan jieshao', *Changsha shizheng*, 10 (January 1951), 35.

Another problem the CCP inherited from the previous government was the unfinished road along the Xiang River. In 1950 plans were drawn up to demolish the remaining buildings and relocate their inhabitants. 3,028 families, comprising 10,395 people eked out a meagre existence along the river. The law abiding were peddlers, water carriers, cart pullers, and engaged in diverse forms of petty commerce. The rest were pickpockets, vagrants, beggars, and prostitutes. Cadres described how the work would benefit Changsha in much the same terms as Nationalist officials. They claimed the project would bring 'long term benefit to all the people in the city and is the prerequisite for the construction of a new Changsha'.[51] They also argued that the straw shacks blocked transport, were a blight on the city's appearance, and posed both a fire and health hazard. Given that work along the road had already begun, it made sense to continue to demolish shacks south of Xiangchun Road and to the north of Xihu Road. Urban residents were to be rehoused, while refugees from the countryside were to be given money to return home. A committee for the demolition of straw shacks comprising the Public Security Bureau, the housing and land office, the construction office, and other *danwei* managed work teams of local cadres, who were responsible for surveying, registration, and relocation. Everyone was to be given 60 *jin* of rice, enough for about two months, and the provincial government provided money for compensation.[52]

Demolition took place between 6 and 18 December, and 3,282 households were relocated, a total of 12,123 people. Of these, 2,466 returned to their homes in the countryside, 9,875 left of their own accord, presumably also to return to their homes, 673 were given vacant land to build housing, and the remaining 109 were placed in government accommodation. In total, 239,362.5 *jin* of rice was distributed along with 591,000 RMB.[53] The work did not proceed without problems. 'Bad elements' tried to disrupt demolition, claiming that the road would not be completed this year, or that the government would not distribute any food or money to assist with relocation. Cadres did not elaborate on who those trying to disrupt the work might be but in the immediate post-revolutionary period they were likely to be enemy agents, people who had worked under the Nationalist Government, former enemy soldiers, and bandits, all of whom were labelled counterrevolutionary.[54] Many people refused to move, and cadres had to go house to house explaining the process of relocation. It was often necessary to return to the same house several times to reassure the occupants, and there were a few cases of outright opposition. For example, on the evening of 2 December, a hairdresser named Xu organised a petition of residents, apparently to stop the work. There were arrests that evening, but Xu escaped the next morning.[55]

[51] 'Changsha shi yanjiang penghu chaiqian gongzuo zongjie', CMA 59-1-50.

[52] 'Changsha shi yanjiang penghu chaiqian jihua caoan', (September 1950), CMA 59-1-50.

[53] 'Changsha shi yanjiang penghu chaiqian gongzuo zongjie'.

[54] Dutton, *Policing China*, p. 140.

[55] 'Changsha shi yanjiang penghu chaiqian gongzuo zongjie'.

Work team reports provide more detail on the problems cadres faced. Registration was often difficult, since residents were at work during the day, and cadres had to visit them in the evening. Those who had bought land before 1949 felt it was unfair that they were being asked to move. Others were suspicious of the new government because of their experiences under the Nationalists.[56] The solution to these issues was more consultation with residents, but cadres found it difficult to find time to hold large public meetings, and often had to visit individual families several times. Some residents understood that demolition and relocation would lead to improvements in the city and were happy with the public housing or the land that was promised them. However, others did not understand why it was necessary to build the road. As the deadline for the work approached the most obstinate, who refused to move and who the work team thought were Nationalist spies, were thrown out of their homes. Then, during the final two days of the period for demolition, the remaining houses were simply torn down.[57] In the end, the new regime also had to resort to force, even as it trumpeted its popular support for urban reconstruction.

Despite these problems, cadres reported that most people were satisfied with their work, and they compared the CCP relocation programme favourably with that undertaken by the KMT, saying the people reported, 'that in the past the Nationalists had used fire, guns, and bullets many times … This time, the People's Government had not used violence, but within a short time all the shacks had been demolished'.[58] The claim that KMT officials had used force to relocate shack-dwellers warrants some investigation. In the archival record prior to 1949, it was not until 1947 that Yin Jiesheng claimed that force was being used to evict people from their homes, and so it is likely that cadre reports were correct. However, it is also possible that there was some exaggeration, since neither Yin nor anyone else complaining about the KMT mentioned guns or fire. Looking at the evidence from before 1949, it is safe to assume that officials in the previous regime used force to demolish housing, but it is not possible to be absolutely sure that the KMT government resorted to arson or threatened people with guns. Since the CCP conquest of the city was inextricably linked to its reconstruction, then any efforts of the previous regime had to be duly delegitimised.

As under the KMT, demolition was to create space for road construction, and this revealed further contradictions in the distinctions that cadres were trying to make with the previous government. The road was planned to run just over 7km long and 60m wide, with an embankment along the river that would prevent flooding and serve as a pleasant walk for people in the city. Between July and October 1950, some 5,000 of Changsha's jobless were employed using work relief to lay crushed stone to a width of 40m along the first section of 4.6km from Xihuqiao in

[56] 'Chaiqian yanjiang penghu gongzuo zongjie, dier gongzu', CMA, 59-1-50.
[57] 'Diyizu gongzuo zongjie', CMA, 59-1-50.
[58] 'Changsha shi yanjiang penghu chaiqian gongzuo zongjie'.

the south to Xinhekou in the north. The work stopped in November to allow for the demolition of the remaining straw shacks but after it resumed in 1951 there were problems. Expensive contractors replaced labourers employed through work relief, and the riverbank was unstable, as two docks fell into the river, blocking shipping. Dredging, flood defences, and drainage pipes all required materials, labour, and time, and so the road was not yet complete. However, cadres believed that, in time and, 'with the support of more people in the city, this great historical responsibility would gradually be completed, and this wide road would be a foundation on which an industrialised new Changsha would develop'.[59]

However, just a few months later cadres took a very different view of the road. By now, the national discourse had moved on as the country had completed its recovery and was embarking on a new phase of its development, that of planned socialist construction.[60] In Changsha, this new phase of socialist transformation brought with it criticism of the immediate past. In 1953, the former secretary of the Changsha Municipal Committee wrote that construction had advanced 'blindly and rashly'. Some projects, such as the provision of running water and the establishment of bus services, were necessary. However, when it came to road repair, 'Other than the relocation of over 3,000 families, over 1,800 buildings were also demolished. Of these over 400 have yet to be rebuilt. This has an impact on the production and livelihood of some of the poorest people and increases the number of homeless in the city'.[61] Given that around 3,000 families were relocated in November 1950, it is almost certain that this criticism was directed at the road along the river. The main reason for 'blind and rash' construction was that cadres had not understood the complexities of urban development in their pursuit of a 'Great Changsha', choosing instead to focus on transforming the city's appearance, rather than meeting the needs of production.[62]

These sentiments matched those in Beijing. In July, the central government published the 'Points about Several Problems Concerning Urban Construction', which subordinated urban development to industrial growth, and highlighted the importance of master plans to solve the problem of *danwei* haphazardly constructing buildings themselves.[63] These guidelines coincided with the first Five-Year Plan, which, drawing on Soviet experience, set the course for urban development.[64]

[59] Jianshe ju, 'guanyu xiujian yanjiang dadao de gongzuo', *Changsha Shizheng* 10 (1 November 1951), 36.

[60] Cao and Chu, *Dangdai Zhongguo de chengshi jianshe*, p. 36.

[61] 'Changsha shi chengshi jianshe gongzuo you mangmu maojin qingxiang', *Neibu cankao*, (20 June 1953).

[62] 'Changsha shi chengshi jianshe gongzuo you mangmu maojin qingxiang'.

[63] 'Zhonggong zhongyang guanyu chengshi jianshe zhong jige wenti de zhishi', *Douzheng* Vol. 228 (1953), 20–2.

[64] In English see for example: Duanfang Lu, *Remaking Chinese Urban Form Modernity, Scarcity and Space, 1949–2005*. Abingdon: Routledge, 2006; David Bray, *Social Space and Governance in Urban China The Danwei System from Origins to Reform*. Stanford: Stanford University Press, 2005; In Chinese see for example: Li Hao, *Bada zhongdian chengshi guihua xin Zhongguo chengli chuqi de chengshi guihua lishi yanjiu*. Beijing: Zhongguo jianzhu gongye chubanshe, 2016.

This new phase in the construction of Maoist cities gave cadres in Changsha license to criticise the 'chaotic' reconstruction of the city since 1949.[65] During the first Five-Year Plan in Changsha, between 1953 and 1955, the total amount of money spent on urban construction remained between 5 and 10 per cent of municipal expenditure, about the same proportion as in the years prior to this. One way to reduce the chaotic construction was to consolidate multiple small building firms into provincial and city construction companies, and together they became responsible for over half of all projects in Changsha. A host of problems remained, including continuing waste and issues with supply of raw materials, especially concrete.[66] Construction increased under the second Five-Year Plan, and in 1958, as the GLF was in full swing, it reportedly doubled over the previous year. Now, it was the responsibility of the Changsha Municipal Committee to ensure that there was more construction, that buildings were put up faster, but also that it was of better quality and delivered at a lower cost.[67]

Conclusion

'More, faster, better, more economical' was the slogan that accompanied urban planning and construction in the 1950s and carried with it an implicit criticism of the wastefulness of the past. In February 1956, in a speech to the National Infrastructure Conference, Wan Li, the head of the Urban Construction Bureau, made this point explicitly. He accused planners of serious problems, including an over-emphasis on the appearance of buildings instead of their use or value. Along with this went a tendency to look at the distant future and ignore how cities should be repaired and constructed to support industrial development.[68] Wan Li's criticism of past mistakes points to the continuing debate on socialist urban construction. This debate was not confined to urban planning or infrastructure, as Jennifer Altehenger's chapter in this volume on how standards for furniture design ran into problems such as supply of raw materials, and Fabio Lanza's on welfare in urban communes demonstrate. Ideals of what it meant to create socialist cities in Maoist China continually floundered on the shores of the gritty realities of constructing them, and this is before we consider that such ideals themselves often changed because of political campaigns. Some elements, most notably that cities should be industrial, remained constant throughout the Maoist era, but others were reframed, de-emphasised, or forgotten as they slipped into the past.

[65] 'Sannian lai Changsha jibenjianshe gongzuo qingkuang baogao, 1953–1955'; 'Changsha shi yijiuwusan nian jibenjianshe gongzuo zongjie', CMA, 44-1-2.
[66] 'Sannian lai Changsha jibenjianshe gongzuo qingkuang baogao, 1953–1955',
[67] *Changsha jiefang shi nian*, p. 35.
[68] Wan Li, 'guanyu chengshi jianshe gongzuo de baogao', in *Wan Li lun chengshi jianshe*, eds. Zhou Ganzhi, Chu Chuanheng. Beijing: Zhongguo chengshi chubanshe, 1994, p. 3.

Maoist China was never made but rather was constantly being remade. This meant that the party endlessly recast the past in the service of the present, and one of the distinctions that was easiest to make was between the new and the old society. In this chapter I have argued that in studying how the CCP sought to make Maoism in the bricks and mortar of Chinese cities, we have not paid enough attention to how the KMT worked to reconstruct them. The years after Japanese surrender were certainly chaotic and the Nationalist Government was incompetent and repressive. However, the revival of China was not all of the CCP's making. KMT officials, for all their faults, were beginning to reconstruct China in the wake of that war. Their efforts, stymied by economic crisis and the Civil War, created a foundation on which Maoism was made. The historical trajectory of post-war reconstruction and socialist construction differed from city to city, and there are many stories of urban revolution that have yet to be told.

Looking once more at the immediate post-war years also challenges us to think more deeply about how the legacies of war have shaped China beyond their contribution to nationalism, which can still be seen today. WWII cast a long shadow over much of the second half of the twentieth century, but in China not enough work has yet been done to shine some light into the dark recesses, to tease out where physical destruction and societal trauma remained significant factors in determining the course of modern Chinese history long after the sounds of fighting had died away.

2

Mediating Disputes, Making *Minzu*: Minoritisation on an Ethnocultural Frontier of Early Maoist China

BENNO WEINER

IN 1953, THE feature film *Gold and Silver Grasslands* (*Jinyintan*) debuted in Chinese theatres.[1] [2] Filmed on location on the high-altitude plateau of Qinghai Province, *Gold and Silver Grasslands* is a Romeo and Juliet story with a Maoist twist. It tells the tale of two Tibetan chiefdoms (*buluo*) locked in a violent, decades-long feud over a strip of pastureland not far from Qinghai Lake. Viewers quickly learn that the conflict first had been instigated and then enflamed by the commander of the local Kuomintang (KMT) garrison, Liu Huimin. Most crucially, with the Red Army approaching Qinghai on its legendary Long March, Liu promises each side ownership of the contested grassland in exchange for 1,500 horses to support the KMT's anti-Communist blockade. A review published in the magazine *Popular Cinema* (*Dazhong Dianying*) described the consequences of Liu's duplicity: 'People were killed or injured and production was damaged; cattle and sheep were deprived of water and grass, [livestock] epidemics spread; people were forced from their homes, and families were ruined.'[3] Not until the Communists' triumphant return to Qinghai 15 years later was the feud resolved and national unity purportedly restored.

[1] I would like to thank the participants of the workshop *How Maoism was Made* for their helpful comments on earlier drafts of this chapter and in particular to co-organisers and editors Aaron Moore and Jennifer Altehenger.

[2] The title has been translated previously as *Gold and Silver River Beach* (Barnett) and *The Gold and Silver Sandbank* (Lu). *Tan*, however, is a transliteration of the Tibetan word *thang* which is often used as a toponym to describe a flatland or plain suitable for grazing.

[3] Jiang Shan, 'Yibu fanyang woguo minzu shenghuo de yingpian: *Jinyintan*', *Dazhong Dianying*, no. 18 (1953), 13.

Gold and Silver Grasslands is fictional, but based on actual events, or actual events as viewed through the Chinese Communist Party (CCP)'s prescriptive lens of what ailed its ethnocultural borderlands in the years leading up to and immediately following the establishment of the People's Republic of China (PRC). As explained in *Guangming Ribao* (*Enlightenment Daily*) soon after the film's release:

> Although *Gold and Silver Grasslands* depicts the story of two Tibetan tribes, its significance is broader. Blood feuds and intra-nationality disunity exist in practically all minority nationality areas. Because of this … mediating feuds and disputes and strengthening nationality unity has become extremely important political work.[4]

The film was inspired by the resolution of a decades-old grassland dispute (*caoyuan jiufen*) 300 kilometres to the southeast of Qinghai Lake that had pitted the Gyelwo chiefdom from Qinghai's Rebgong region (present-day Tongren County) against the Gengya, a chiefdom located in the grasslands north of Labrang Monastery on the Gansu side of the provincial boundary.[5] The CCP-mediated resolution of the Gyelwo–Gengya dispute in summer 1951 was not only touted in the national media, but for several years after was repeatedly cited as one of the party's notable achievements in nationality work during the first years of the PRC.[6]

The CCP, then as now, considered the near-territorial entirety of the former Qing Empire to be a single, historical, multi-nationality state, and its diverse inhabitants to form one 'big family' consisting of the Han majority and an as-yet-unknown number of 'minority nationalities'.[7] If so, it was a family in desperate need of an intervention. Party leaders openly acknowledged that relations between nationality groups often were marked by deep distrust and longstanding grievances that frequently erupted into bouts of inter-community conflict. For most of the 1950s, they insisted that these rifts were disproportionately the fault of the Han majority – that they were caused by what was termed 'great nationality chauvinism' (*da minzu zhuyi*) or more commonly 'great Han chauvinism' (*da Hanzu zhuyi*). The claim was that centuries of discrimination and exploitation committed by the Han majority had driven a wedge between ethnocultural communities.[8] As in *Gold and Silver Grasslands*, however, party operatives quickly discovered that many of the most intractable and destructive disputes involved members of a single nationality group. In fact, during the early 1950s the familiar

[4] Li Youyi, 'Kanle dianying *Jinyintan* yihou', *Guangming Ribao*, 6 October 1953, p. 3.

[5] Huangnan Zhou Zhengxie, 'Minzu tuanjie hao, caoyuan qixiang xin: Jiefang chuqi Huangnan diqu tiaojie minzu jiufen jishi', *Qinghai wenshi ziliao xuanji*, 12 (1984), 63–7.

[6] For example, *Guangming Ribao*, 8 August 1951, p. 2; *Renmin Ribao*, 20 January 1952, p. 2; *Renmin Ribao*, 23 March 1957, p. 13.

[7] The major exception was what was then the Mongolian People's Republic.

[8] See Benno Weiner, 'This Absolutely is not a Hui Rebellion! The Ethnopolitics of Great Nationality Chauvinism in Early Maoist China', *Twentieth Century China*, 48 (2023), 208–29.

phrase 'nationality unity' (*minzu tuanjie*) was employed in the context of efforts to consolidate unity *within* a nationality at least as often as it referred to establishing unity between nationalities.

It therefore was no accident that one of New China's earliest films to feature non-Han people depicted a feud between two groups of Tibetans and its resolution under the guidance of the Communist Party. Written by the veteran Communist screenwriter and cultural work cadre Lin Yi (1914–2004) and directed by emerging filmmaker Ling Zifeng (1917–99), *Gold and Silver Grasslands* was contemporaneously celebrated as the second entrant in an important genre of PRC filmmaking soon dubbed 'minority nationalities film' (*shaoshu minzu ticai yingpian*).[9] Historians of Chinese film have attributed the genre's popularity among both filmmakers and audiences to a 'search for the exotic' while imparting the message that the salvation of non-Han people depended on 'simultaneous Han Chinese and socialist self-styled benevolence'.[10] Yet, film historian Xiaoning Lu insists that these movies also had a generative function. She writes, 'this film genre ultimately aims to cultivate socialist fraternity and to shape socialist subjectivity that transcends any kind of ethnic boundaries.'[11] As Lu notes, however, before those boundaries could be transcended, they first had to be constructed. In other words, minority nationalities film was not solely a genre in which often exoticised non-Han people could be saved and civilised by their Han elder brothers – represented by the CCP. It was also a key component of the early Maoist state's 'political endeavour to construct ethnicity'.[12] Both in film and on the ground, before a pan-nationality 'patriotic consciousness' (*aiguo juewu*) could emerge, parochial sub-ethnic loyalties would need to be replaced by *minzu*-based identities as defined by the CCP.

This chapter uses the Gyelwo–Gengya feud as an entry point through which to explore early efforts by the CCP to make *minzu* (nationalities), and specifically *shaoshu minzu* (minority nationalities), out of the disparate people that inhabit the region known to Tibetan speakers as Amdo.[13] Minoritisation, here, does not refer to demographic transformation, but, following Ilyse Morgenstein Fuerst, to a 'process through which a group, often formerly dominant or prominent, is stripped

[9] Xu Wen, '*Jinyintan* shi yibu zenmeyang de yingpian?' *Dazhong Dianying*, no. 18 (1953), 10–11. For a history of the genre, see Rao Shuguang, *Zhongguo shaoshu minzu dianying shi*. Beijing: Zhongguo dianying chubanshe, 2011. On films and television shows produced in the PRC that take Tibet/ans as their subject, see Robert Barnett, 'Close Encounters of the Filmic Kind: Visualising the Chinese Arrival in Tibet', in *Conflicting Memories: Tibetan History under Mao Retold*, eds. Robert Barnett, Benno Weiner, and Françoise Robin. Leiden: Brill, 2020, pp. 141–203.

[10] Paul Clark, *Chinese Cinema: Culture and Politics since 1949*. Cambridge: Cambridge University Press, 1987, p. 96; Chris Berry and Mary Ann Farquhar, *China on Screen: Cinema and Nation*. New York: Columbia University Press, 2006, p. 182.

[11] Xiaoning Lu, *Moulding the Socialist Subject: Cinema and Chinese Modernity (1949–1966)*. Leiden: Brill, 2020, p. 95.

[12] Lu, *Moulding the Socialist Subject*, p. 72.

[13] The neologism *minzu* has different possible meanings in different ethnopolitical contexts. However, at the time the CCP purposefully used it as a gloss for the Stalinist notion of nationality.

of agency, power, or prestige.'[14] In 1950s Amdo, 'mediation of disputes' (*tiaojie jiufen*) was an integral part of Maoist processes of state- and nation-making, and thus minoritisation. It was an official work category that regularly appeared in ambitious start-of-the-year work plans – alongside headings such as social welfare, public security, and civil administration – and in end-of-the-year reports where the number of disputes allegedly resolved often stood in contrast to the local leadership's admitted failure to meet various production targets and other setbacks. Throughout the chapter I invoke the Gyelwo–Gengya dispute's cinematic avatar, *Gold and Silver Grasslands*, not so much to illustrate the depiction of minoritised people in socialist-era Chinese film – a topic that has been productively examined by several specialists[15] – but as a device that reveals the ethnopolitical anxieties and policy priorities of early-PRC state builders as they sought to make a complex ethnocultural frontier legible and governable. As I show, the triumphalism that would accompany the official resolution of the Gyelwo–Gengya feud and that is depicted in its on-screen adaption masks a drawn-out process of dispute mediation and *minzu* making in 1950s Amdo that was marked more by failure than success. In the end, minoritisation along the Sino-Tibetan frontier would be achieved not through acts of benevolent paternalism as represented by CCP-sponsored dispute mediation. It instead was accomplished largely through state violence.

Like brothers in an intimate and harmonious family

Like majorities, minorities do not just exist. Janet Klein argues that 'it is essential for us to critically reconsider our use of the term "minority," to see minorityhood *as historically and socially constructed as we recognize nationhood to be*, and to understand the specific links between them.'[16] Klein, a historian of the Ottoman Empire and its successors, is particularly concerned with the 'unique brand of repression and mass violence' that has often accompanied processes of minoritisation, 'when minorities—now conceived as such—came to be regarded as threats to the territorial integrity and sovereignty of "the nation" and to the imagined privilege and power of the dominant … group, now envisioned as the "majority," or the *real* citizen.'[17] By promising non-Han people political autonomy and equality in a multinationality state, the CCP codified ethnocultural difference as a legally protected category. Yet, even in states with pluralistic pretensions like the early PRC that explicitly

[14] Ilyse R. Morgenstein Fuerst, 'Minoritization, Racialization, and Islam in Asia', in *Routledge Handbook on Islam in Asia*, ed. Chiara Formichi. London: Routledge, 2021, pp. 16–30 at 17.

[15] In addition to the scholars cited above, see Peng Hai, 'Habitus of the Minor: The Visuality of non-Han Peoples in Modern China (1930s–2010s)' (Ph.D dissertation, Harvard University, 2023).

[16] Janet Klein, 'Making Minorities in the Eurasian Borderlands: A Comparative Perspective from the Russian and Ottoman Empires', in *Empire and Belonging in the Eurasian Borderlands*, eds. Krista A. Goff and Lewis H. Siegelbaum. Ithaca: Cornell University Press, 2019, pp. 17–31 at 19. Italics original.

[17] Klein, 'Making Minorities', pp. 17–18. Italics original.

reject ethnonationalist foundations and instead commit to protecting and promoting the interests of newly minoritised people, those made into minorities often become 'marked citizens' and their homelands 'marked territories' rendering these communities vulnerable to the 'repression and mass violence' Klein warns of.[18]

The majority of Amdo lies in present-day Qinghai Province, while the remainders spill into southern Gansu and northern Sichuan. Although Amdo is often considered one of the three major ethnolinguistic regions of Tibet, it also serves as a 'contact zone' or 'shatter zone' in which the Tibetan, Chinese, Mongol, and Central Asian worlds collide. Under the Qing Empire, which conquered Amdo in the eighteenth century, there were no *shaoshu minzu* – no 'minorities'.[19] The Qing managed separate and unequal subject populations through a variety of governing practices that ranged from more bureaucratic to more paternalistic and fluctuated between some that prized acculturation and others that reenforced difference. Like much of Qing Inner Asia, Amdo tended toward the latter. Direct authority over Amdo's Tibetan and Mongol communities historically was most often in the hands of monastic estates (T. *labrang*) and/or hereditary secular chieftains. Where representatives of larger states sought to impose their rule over the region – be it the Manchu Qing, the Muslim 'Ma family warlord' regime that controlled Amdo through much of the Republican period, or the CCP in the early years of the PRC – they most often did so indirectly through these elite intermediaries.

Most troublesome for all three states, and the focus of this chapter, was Amdo's vast pastoral regions. 'Incomplete statistics' cited in a 1952 speech by Qinghai's First Party Secretary, Zhang Zhongliang, emphasised that Qinghai's pastoral population was divided between 315 Tibetan tribes and Mongol banners ruled by 368 headmen. In addition, there existed 695 reincarnate lamas (C. *huofo*; T. *trülku*) and other monastic leaders, most of whom would have been associated with one or more of the region's 249 'lama temples'.[20] While the figures themselves are certainly inaccurate, Zhang's point is clear: the grassland system was 'backward', 'tribal', and antithetical to the CCP's vision of a unified, modern, and socialist multinationality-state.[21]

[18] Janet Klein, 'The Kurds and the Territorialization of Minorityhood', *Journal of Contemporary Iraq & the Arab World*, 14 (2020), 13–30. Klein is building on the work of Gyanendra Pandey, *Routine Violence: Nations, Fragments, Histories*. Stanford: Stanford University Press, 2006, chapter 6.

[19] As in the late-Ottoman Empire, processes of minoritisation may have begun during the last decades of the Qing. For example, see Hannah Theaker, 'Old Rebellions, New Minorities: Ma Family Leaders and Debates over Communal Representation following the Xinhai Rebellion, 1911', *Global Intellectual History*, 7 (2021), 1016–36.

[20] 'Zhang Zhongliang tongzhi zai muyequ gongzuo huiyi shangde zongjie baogao', in *Qinghai shengzhi*, ed. Qinghaisheng Difangzhi Bianji Weiyuanhui, v. 81 *Fuluzhi* (hereafter *QSFL*). Xining: Qinghai renmin chunbanshe, 2003, pp. 567–75 at 567.

[21] On debates among early Soviet social scientists, theorists, and planners over the characteristics of pastoral societies and their position within Marxist paradigms of social evolution, see Sarah Cameron, *The Hungry Steppe: Famine, Violence, and the Making of Soviet Kazakhstan*. Ithaca: Cornell University Press, 2018, 60–8.

As alluded to in *Gold and Silver Grasslands*, prior to 'liberation' the Communist Party's main experience in the region had been its harrowing traversal of the great grasslands of northern Sichuan and southern Gansu during the Long March. When it returned a decade and a half later, the regional leadership freely admitted that the CCP had no pre-existing institutional presence in the area and few friends on which to rely.[22] Who and what, then, did party leaders expect to find when they arrived in Amdo? Focusing on the southwestern province of Yunnan, Thomas Mullaney writes, 'having committed themselves to a concept of China as a multi-minzu country, ... the Communists unwittingly stumbled upon a remarkably complex problem: who were they?'[23] As Mullaney shows, this 'ethnotaxonomic volatility' persisted well into the early years of the PRC.[24] In January 1952, for example, vice-director of the state Nationalities Affairs Commission (NAC), Liu Geping, admitted that officials were still far from clear how many minority groups existed within China.[25]

In contrast to Yunnan, when Communist soldiers and cadres poured across Qinghai's borders in late summer and autumn of 1949, they did so with a relative fixed notion of the province's *minzu* make-up. This comparative ethnotaxonomic certainty seems to have been a partial outgrowth of the more general scrutiny China's northwest had come under in the years following Japan's occupation of Manchuria. Speaking of a 'frontier crisis' (*bianjiang weiji*) and fearing that the northwest might be next, during the 1930s and 1940s Chinese political elites and intellectuals endeavoured to bind the region closer to the state and nation through administrative regularisation, investment in infrastructure, resource extraction, the dissemination of nationalist propaganda, and, perhaps most successfully, the pursuit of ethnographic, topographic, and geological knowledge.[26] A disproportionate amount of these energies were directed toward Qinghai, which had only been established as a province in 1929. Three years later, the inaugural issue of the journal *New Qinghai* announced to its readers that the new province was home to Mongols, Tibetans, Han, and Hui.[27] A survey of Qinghai published in 1934 by the New Asia Society added the Tu (Monguors).[28] By the end

[22] Qinghai Shengwei, 'Guanyu dangqian gongzuo zhishi', in *Jiefang Qinghai shiliao xuanbian* (hereafter *JQSX*), eds. Zhongguo Renmin Jiefangjun Qinghaisheng Junqu Zhengzhibu and Zhongguo Qinghai Shengwei Dangshi Ziliao Zhengji Weiyuanhui. Xining: neibu ziliao, 1990, pp. 76–8.
[23] Thomas S. Mullaney, *Coming to Terms with the Nation: Ethnic Classification in Modern China*. Berkeley: University of California Press, 2011, p. 31.
[24] Mullaney, *Coming to Terms*, p. 2.
[25] *Renmin Ribao*, 20 January 1952, p. 2.
[26] Zhihong Chen, 'Stretching the Skin of the Nation: Chinese Intellectuals, the State, and the Frontiers in the Nanjing Decade (1927–1937)' (Ph.D dissertation, University of Oregon, 2008); Jeremy Tai, 'Opening up the Northwest: Reimagining Xi'an and the Modern Chinese Frontier' (Ph.D dissertation, University of California Santa Cruz, 2015); Andres Rodriguez, *Frontier Fieldwork: Building a Nation in China's Borderlands, 1914–1945*. Vancouver: University of British Columbia Press, 2022.
[27] 'Fakan ci', *Xin Qinghai* 1, no. 1 (1932).
[28] Qinghai Sheng Minzhengting (ed.), *Zuijin zhi Qinghai*. Nanjing: Xin Yaxiya shehui, 1934.

of the decade, Salar and Kazakh were regularly included in overviews of Qinghai's ethnocultural landscape.[29]

After its fall to the Communists in 1949, Qinghai's new leaders announced that its people had been given a 'new life' and a chance to build a 'new Qinghai'.[30] However, the 'earth-shaking changes' distinguishing the revolutionary regime from its predecessor evidently did not extend to the taxonomies used to categorise Qinghai's population.[31] For example, when on 1 January 1950 Qinghai's leadership gathered to officially inaugurate the provincial People's Government, Chairman Zhao Shoushan 'pledged' 'to unite Qinghai's Tibetan, Han, Hui, Mongol, Tu, Salar and Kazakh people'.[32] Later that month, military chief Liao Hansheng confidently proclaimed, 'Qinghai's seven nationalities are not divided by language, are not divided by region, are not divided by religious belief, and are not divided by customs, but are united together under the flag of Mao Zedong. They are like brothers in an intimate and harmonious family.'[33]

Liao was well-aware that the fraternal feelings he pointed to were far more aspirational than they were actual. More fundamentally, despite Liao's assured enumeration of Qinghai's seven *minzu*, it was becoming clear to the regional leadership that these categories had less local currency than imagined. The problem in Qinghai was not so much figuring out who people were as it was making lived reality accord to the party's presumptive ethnopolitical blueprint.[34]

Unite with all who can be united

As was expected of all cultural production during the Maoist period, minority nationalities film was not meant just for the audience's entertainment but also for its edification. If there had been any doubts about the leadership's attention to and expectations for the genre, they were quieted by the fate of the predecessor to *Gold and Silver Grasslands*. *Inner Mongolian Spring* (Neimenggu chunguang) had been released in April 1950 to great fanfare and rave reviews. Within weeks, however, the film came under high-level criticism for 'political defects' in its depiction of the Mongol nobility and Buddhist leadership, but not for the reasons that one might

[29] See for example the four-part essay by Gao Tian, 'Jianghe yuantou de jian'ermen', that ran in *Shishi Xinbao* (*The China Times*) from 25–9 November 1939.

[30] *JQSX*, pp. 40, 56.

[31] Liao Hansheng, 'Qinghai jiefang de lishi huiyi', in *Jiefang Qinghai*, eds. Qinghai Shengwei Dangshi Ziliao Zhengji Weiyuanhui and Qinghai Junqu Zhengzhibu. Xining: Qinghai renmin chubanshe, 1987, pp. 26–36 at 26.

[32] *QSFL*, p. 490. In the months just before and immediately following Qinghai's September 'liberation', at times Party and military leaders only referred to five *minzu*: Han, Hui, Tibetans, Mongols, and Tu.

[33] Zhang Bo, 'Jiefang chu Qinghai gezu renmin lianyihui shengkuang jishu', in *JQSX*, pp. 215–23 at 223.

[34] Mullaney shows that the CCP in Yunnan actively constructed ethnic identities as well. My point is that in Qinghai the categories themselves were relatively stable.

expect. It was not that the filmmakers had created too sympathetic a portrayal of Mongol elites. It was because they had painted these figures as irredeemable class enemies. At a meeting called to determine the film's fate, Premier Zhou Enlai explained to the roomful of leading cultural figures that included Mao Dun, Guo Murou, Lao She, Ding Ling, Deng Tuo, Jiang Qing, and more, 'during the current moment our domestic enemies are the reactionaries led by Chiang Kaishek … they are not the nobility and lamas. The nobility and lamas are the main [targets] for us to win over.'[35]

The film was therefore pulled from theatres. When it was rereleased the following year, it not only had a new name, *Victory of the People of Inner Mongolia* (Neimenggu renmin de shengli), but also new characters, new scenes, and a new lesson.[36] In a self-criticism published in the *People's Daily*, screenwriter Wang Zhenzhi admitted that the original version mistakenly had focused on two, simultaneous struggles: one the struggle against Han reactionaries led by Chiang Kaishek (nationality struggle), the other against internal class enemies from among the traditional elite (class struggle).[37] In *Inner Mongolian Spring*, these two groups – the Mongol 'feudal' leadership and KMT 'bandits' – had conspired to exploit the Mongol masses. In its revised version, the focus turned to liberating *all* Mongols – regardless of class – from the clutches of KMT reactionaries.

The decision to deemphasise class struggle and instead form a 'broad-based United Front [*tongyi zhanxian*] by winning over and uniting with all local [minority] nationality and religious elites who can be united' was a practical solution for the specific historical problem caused by Han chauvinism.[38] At the 1950 opening of Qinghai's 'Unity and Friendship Conference', Liao Hansheng had insisted that the 'mutual suspicion' and even 'hatred' that existed between the nationalities was a direct consequence of the 'cruel nationality exploitation and slaughter' minority peoples in Qinghai had historically experienced under a succession of feudal and reactionary regimes.[39] While the 'narrow nationalism' (*xia'ai minzu zhuyi*) or 'local nationalism' (*difang zhuyi*) displayed by non-Han people was also considered an obstacle to nationality unity, party leaders made clear that this was a parochial response to the oppression committed over many years by the Han majority. Because 'local nationalism is the product of past nationality exploitation', Qinghai's United Front Work Department (UFWD) head Zhou Renshan declared,

[35] Quoted in Rao, *Zhongguo shaoshu minzu dianying shi*, pp. 27–8.

[36] Rao, *Zhongguo shaoshu minzu dianying shi*, pp. 26–35; Sun Lifeng, 'Cong *Neimengu Chunguang* dao *Neimenggu Renmin de Shengli*', *Minzu wenxue yanjiu*, no. 5 (2017), 43–52 at 48–9.

[37] Wang Zhenzhi, '*Neimeng Chunguang* de jiantao', *Renmin Ribao*, 28 May 1950, p. 5.

[38] Dangdai Zhongguo Congshu Bianjibu (ed.), *Dangdai Zhongguo de minzu gongzuo*, v.1. Beijing: Dangdai Zhongguo chubanshe, 1993, pp. 71–2.

[39] Liao Hansheng, 'Guanqie shixing minzu pingdeng he minzu tuanjie wei Qinghaisheng renmin zhengfu de jiben zhengce', in *JQSX*, p. 105. In truth, prior to the rise of Muslim power in the late-nineteenth century, Mongol and Tibetan monastic and secular polities had long held the lion's share of political, military, and economic authority across the Amdo region.

it would disappear only *after* its cause had been eradicated.[40] As Wang Zhenzhi, the chastised screenwriter of *Inner Mongolian Spring*, wrote in the *People's Daily*, 'The immediate enemy of all of the people of China is the great Han chauvinism of the KMT reactionary clique, this is a historical truth that cannot be doubted.'[41]

Victory of the People of Inner Mongolia would serve as a template for the next several minority nationalities films. All told the story of an intra-nationality feud instigated by KMT reactionaries who had deceived non-Han elites into harming their own people and, according to film historian Rao Shuguang, 'all revolved around winning over the upper stratum of fraternal nationalities'.[42] It is therefore no surprise that in *Gold and Silver Grasslands* the warring chieftains were not depicted as venal villains, but instead as victims of KMT trickery and therefore were redeemable. Yet, in a telling critique of her script published in the *Guangming Ribao*, screenwriter Lin Yi admitted that her depiction of the chieftain Sonam Gyel was more fleshed out and successful than her relatively two-dimensional, 'idealised' portrayal of the film's putative protagonist, the Tibetan 'advanced element' Dolung, who, she admitted, functioned more like a 'prop' than a 'leading character'. Among the reasons, Lin suggested, was that when she had gone to the Qinghai grasslands she spent more time amidst the upper strata than she had among the masses.[43]

Li's experience is anecdotal but instructive. When People's Liberation Army (PLA) troops and CCP cadres swept into Amdo in 1949, they did not primarily seek out progressive members of the masses because, by and large, these people did not exist. Unlike the stories transmitted through *Gold and Silver Grasslands* and other minority nationalities films of the early-to-mid 1950s, cadres did not ally with progressives to win over hoodwinked elites. Quite the opposite, in Amdo they went to sometimes extraordinary lengths to recruit the traditional leadership – including in many cases figures who had colluded with the Ma warlords, previously joined the KMT, or even taken up arms against the new regime – into its patriotic United Front.

Referred to by Mao Zedong as one of the 'three magic weapons' (*sanda fabao*) of the revolution, the United Front is a concept that allowed Communists to ally on a temporary basis with non-Communist forces in order to advance socialist goals. Most commonly associated in China with the two United Fronts struck between the CCP and KMT first in the 1920s and again during the anti-Japanese War, by 1949 it had become the theoretical justification and bureaucratic instrument for bringing nonproletarian elements into the political process in preparation for the final transition to socialism. In Amdo, where the party's reach into non-Han communities

[40] Zhou Renshan, 'Jixu kaizhan minzu quyu zizhi yundong', in *Minzu zongjiao gongzuo wenjian huiji* (hereafter *MGWH*), ed. Zhonggong Qinghai Tongzhanbu [Xining], n.p., 1959, pp. 637–56 at 648–9.

[41] Wang, '*Neimeng Chunguang* de jiantao', p. 5.

[42] Rao, *Zhongguo Shaoshu minzu dianying shi*, pp. 39, 43. See also Sun, 'Cong *Neimenggu Chunguang*', p. 43.

[43] Lin Yi, 'Guanyu *Jinyintan* de liangge renwu', *Guangming Ribao*, 22 May 1954, p. 3. Similar critiques can be found in Xu, *Jinyintang*, p. 11.

was so limited and the influence of traditional elites was imagined to be so deep, this meant welcoming almost any influential non-Han figure into its United Front. These were often astute political actors who had years of experience and even longer historical memories navigating complex frontier dynamics. They were also the same people who had long served as intermediaries between local Amdo communities and larger powers.

To rule over vast territories and diverse groups of people, the Qing, like other successful imperial formations, had both tolerated and relied on indirect rule through these types of elite intermediaries who in exchange for expressions of loyalty received certain benefits for themselves and their communities. The United Front capitalised on and to a degree replicated these imperial traditions and useful comparisons can certainly be made between the CCP's policies among non-Han people and colonial practices in other times and places. Yet, as Adeeb Khalid argues for the Soviet Union in Central Asia, the transformative agenda and participatory politics of the PRC only makes sense if it is understood as a fundamentally different type of state than a traditional empire. Karen Barkey writes, 'the imperial state does not have complete monopoly of power in the territory under control. It shares control with a variety of intermediate organisations and with local elites, religious and local governing bodies, and numerous other privileged institutions.'[44] By contrast, Khalid contends, 'Modern mobilisational states have instead sought to cut through layers of intermediaries and to deal directly with their citizens'.[45]

For the CCP, the United Front was just such a mechanism. By harnessing the charismatic authority of religious leaders and secular headmen in support of programmes that promised autonomy, equality, and material prosperity, policy makers predicted that propaganda could be spread, activists cultivated, fraternalism nurtured, and the benefits of cooperative production practices demonstrated. In the process, both class awareness and 'patriotic consciousness' would slowly rise while faith in hereditary headmen and monastic leaders would dissipate and eventually disappear. At that point, the masses themselves would demand what the party called 'democratic reforms' (full political integration) and socialist transformation (collectivisation) and the period of the United Front would come to its peaceful conclusion.[46]

From tribal divisions to democratic unity

The cornerstone upon which the United Front rested in Qinghai was nationality autonomy. At a 1951 pastoral work conference, UFWD Director Zhou Renshan called it a 'powerful weapon' with the ability to 'eliminate inter-nationality estrangement and

[44] Karen Barkey, *Empire of Difference: The Ottomans in Comparative Perspective*. Cambridge: Cambridge University Press, 2008, p. 10.

[45] Adeeb Khalid, 'Backwardness and the Quest for Civilization: Early Soviet Central Asia in Comparative Perspective', *Slavic Review*, 65 (2006), 231–51 at 233.

[46] Zhou, 'Jixu kaizhan minzu quyu', pp. 637–56.

unify each nationality'. Nationality autonomy was a new, intrusive, 'anticipatory form of territorialization' that not only threatened to replace the overlapping and often diffuse spheres of authority that had long governed sociopolitical life in Amdo, but also promised to make minorities out of its inhabitants.[47] It was, Zhou emphasised, the transformative institution through which 'tribal division would become democratic unity'.[48]

Nonetheless, it was easier to issue proclamations in the provincial capital than it was to enact the party's agenda on Qinghai's vast grasslands. In the newly formed Zeku (T. Tsékhok) Tibetan Autonomous County, for example, an often-frustrated party and state apparatus consistently struggled to fulfil work plans, meet production quotas, build administrative capacity, or implement almost any other task. Just finding materials with which to build Zeku's administrative centre was difficult enough, and that was only after the CCP work team managed to cajole the area's divided indigenous leadership – representing 10 often warring chiefdoms – to agree on where to locate the county seat.[49]

Like the headmen from *Gold and Silver Grasslands*, when meeting with CCP representatives Zeku's chieftains and lamas would say all the right things about promoting nationality unity. Off screen, as it were, these same figures often failed to follow through on their commitments. For instance, headmen were meant to set an example for the masses by honestly paying 'patriotic animal taxes'. Investigators instead found that they often underreported their own herd sizes and allowed others to do the same. Despite efforts to reduce inter-community tensions by regulating grassland use, a 1956 report charged that when rules were transgressed, 'it is always caused by the headmen'.[50] Yet each setback was blamed not on deficiencies in the overall policies sent down from the provincial capital, or even on obstruction by 'feudal' elites, but on poor work style and political consciousness among grassroots Han cadres, many of whom stubbornly harboured chauvinistic attitudes toward the Tibetans they now worked among. Even in the midst of the 1957 Anti-Rightist Campaign, Qinghai's leaders continued to condemn Han chauvinism within its cadre force.[51] By contrast, it ordered criticism of pastoral elites limited to 'talk but no struggle'.[52] As explained several years earlier by the CCP's Northwest Bureau, 'The reason is simple: it is because in the past under the rule of Han chauvinism, [upper strata elements] played the leading role in opposing nationality exploitation.'[53] For the foreseeable future, it seemed, turning Amdo's 'tribal' people into Tibetans, and then Tibetans into socialist citizens, would continue to require the

[47] Donald S. Sutton, 'Territorialization and Ethnic Control in China's Borderlands: Aba Prefecture in the People's Republic', unpublished manuscript.

[48] Zhou, 'Jixu kaizhan minzu quyu', pp. 637, 647–8.

[49] Benno Weiner, *Chinese Revolution on the Tibetan Frontier*. Ithaca: Cornell University Press, 2020, chapters 3–4.

[50] Zeku County Party Committee Archives (hereafter ZCPC), folder 11, folios 33–4.

[51] Weiner, *Chinese Revolution*, chapter 5.

[52] ZCPC, folder 16, folio 23.

[53] Xibeiju Tongzhanbu, Minwei Dangzu, 'Guanyu pingxi Angla, Xiji deng shaoshu minzu diqu panluan de baogao', in *JQSX*, pp. 300–10 at 307.

mediation of traditional elites. Class struggle could not begin until nationality struggle – the struggle against Han chauvinism – was complete.

Nothing tested Director Zhou's assertion that nationality autonomy would 'cause a gradual reduction in tribalism and put it on a path toward disappearing naturally' quite like grassland disputes.[54] By 1949, these seemingly ubiquitous feuds had long been legend among those with experience living in or travelling through Amdo's grasslands. The missionary-cum-ethnographer Robert Ekvall, for example, chalked up the high incidence of inter-community violence to a combination of weak and fragmented political control, the 'basic mobility [of pastoralists] and the considerable evasive capability that mobility conferred', and what he referred to as 'the primal law of retaliation'.[55] Once begun, the retributive violence could last years, or even generations.

The Gyelwo–Gengya dispute was just one of probably thousands, large and small, that party operatives identified in the years following 1949. For instance, early reports out of Zeku indicate that nearly all of its chiefdoms were engaged in multiple feuds with rivals both within and across the new county's invisible boundaries. Perhaps even more common were conflicts between sub-units of a single chiefdom.[56] As a result, one of the party work team's first acts after entering the Zeku region in September 1953 was to hold a meeting with local headmen in order to find ways 'to restore social order, build the new administration, develop the economy, improve peoples' livelihoods, and handle all types of disputes between the people.'[57] Soon after, 'mediation committees' led by local headmen were formed in eight of the county's 10 chiefdoms and within weeks more than 130 'internal disputes' purportedly had been resolved.[58] A year later, the number of 'historical disputes' successfully mediated within Zeku had risen to 315.[59]

Zeku was not an outlier. For instance, in 1951 the counties that neighbour Zeku to its west and east reported the resolution of over 300 and over 260 disputes respectively. These were said to be among the more than 3,000 grassland, water, border, and domestic disputes already resolved across the Northwest.[60] At least on paper, provincial leaders seemed correct when they insisted that where nationality autonomous governments were established, 'inter-tribal disputes greatly lessen, looting and armed feuds basically disappear, and when disputes do arise, they are easily settled'.[61]

[54] Zhou, 'Jixu kaizhan minzu quyu', p. 648.

[55] Robert B. Ekvall, 'Peace and War among the Tibetan Nomads', *American Anthropologist*, 66 (1964), 1119–48 at 1122–3.

[56] Zeku County People's Government Archives (hereafter ZCPG), folder 2, folios 68–70.

[57] Zeku Xianzhi Bianji Weiyuanhui (ed.), *Zeku xianzhi*. Beijing: Zhongguo xian zhen nianjian chubanshe, 2005), p. 305.

[58] ZCPC, folder 1, folios 10–11.

[59] ZCPG, folder 15, folio 69.

[60] *Renmin Ribao*, 10 February 1952, p. 3.

[61] Chen Sigong, 'Qinghai jingxing minzu quyu zizhi de jingyan ji jinhou yijian', in *MGWH*, pp. 656–61 at 656. See also 'Qinghai sheng muqu tongyi zhanxian gongzuo qingkuang', *Neibu cankao*, 153 (8 July 1952), 76–9.

Upper levels made clear that dispute mediation was a political task that offered the CCP opportunity to build support among non-Han people. By the same token, incorrect implementation of mediation work would harm the party's reputation. Nevertheless, considering the attention and importance given by local, provincial, and even national leaders to inter-community disputes, instructions for resolving them were often big on platitudes but surprisingly thin on specifics. Pointing to the complex nature of many 'nationality disputes', Qinghai's party secretary Zhang Zhongliang told subordinates, 'We must be proficient at analysing [conditions], grasp how various kinds of contradictions are connected, concentrate strength, resolve the problem that is most pressing and for the masses easiest to understand, and then move outward to solve surrounding issues.'[62] Using similarly circumspect language, other missives directed cadres to proceed 'according to the principles of fairness, unity and mutual aid',[63] 'to take the present situation as the starting point, pay attention to history, show consideration for the overall situation, [and] benefit production and unity',[64] and, most commonly, to promote 'mutual understanding and mutual concessions'.[65]

In theory, this type of cooperation was now possible because the exploitative conditions that had historically produced conflict were in the process of being eliminated. Using language that sounds suspiciously as if it came from the screenplay for *Gold and Silver Grasslands*, a headman in Zeku explained, 'In the past the bandit Ma Bufang gave us guns, bullets and horses to rob people. And we got rewarded for sharing with him. In this way he sowed discord [and harmed] our unity.'[66] Unlike in the movies, however, it was the old feudal class that was expected to take the lead in repairing the damage by negotiating in good faith and making concessions for the benefit of the greater good. In Zeku, one dispute was purportedly resolved when a headman agreed to temporarily loan a piece of grassland to a rival. Another was diffused when the area's largest chiefdom ceded a winter pasture to a cross-border neighbour.[67]

The herders unite hand in hand

For CCP leaders, reliance on indigenous elites was necessary because of their considerable influence over the 'masses'. Conflict mediation, however, also was one of the primary functions that hereditary lay elites and important religious figures

[62] Zhang Zhongliang, 'Ba dang de minzu guanche dao gexian gongzuo zhong qu', in *JQSX*, pp. 141–54 at 147–8.
[63] ZCPG, folder 4, folio 109.
[64] Zeku Xianzhi Bianji Weiyuanhui, *Zeku Xianzhi*, p. 364.
[65] ZCPG, folder 15, folio 69; ZCPC, folder 9, folio 10.
[66] ZCPG, folder 4, folios 121–2. Ma Bufang (1903–75) was the last and most infamous of the three men who would lead the Qinghai Ma clique during the Republican period.
[67] ZCPC, folder 6, folio 79; ZCPG, folder 5, folio 78; ZCPG, folder 15, folios 62–3.

traditionally played across the region. When a dispute was internal to a chiefdom, resolution could be relatively straightforward. In those cases, Fernanda Pirie writes, 'mediation is carried out by the *gowa* [chieftain] and there is considerable social pressure on the disputants to agree to a settlement'.[68] This may help explain how Zeku County was able to settle more than 130 'internal disputes' over a few weeks in autumn 1953. In comparison, larger disputes between chiefdoms commonly demanded complex mediation processes in which outside intermediaries (T. *zowa*) played vital roles. These were often headmen of other chiefdoms, but, Ekvall adds, 'the most preferred of all mediators were lamas and other high ecclesiastic dignitaries'.[69]

Among the few, frustratingly brief, descriptions of the mechanics of mediating grassland disputes in the early 1950s is the resolution of the Gyelwo–Gengya feud. At issue was Saiqinggou, a roughly 75km^2 stretch of grassland that fell along the boundary between Qinghai and Gansu provinces. According to most accounts, the conflict began around 1915. By 1950, 86 people had been killed, more than 70 injured, and a large number of livestock lost in fighting that seemingly broke out each summer.[70] Much like KMT commander Liu Huimin in *Gold and Silver Grasslands*, PRC sources add that Ma Bufang had inflamed the feud by giving weapons to the Gyelwo while recognising both chiefdoms' ownership over the disputed valley.[71]

Details remain sketchy. However, Trashi Namgyel (1923–97), the last Rongwo *nangso* (a hereditary pan-chiefdom position unique to this part of Amdo), reports that in July 1950 35 members of the area's traditional elite along with representatives of the state began to negotiate a permanent end to the feud. Investigations were first conducted on site at Saiqinggou and then a smaller group met in Lanzhou under the direct supervision of Northwest NAC director Wang Feng. In November, an agreement was reached in principle to divide the grassland near its midway point.[72] According to a reporter from the *Gansu Daily*, on 7 June 1951 hundreds of members of both chiefdoms 'braved rain, snow, and hail' to warmly welcome regional, provincial, and county representatives to the Saiqing grasslands. Over the next month, state representatives surveyed the rugged, hilly land, mediated between the headmen who had settled into separate encampments, provided medical care to

[68] Fernanda Pirie, 'Feuding, Mediation and the Negotiation of Authority among the Nomads of Eastern Tibet', *Max Plank Institute for Social Anthropology Working Papers*, no. 72 (2005), 14.

[69] Ekvall, 'Peace and War', 1141.

[70] *Guangming Ribao*, 8 August 1951, p. 2; Xibei Junzheng Weiyuanhui Minzu Shiwu Weiyuanhui, 'Huzhu hurang, jiaqiang tuanjie Zangzu Ganjia, Jiawu buluo caoshan jiufen tiaojie huajie xuanchuan gangyao', *Huangnan wenshi ziliao*, 5 (2001), 8–13 at 9.

[71] Ruan Hai, 'Huigu tiaojie Tongren Jiawu buluo yu Xiahe Ganjia buluo lishi yiliu caoshan jiufen zhong guanche zhixing minzu zhengce de jingguo', *Huangnan wenshi ziliao*, 5 (2001), 1–7 at 3.

[72] Zhaxi Anjia [Trashi Namgyel], 'Yi wo jiefang qianhou de jingli he gongzuo pianduan', *Huangnan wenshi ziliao*, 2 (1991), 246; Ruan, 'Huigu tiaojie Tongren Jiawu buluo', p. 4.

both people and livestock, drew topographical maps, and eventually determined a suitable division between the Gyelwo and Gengya chiefdoms.[73]

Finally, on 1 July – 'a rare sunny day at Saiqinggou' – a celebration was held to coincide with the 30th anniversary of the founding of the CCP. On this 'doubly joyous occasion', in the presence of delegations from surrounding chiefdoms and beneath two red flags perched atop opposing mountains marking the line of demarcation, the two sides signed a 'Pact of Unity and Patriotism'. Afterward, a 'unity meal' was held at which leaders of the two chiefdoms 'clinked glasses and shook hands for the first time in over thirty years'. The Gyelwo headman Dorjé (1899–1974) is reported to have donated 100 sheep and two machine guns to the volunteers fighting in Korea and presented Wang Feng with a ceremonial scarf (*khata*) and a letter expressing his gratitude to the party and Chairman Mao. His Gengya counterpart offered a silk banner embroidered in both Tibetan and Chinese with the words, 'The Chinese Communist Party is the great benefactor of the Tibetan people!'[74]

The symbolic value of the resolution of the Gyelwo–Gengya dispute was immediately seized upon by propagandists and policy makers who championed it as a great victory for nationality unity. When *Gold and Silver Grasslands* was released two years later, accompanying articles declared the fictionalised account an 'accurate' portrayal of how the Communist Party had eliminated nationality exploitation and helped minorities mediate feuds, making 'the PRC a big family based on the friendship and cooperation of each nationality'.[75] Decades later, when the history of Amdo's 'early-liberation period' was written, its importance was recycled as a near-parabolic example of the correct implementation of the CCP's nationality policies and thus as a repudiation of the Cultural Revolution's unfettered Han chauvinism and a model for the post-Mao period.[76] A Han cadre who had attended the celebration later recalled attendees joyously singing:

> On the bright green Saiqing grassland,
> The *kelzang* flower emits a refreshing fragrance,
> The Gyelwo and Gengya herders unite hand in hand,
> Presenting each other with *khata* and wishes of good fortune and success,
> The one-hundred-year enemies have become 'in-laws.'[77]

At the time, he continued, 'many people said things like, "If such a large problem can be settled, what nationality dispute cannot be resolved?" '

[73] Ya Lin, 'Tuanjie de qizhi piaoyang zai caoyuan shang', *Guangming Ribao*, 8 August 1951, p. 2.

[74] Ya, 'Tuanjie de qizhi piaoyang', p. 2.

[75] Jiang, 'Yibu fanyang woguo minzu shenghuo', pp. 12–13.

[76] See Benno Weiner, 'The Aporia of Re-Remembering: Amdo's "Early-Liberation Period" in the Qinghai *Wenshi Ziliao*', in *Conflicting Memories: Tibetan History under Mao Retold*, eds. Robert Barnett, Benno Weiner, and François Robin. Leiden: Brill, 2020, pp. 41–77.

[77] Ruan, 'Huigu tiaojie Tongren Jiawu buluo', 6. The *kelzang* (C. *gesang*) is a usually purplish flower found on the Amdo grasslands that is said to bring happiness.

What indeed? While the end of the Gyelwo–Gengya dispute was hailed as a historic victory for nationality unity, it is also an example of the pyrrhic nature of so many of the CCP's 'successes' in 1950s Amdo. On the surface, what little we know of the CCP's mediation tactics resembled those traditionally used to resolve grassland disputes as described by Ekvall. Most obvious was the reliance on elite intermediaries. Similarities also included the formation of 'mediation committees', the responsibility of the host to act as guarantor of the peace, the division of factions into what the party called 'small groups' so that the mediator could work among each separately, a determination not to accord blame to either side, the employment of 'lengthy and sententious speech-making', and the accompaniment of resolution with shared meals and celebration.[78]

Perhaps most importantly, both the CCP and traditional mediators were meant to operate through what Ekvall called 'consensus making' and the CCP referred to as 'consultation and persuasion'. However, herein lies a major difference: according to Ekvall, in the pre-1949 period a resolution was 'never imposed, and the parties ... retained the right of final choice'.[79] By contrast, the CCP put tremendous pressure on both cadres and local elites to produce results. At first glance, the hundreds of disputes that under the aegis of the party were purportedly resolved across Amdo each year suggest that its leadership's diagnosis had been correct, that the feudal and reactionary rule of past regimes had been the primary driver of intra-nationality disunity. Yet, throughout the 1950s new conflicts seemed to always spring up and an alarming number of feuds that were thought to have been settled reignited.[80] By 1956, internal documents quietly began to place blame at the feet of the 'nationality ruling class' (*minzu tongzhi jieji*) who in order to 'maintain their dominant position ... create conflict among the people and even armed fighting'.[81] Albeit initially behind closed doors, disunity among and between *minzu* was no longer solely ascribed to the legacy of great nationality exploitation. Increasingly the principal cause of intra-nationality conflict was being attributed to something both more familiar and more nefarious to the Han cadres sent to work in the Amdo grasslands: class struggle.

The consequences of this shift would not fully manifest until the United Front was abandoned at the start of the Great Leap Forward (GLF) in 1958. National leaders like the NAC's Wang Feng, who had overseen the resolution of the Gyelwo–Gengya feud and at one time had loudly excoriated the poisonous influence of 'Han chauvinism', now insisted that it was local nationalism that posed the gravest danger to the 'unity of the motherland'. Wang insisted that some Uyghurs were seeking to establish an 'autonomous republic', some Sino-Muslims a 'Hui-stan' (*Huizu sitan*), some Mongols a federated republic, and – as a prelude to independence – some

[78] Ekvall, 'Peace and War', 1142–7; Pirie, 'Feuding, Mediation', pp. 14–18.

[79] Ekvall, 'Peace and War', 1123.

[80] ZCPG, folder 15, folios 32–3.

[81] ZCPC, folder 6, folios 78–9.

Tibetans wished to combine the plateau into a single 'Greater Tibetan Autonomous Region'.[82] Minority status indeed had become 'regarded as [a threat] to the territorial integrity and sovereignty of "the nation" and to the imagined privilege and power of the dominant ... group.'[83]

In Amdo and surrounding areas, the causal decoupling of local nationalism from Han chauvinism was accompanied by attacks on the region's Buddhist and Islamic leaderships and the coerced collectivisation of pastoral areas. In response, a massive rebellion swept across Amdo in spring and summer 1958. Security forces answered with a brutal counterinsurgency campaign. Tens of thousands of Mongols, Hui, Salar, Tu, Kazakh, and especially Tibetans were arrested; untold numbers were killed. Among them was the vast majority of elite figures who had been so prominently relied upon under the United Front. This included Rongwo nangso Trashi Namgyel and the Gyelwo chieftain Dorjé, both of whom were detained in June 1958. The latter died 16 years later while still a prisoner. Like so many others, Trashi Namgyel only reappeared at the start of the Reform era.[84]

Conclusion: Minoritisation in a time of great changes

For Amdo Tibetans, 1958, not 1949, was 'the time of great changes'.[85] In a fitting bit of symbolism, not only was the film *Gold and Silver Grasslands* pulled from circulation that year, but Jinyintan – the actual place near Qinghai Lake – was literally erased from the map. It had become one of China's primary nuclear research sites and its very existence demanded utmost secrecy.[86] In the film, the fictionalised Tibetan chiefdoms living in the area unite under the leadership of the CCP to share their pastures. In reality, Jinyintan's inhabitants were removed from the Gold and Silver Grasslands at gunpoint. According to a former public security official, 'hundreds died along the way'.[87] By then, it was no longer proper to portray Amdo's traditional elite as victims of past nationality exploitation. They had become class enemies.

Of course, members of minority nationalities were not the only ones persecuted in the wave of political campaigns that washed over China during the Maoist period. Yet, the socialist (and now post-socialist) project in China has long been embedded

[82] Wang Feng, 'Guanyu shaoshu minzu zhong jinxing zhengfeng he shehui zhuyi jiaoyu wenti de baogao', in *Fandui difang minzu zhuyi xuexi ziliao*. [Lanzhou]: Gansu sheng minzu shiwu weiyuanhui, 1958, pp.11–37 at 13.

[83] Klein, 'Making Minorities', pp. 17–18.

[84] See Weiner, *Chinese Revolution*. For a riveting description of the state violence inflicted upon pastoral communities in Amdo, see Naktsang Nuluo, *My Tibetan Childhood: When Ice Shatters Stone*. Durham: Duke University Press, 2014.

[85] Naktsang, *My Tibetan Childhood*, p. 7; Tsering Woeser, *Tibet on Fire: Self-Immolations against Chinese Rule*, trans. Kevin Carrico. London: Verso, 2014, p. 30.

[86] Qiu Ruixian, 'Jinyintan: Wei "liangdan yixing" yinshen 30 nian', *Chuancheng*, no. 10 (2010), 20–1.

[87] Yin Shusheng, 'Jinyintan zhi tong', *Yanhuang Chunqiu*, no. 3 (2010), 41–5 at 45.

within a quest for ethnonational salvation that at various points and places has targeted minoritised peoples.[88] Despite the CCP's initial commitment to its own brand of pluralism and minority rights, the process of minoritisation in Amdo led to the kind of 'repression and mass violence' seen elsewhere. The late-Ottoman Empire and its successor states that Klein analyses are far from perfect analogues. Nonetheless, as Eric Weitz suggests, forced deportations and ethnic pogroms like those that occurred at the end of the Ottoman Empire and minority protections as written into the Chinese state constitution, 'were, and are, two sides of the same coin ... an entirely new way of conceiving of politics focused on discrete populations and the ideal of national homogeneity under the state'.[89] In Amdo by 1958 and in many other contexts since – including among Turkic Muslims in Xinjiang today – being a minority in China has the potential to 'mark' you as particularly dangerous and/or particularly backward and therefore in need of transformation or even elimination.

But what of the party's plan to make *minzu* out of the disparate interests and identities in Amdo and beyond? If in 1949 'Tibetan' (C. Zangzu; T. Bödpa) was not a pivotal part of a shared identity for many people across the Tibetan Plateau, it is now. Yet, nationality autonomy has not been the powerful weapon imagined by Zhou Renshan and his colleagues; it did not become a transformative institution with the ability, borrowing again from Lu, 'to shape socialist subjectivity that transcends any kind of ethnic boundaries'.[90] Instead, it and other legal and less formalised markers of minzu-based difference – including minority nationalities film – along with frequent and in some cases ongoing acts of state violence against non-Han communities, have helped consolidate both minority and majority *minzu* subjectivity in China in part by 'marking' minorityhood in multiple ways while leaving the majority 'unmarked',[91] the 'default ethnicity'.[92]

In Amdo's grasslands, as elsewhere in China, *minzu* consciousness has hardened. But it is not all encompassing. Amdo Tibetans have complicated relationships to the Chinese state and nation, but they also often continue to strongly identify with their local communities. In fact, even as state power transformed Amdo after 1958, grassland conflicts continued to regularly occur. In Zeku, for instance, one dispute exploded into violence on at least five occasions between 1953 and 1986, while another that was thought to have been resolved in the 1950s resurfaced in

[88] Uradyn E. Bulag, 'Ethnic Resistance with Socialist Characteristics', in *Chinese Society: Change, Conflict, Resistance*, eds. Elizabeth J. Perry and Mark Seldon, first edn. London: Routledge, 2000, pp. 178–97.

[89] Quoted in Klein, 'Making Minorities', p. 22.

[90] Lu, *Moulding the Socialist Subject*, p. 95.

[91] Pandey, *Routine Violence*, chapter 6. On Han as an 'unmarked category', see Dru Gladney, *Dislocating China: Reflections on Muslims, Minorities, and Other Subaltern Subjects*. Chicago: University of Chicago Press, 2004, p. 83.

[92] Stevan Harrel, *Ways of being Ethnic in Southwest China*. Seattle: University of Washington Press, 2001, p. 293.

the 1960s, 1980s, and 1990s.[93] As for the Gyelwo and Gengya feud that had been 'settled' to such celebration in 1951, a local headman named Shawo Tsering (1932–2015) would later write, 'the mediated peace was repeatedly violated, and those violations were repeatedly mediated, over and over again, subsiding for a while and then erupting again.'[94]

Beginning in the early 1980s, when livestock was returned to individual herders, boundaries redrawn, and grasslands reallocated and fenced, grassland disputes appear to have spiked across Amdo. Some were new, but others were continuations of feuds that had their origins in the pre-1949 period. And although the state has legal and coercive capacities it did not possess in the early 1950s, authorities often have ceded at least partial authority to lamas and headmen, who, as in the period before the arrival of the Communists, mediate through consensus formation rather than statutory codes and punishments.[95] More recently, in Qinghai and elsewhere on the Tibetan Plateau the state has pursued efforts aimed at pastoral sedentarisation. It remains to be seen what the long-term impacts of the campaign will be for the government's stated goals of poverty alleviation and environmental protection, although initial findings are not encouraging. It is even more difficult to predict what its impact will be on some unstated goals: disrupting longstanding 'social and cultural structures', reducing the risk of armed resistance, and facilitating the full integration of Amdo's pastoral population into the Han-dominated, market-based society.[96]

[93] Zeku Xianzhi Bianji Weiyuanhui, *Zeku Xianzhi*, pp. 394–5.

[94] Xiawu Cailang [Shawo Tsering], *Bazong qianhu cangsan: Xiawu Cailang zishu*, recorded and compiled by Zhao Qingyang, revised and edited by Zhao Shunlu (NP, Huangnanzhou zhengxie wenshi ziliao weiyuanhui, 2010), pp. 134–6. Shawo Tsering is referring to the breakdown of settlements brokered both before and after 1949.

[95] Pirie, 'Feuding, Mediation'; Emily T. Yeh, 'Tibetan Range Wars: Spatial Politics and Authority on the Grasslands of Amdo', *Development and Change*, 34 (2003), 499–523.

[96] Jarmila Ptáčková, *Exile from the Grasslands: Tibetan Herders and Chinese Development Projects*. Seattle: University of Washington Press, 2020.

3

The Political Economy of the Everyday: Theory and Praxis of the Urban Commune Movement

FABIO LANZA

DURING A PRESENTATION on my research on the urban commune movement, I noted in passing that one of my goals was to link a description of the changes in everyday life and street-level practices to an analysis of the nationwide debate on the socialist economy and the Maoist critique of Stalinist economic orthodoxy. One of the attendees, a senior China historian, took exception to that goal, asserting that common people did not care about political or intellectual debates and only reacted to what directly affected their lives and families. Now, it is far from clear that 'common people' in post-1949 China were estranged from political discussions, given how Maoism was configured as a pervasive educational process for the masses and the Party in which both were supposed to contribute to the creation of the ideological discourse. But even leaving that aside, the incredibly intense if seemingly abstract discussions on the Law of Value, the persistence of the commodity form, wages, and bourgeois right that took place in coincidence with the Great Leap was part and parcel of the larger search for the revolutionary praxis that could more rapidly transform the lives of people and forge new (communist) social relationships. This was neither the case of a theoretical position or an ideological argument at the top fuelling a policy change at the bottom, nor that of a political experiment at the street level that needed to be justified and rearticulated at the level of Marxist theory. Rather, as it often is the case for Marxist politics, the two aspects were interdependent, co-determined, and yet always in a state of profound tension. In this chapter, I bring together street-level sources on the urban commune movement with some crucial texts in the debate on the socialist economy to argue that those discussions shaped policies that did indeed affect the everyday of urban residents and that were cited by cadres at all levels to argue for and against specific changes in terms of retribution, labour participation, and welfare (*fuli*). By combining the analysis of experiential praxis with that of theoretical debates I aim at clarifying what were the political stakes in the experience of urban collectivisation, and I also propose a different approach to the history of the Maoist 'everyday'. The everyday is not a space separated from politics, nor is it a space of real, unadulterated, 'true' experience, which exists outside structures of power. The space of everyday life contains both the 'ordinary' struggles of living beings and the more abstract forces of political conflicts, and thus it connects the state and the people. I therefore take the 'everyday' to be the location where political change can be measured, viewed, and understood.[1]

Remuneration and welfare,[2] i.e. what is provided to workers in exchange for their labour, constitute just one element in the purported socialist transformation of the quotidian, but they are crucial for that purpose and, by the end of the 1950s, the situation in Chinese cities looked heavily unbalanced. The institution we rightfully associate with the Maoist (urban) everyday is the *danwei* (work-unit), which had emerged in the 1950s as the basic component in the organisation of society, labour, and, ultimately, life itself. Under planned economy, the *danwei* evolved from a political to an economic unit, and eventually became the conduit by which goods and services were guaranteed and dispensed to an increasingly large sector of the urban population. This process did not necessarily happen by design. The socialist state originally wanted to achieve a system of equal distribution at the city level, but a series of constraints – lack of resources, the resilience of urban structures, and the very administrative and financial functioning of state-owned enterprises (SOEs) – led to a lopsided development by which the growing number of state employees, at various administrative levels, lived a life that was completely dependent on and framed by the *danwei*.[3] By the late 1950s, the system was largely set. Workers belonged to a work-unit, and while that relationship had a darker, controlling side,[4] the *danwei* was also and mainly a provider: food, consumer goods, basic services like clothing, showers, haircuts, and even entertainment were either offered directly or more often indirectly allotted through a complex rationing system.[5] Within the

[1] For a discussion on approaches to the history of the Mao era, including a discussion about the 'grass-roots' and the 'everyday', see the special issue of *positions* I co-edited with Aminda Smith, 'The Maoism of PRC History: Against Dominant Trends in Anglophone Academia', *positions: asia critique* 29: 4 (November 2021).

[2] Here by 'welfare' I mean all the provisions of goods and services that went under the name of '*fuli*', especially in the *danwei* system: housing, medical care, education, childcare, all kinds of food rations, sick leave, entertainment, but also everyday services such as barbers, laundry, tailoring, etc.

[3] David Bray, *Social Space and Governance in Urban China. The Danwei System from Origins to Reform*. Stanford: Stanford University Press, 2005. See also, Fabio Lanza, 'A City of Workers, A City for Workers? Remaking Beijing Urban Space in the Early PRC', in *China: A Historical Geography of the Urban*, eds. Ding Yannan, Maurizio Marinelli, and Zhang Ziaohong. London: Palgrave Macmillan, 2018, 41–66.

[4] *Danwei* employees were monitored through the system of personnel files, *renshi dang'an*.

[5] Joel Andreas has aptly labelled this set of privileges, which included lifetime employment, 'industrial citizenship'. Joel Andreas, *Disenfranchised. The Rise and Fall of Industrial Citizenship in China*. Oxford: Oxford University Press, 2019.

state-owned sector of the economy, under the so-called 'ownership by the whole people' (*quanmin suoyouzhi*), while wages still constituted part of the employees' compensation and accounting at the enterprise and inter-enterprise levels was still measured in monetary terms, money did not seem to function anymore as either the universal medium of exchange or the representation of socially necessary labour embodied in commodities. The massive collectivisation effort of the Great Leap was, among other things, also an attempt to expand those forms of remuneration and distribution to those sectors of society and the economy which were not (yet) under 'ownership by the whole people'. Accordingly, the coeval debate on the socialist economy focused precisely on how such forms actually functioned, on which of them were appropriate to the socialist economy, and on how they should be transformed so as to guarantee the vaunted transition into communism.

In fact, while it was potentially the aspirational and normative model for the socialist everyday, life within the SOE *danwei* was not synonym for life in Maoist China or even life in a Chinese city. Besides the overwhelming majority of Chinese citizens who lived in the countryside and had little or no access to welfare or services, a sizeable sector of the urban population was also systematically excluded from those benefits: the unemployed or those employed in collectively or privately owned enterprises, which were, on average, much more likely to be female. As Nara Dillon has cogently argued, by the end of the first Five-Year Plan, the limits of the Stalinist development strategy became evident in that industrialisation seemed incapable to solve the issue of unemployment. Fearful that expansion of welfare outside the SOEs would outstrip resources, the state explicitly restricted and rationed services.[6] By 1957, urban life was experienced in radically different ways, depending on whether one was inside or outside the state-owned *danwei* (and few very large collectively owned enterprises). Despite the massive reduction of private businesses that took place in the late 1950s, the excluded still relied on commercial venues and on transactions mediated by money for much needed products and services. This separation was inscribed in urban space, as the *danwei* came to constitute the building bloc of Chinese cities; the people left out of socialist welfare therefore tended to live in the interstitial spaces between large compounds, or in urban areas that had been historically resilient to massive industrialisation or rebuilding, like central Beijing, where most of my archival sources come from.

Forging a communal everyday

The establishment of urban people's communes (*chengshi renmin gongshe*), which cropped up in major and minor centres between 1958 and 1962, pushed collectivisation to the entirety of urban society, or at least as close to that extent

[6] Nara Dillon, *Radical Inequalities. China's Revolutionary Welfare State in Comparative Perspective*. Cambridge, MA: Harvard University Press, 2015. Carl Riskin, *China's Political Economy. The Quest for Development since 1949*. Oxford: Oxford University Press, 1987.

as possible.[7] One of the first actions was placing private enterprises and service providers under the authority of sub-district committees, cooperatives, and eventually communes. The sources provide scattered but vivid evidence both of the resilience of these businesses through the 1950s and of their ultimate collectivisation. A 1960 essay analysing the cases of three major cities in Sichuan (Chengdu, Chongqing, and Zigong) describes an urban texture dotted with small private enterprises, private shops, peddlers, artisans, but also private clinics, and even shamans and quacks (*wupo, jianghu yisheng*). The authors candidly admitted that these businesses responded to the actual needs of the urban population and, while they had 'capitalist tendencies' and allegedly engaged in illegal activities, they could not be summarily eliminated. The solution was then to integrate these enterprises in the urban communes, allowing the new organisations to take over private materials and means of production, which were not insignificant: one commune was reported to have a starting capital of ¥24,620, 70.93 per cent of which was originally privately owned.[8] In Beijing, the Fusuijing commune (Xicheng district) created its lending library in 1960 by absorbing the four existing lending centres, through a seemingly more negotiated process. They convened meetings, set prices for books and shelves, and hired some of the owners as managers. The booklenders tried to wrestle specific concessions from the commune administration, one of them even requesting an urban *hukou* for her son. But, in the end, books, stores, and furniture ended up under the management of the commune and as property of its service centres.[9] Once private enterprises were turned over, they became an integral part of the commune capital and collective ownership.[10]

Yet, while the Great Leap expanded collective life outside the confines of the *danwei*, urban communes did not and could not replicate the everyday of the work-unit. The commune was a different organisation, with a different articulation between production and welfare, and different forms of belonging and ownership. It was the form adopted to specifically address the impossibility of extending a state-funded and more uniform welfare system to urban centres. It was, by definition, a

[7] Urban collectivization during the Great Leap started at the same time as its rural counterpart but proceeded with a different timeline. The establishment of communes was halted in China's five major cities in late 1958 and it did not start again until April 1960, so that the urban campaign had a significant resurgence while the countryside was suffering the disasters of the Great Leap famine. Li Duanxiang, *Chengshi renmin gongshe yanjiu*. Changsha: Hunan Renmin Chubanshe: 2006.

[8] Wang Yongshi, Zhao Guoliang, Wang Fangtian, Guo Shaoxiang, Li Bicheng, Xiao Deyu, 'Guanyu chengshi renmin gongshe suoyouzhi wenti de chubu tantao', *Caijing kexue*, Issue 04 (1960): 21–33.

[9] Beijing Municipal Archive (hereafter BMA) 001-024-00110: 'Fusuijing renmin gongshe dangwei guanyu tushu zulinye gaizao de qingkuang baogao' (June 1960). Three of the four lending centres were privately owned, and one was a cooperative store. They were all reported to be quite profitable.

[10] One extreme case was that of two Beijing communes, Deshengmenwai and Yuetan, whose administration appropriated local mosques and converted them into showers and nurseries. In 1962, hundreds of residents wrote to the district to have their mosques returned to their original use. BMA 001-028-00036, 'Dewai, yuetan gongshe huimin qunzhong lianming xiexin yaoqiu tuihui bei gongshe zhanyong de qingzhensi' (21 July 1962).

transitional form, or rather the form best suited for the socialist transition, under which the vestiges of the capitalist era were supposed to be replaced by always more progressive structures. As the saying went, 'communism is heaven, the commune is the bridge to heaven'.[11] Therefore, commune *fuli* and remuneration were managed in complex and uneven ways; even before 1961–2, when most services were eliminated or massively reduced, welfare varied from commune to commune, and, more surprisingly, within the same commune. Who had access to what services and how these services could be provided were contested and vital issues.

Those communal canteens, service centres, and childcare facilities that were cited as models for efficiency and profitability provided very much needed services largely by deploying cheap labour and by establishing connections with the fully socialised and collectivised sectors of the economy. A celebrated service centre in Lanzhou, Gansu Province, had started by delivering boiled water and doing laundry but had quickly expanded into hairstyling, cleaning, helping the pregnant and the sick. They helped with funerals and weddings, from providing the bridal chamber to renting blankets and hosting guests. What could not be done through commune labour was achieved by setting up agreements with a series of non-commune *danwei*: the workers' hospital, the city traditional medicine clinic, the cinema, the theatre, the railway ticket office, the *sanlunche* (pedicab) station.[12] In Beijing, the Chongwen district set rules for the commune service centres: they should coordinate with the state-owned trade and service bureaus and be located in the interstitial areas where state-owned services were sparse or non-existent. Provisions of raw materials as well as prices should be set in coordination with city and state agencies, but prices should in general be lower than in the state-owned facilities.[13] However, how the connections between the state-owned *danwei* and the communes were practically negotiated is far from transparent, especially in that this implied merging an economic structure based on rations and one still predominantly based on money. In some cases, this led to a convoluted system of exchange between coupons and money. Canteen employees in Lanzhou, for example, were supposed to provide coupons in exchange for the rations of flour and vegetables they took out of storage, while the canteen manager had to repay the coupons he received fully with cash.[14]

Even more stubbornly intractable was the problem of who should benefit from the new services provided and at what price. And that became especially urgent

[11] According to Roderick MacFarquhar, the ditty was attributed to Kang Sheng. Roderick MacFarquhar, *The Origins of the Cultural Revolution. 2: The Great Leap Forward 1958–1960*. New York: Columbia University Press, 1983, 103.

[12] Zhonggong Gansu shengwei caimaobu chengshi renmin gongshe diaochazu, 'Yige shenshou qunzhong huanying de wanshi fuwubu', *Caijing yanjiu*, Issue 08 (1960), 37, 30.

[13] BMA 001-006-01864, 'Zhonggong Chongwenqu bangongshi zhengli, chengshi renmin gongshe diaocha cailiao zhiliu — guanyu gongshe de fuwu shiye' (28 August 1961).

[14] Zhonggong Gansu shengwei caimaobu chengshi renmin gongshe diaochazu, 'Banhao chengshi jumin shitang de yixiang zhongyao cuoshi: Lanzhoushi Baiyinlu renmin gongshe shiban shitang guanlizhan de jingyan jieshao', *Caijing yanjiu*, Issue 08 (1960), 38–9.

by 1960–1, when the welfare sector of many urban communes showed evident signs of crisis, and decisions had to be made about whether and how to finance welfare. As for who could access the services, early commune regulations painstakingly detailed how to assess fees for families where one spouse was a commune employee and the other was employed somewhere else, as well as for contract workers and apprentices.[15] A survey of the service centres in Beijing's Chongwen district revealed that they oscillated between two models: they were supposed to be a 'collective welfare' project (*jiti fuli shiye*), providing services to commune employees, yet they were not profitable and required large subsidies from the commune administration. The Qianmen commune, for example, had given its 16 centres over ¥6,200 in the first half of the year. In practice, they had been run as service enterprises (*fuwuxing qiye*), open to all residents (*jumin*) inside and outside the commune, with independent accounting and full responsibility for profit and losses. They belonged to 'collective ownership' and, according to the bureaucrats' assessment, that was the only way in which they could be self-sustaining.[16] Another survey, this one in Xuanwu district (1961), portrays a similar situation. The Chaoyangmen commune had 13 canteens and was losing ¥1,292 just in the month of February, equivalent to a subsidy of ¥1 per customer. Therefore, in March, they changed the financial structure, reduced the number of employees, and opened services to non-commune members, at a fee – ¥0.5 for students, ¥1.5 for workers – and all the canteens immediately turned a profit. Other communes tried to stay faithful to the mission of providing welfare at almost no cost to their members, but, in all cases, this required some form of subsidy, and subsidies affected capital accumulation and reinvestment in expanding production, which was the other main goal of the Great Leap.[17]

The cost of welfare

Even this brief description of how welfare worked within what was the most radical experiment in urban collectivisation under Maoism reveals how riddled with contradictions that experiment was. The contradictions are not simply the expression of a tension between the revolutionary goal of massive social change and a putative reality that resisted being so fundamentally manipulated. Rather, the contradictions were intrinsic to the Maoist project itself, in that some of the very organisations and practices informed and shaped by the desire to move rapidly towards that communist horizon produced effects that made that transition ever more distant. Those

[15] BMA 001-114-00566 1, 'Zhonggong Beijing shiwei guanyu renmin gongshe de guanli tizhi he ruogan zhengce de guiding (caogao)' (18 March 1959).

[16] BMA 001-006-01864, 'Zhonggong Chongwenqu bangongshi zhengli, chengshi renmin gongshe diaocha cailiao zhiliu — guanyu gongshe de fuwu shiye'.

[17] BMA 084-003-00072, 'Dang'an biaoti: chengshi gongshe tuoersuo, shitang gongye, dongyuan huanxiang deng gongzuo de diaocha cailiao' (30 June 1961).

tensions became evident when parts of the communist state castigated officers and activists for promoting policies of excessive egalitarianism, pushed for a continuing reliance on the money economy, gladly accepted and sometimes actively fostered social differences. And the experiment of urban collectivisation was too brief and short-lived to allow for a meaningful search for a solution to those complex tensions.

Welfare provisions proved to be one of the contested grounds, with divergent opinions about how these benefits should be financed under collectivisation. One of the crucial goals of the urban commune experiment was to provide welfare to non-*danwei* urbanites (*jumin*) *without affecting the state coffers*, and therefore welfare was supposed to be managed and financed within the commune budget. That, however, did not answer the question of who should pay for it, that is whether welfare should be financed by deduction from overall commune revenues or commune members should pay for it as individuals, thanks to their newly available income. Several sources from Beijing vocally argued for the continuing need to subsidise canteens and childcare centres, in order to make those services economically viable to the majority of residents, and especially the poorer among them. At the Shijingshan commune, they cited a case of a female worker, with three children, earning around ¥22 per month in the new commune factory, but spending ¥18 in fees for mess halls and childcare, thus dissipating any economic advantage promised by the new system.[18] At the Chunshu commune, another newly employed female worker had a large family (four children) and a very tight budget: the family earned ¥84 per month, vis-a-vis an average monthly need estimated at ¥15 per person. She therefore did not dare to eat at the canteen or send her kids to childcare, and she also kept mending and washing the family clothes.[19] These and other cases were upheld to justify a call for expanded subsidies, without which most commune members could not afford the services now available to them, and therefore could not fully enjoy the benefits of commune membership. Cadres, mostly at the local level, voiced support for providing health care to female workers, as well as for reducing prices at the service centres, financing it with commune reserve capital.[20] This was because the communes, at least in most of the cases cited, had proven to be quite profitable and had indeed accumulated some capital. At Fusuijing, 60 per cent of factory profits were transferred to the commune (40 per cent stayed at the factory level and was used for improvements and expansion), but that capital was almost entirely reinvested into production, with only a small portion devoted to

[18] BMA 001-006-01493 'Shijingshan Zhongsu youhao renmin gongshe zuzhi zhigong jiashu canjia shengchan laodong qude jingyan'.

[19] BMA 001-028-00029 'Chunshu gongshe fuli daiyu qingkuang' (1 September 1961).

[20] BMA 101-001-00782 'Zhigong jiashu chengle shehui caifu de chuangzaozhe: Shijingshan Zhongsu youhao renmin gongshe jieshao'. Note that women who were involved in 'scattered production' (doing finishing work at home or in very small workshops) had no *fuli* at all: BMA 001-028-00031 'Erlonglu renmin gongshe guanyu jizhong shengchan he fensan shengchan qingkuang diaocha' (15 May 1961).

welfare or salary increases.[21] In 1959, Shijingshan had an overall profit of ¥530,000 and allocated less than half of it (¥230,000) to subsidies and welfare;[22] at Chunshu, welfare subsidies counted for 9 per cent of the total salaries paid in the commune, each worker receiving on average ¥1.33 a month, clearly not a significant contribution.[23] Other reports provided comparable accounting figures, yet they came to the opposite conclusion about financing welfare. In Xinjiekou, the commune had a surplus of ¥1,280,000 but had to give ¥249,000 to the service sector; at the Erlonglu commune 20 per cent of the profits went to welfare subsidies and at Chunshu it was 29.8 per cent.[24] These percentages do not seem particularly high or financially threatening, especially if one considers that welfare provisions were essential not only for the radical transformation of social life in the cities but also for the very success of the productive push of the Great Leap. It was impossible to move along the path to communism, in terms of reduction of inequalities, women's liberation, socialisation of domestic labour, and mass participation in culture, politics, and education without a massive expansion of widely accessible welfare services. Yet, despite many dissenting voices, by 1960–1 the Chinese Communist Party (CCP) adopted the policy that communes should not keep throwing money in the 'bottomless pit' (*wudidong*) of welfare.[25] Rather, canteens, childcare centres, and service stations should be run as self-standing enterprises, fully responsible for profits and losses, even if that meant increasing fees and making those services unaffordable for workers and residents. By then, in those commune enterprises that were left, workers' compensation reverted to a mainly monetary form, with wages paid according to time, productivity, and sheer production output, often calculated on a piece-rate basis.

The people and the collective

The contradictions and tensions in the day-to-day running of urban communes and in the administration of collective welfare found a theoretical echo in the debate on the socialist economy waged from the very top of the Party. I cannot do justice here to a discussion that spanned many months and several publications so, at the risk of oversimplification, I will briefly sketch here two main trends of that debate, which are directly connected to the issues discussed in this piece. The first one is the question of ownership and the correlation of various forms of ownership with

[21] Beijing Daxue jingjixi chengshi renmin gongshe diaochazu/ Zhou Zhenhua, 'Fusuijing renmin gongshe diaocha baogao', *Beijing Daxue Xuebao: Zhexue shehui kexue ban*, Issue 04 (1960), 91–108.
[22] BMA 001-006-01493 Zhonggong Beijing shiwei bangongting, 'Guanyu Shijingshan Zhongsu youhao renmin gongshe zizhi zhigong jiashu canjia shengchan laodong de baogao' (28 December 1959).
[23] BMA 001-028-00029 'Chunshu gongshe fuli daiyu qingkuang'.
[24] BMA 084-003-00072 'Dang'an biaoti: chengshi gongshe tuoersuo, shitang gongye, dongyuan huanxiang deng gongzuo de diaocha cailiao'.
[25] BMA 084-003-00072 'Dang'an biaoti'.

the process of transition to communism and the establishment of certain structures of compensation, distribution, and circulation. Socialism was supposed to be a transitory stage, during which a system based on exchange of commodities of certain value (expressed in prices and money) and on remuneration according to labour (expressed in wages) would be progressively replaced by one in which products (not commodities) would be distributed (not sold) and labour would be provided according to the principle 'from each according to their abilities, to each according to their needs'. Postulating a particular social formation as transitional obviously raises a series of questions, concerning primarily how long that transition is supposed to take, how to identify the best path to realise that transition, and which socioeconomic indicators can best provide evidence that the transition is indeed moving along.

The background of this discussion was the publication in the Soviet Union of a new textbook on socialist political economy and Stalin's treatises that accompanied it, conferring it supreme imprimatur.[26] In his analysis, Stalin, following the principle that 'the relations of production *must necessarily conform* with the character of the productive forces', argued that, even in the relatively advanced Soviet Union, while capitalism had been eliminated, two different forms of ownership still persisted: state ownership, which included the modern industrial sector, of whose products the Soviet state could fully dispose; and collective ownership, which comprised of the entire rural sector, whose products could only enter into circulation in the form of commodities. Hence, Stalin concluded, a commodity economy, operating under the Law of Value, with a system of purchase and sale regulated by accounting, currency, productivity, and efficiency, would continue to exist under socialism as long as there is collective ownership. Stalin was notably evasive on how any further evolution could occur, and what specific, active measures would help pave the way and potentially hasten the transition towards communism. Mao and others in the CCP seemed much more concerned about the process, pace, and even the possibility of that transition. In China, like the Soviet Union, the state-owned sectors were under 'ownership by the whole people', which implied full integration in the planned economy, including distribution of goods, resources, and welfare, as well as regulated salaries and prices. All forms of rural coops were instead under a different form of ownership system, 'collective ownership' (*jiti suoyouzhi*), and therefore they were always fully responsible for their profits and losses, and could provide for the overall welfare of their members only on the basis of their own available resources. The economic and social transformations produced by the Great Leap, including the creation of large communes, were believed to bring the transition into 'ownership by the whole people' much closer in time. Mao expressed his dissatisfaction with Stalin's seemingly static assessment both at the Beidahe and the Zhengzhou conferences (in August and November 1958, respectively). In Zhengzhou, he chastised the Soviets for

[26] J.V. Stalin, *Economic Problems of Socialism in the U.S.S.R.* Beijing: Foreign Language Press, 1972.

bragging about entering communism, 'but you only hear a noise on the staircase, you don't see anyone coming down'. Stalin, according to Mao, 'did not promote the elements of communism at all', meaning those economic features that prefigured the future society, such as the direct distribution of products through a supply system. Under those circumstances, the transition to communism was hardly possible.[27] To that Mao compared the initial achievements of the Great Leap, which was bringing forth two major transitions, one concerning ownership, one concerning distribution. He toyed with the timeline for moving communes into 'the system of communist ownership – almost the same as in factories – that is, public ownership of all eating, clothing, and housing'.[28] 'Three, four, five, six years or a bit more—isn't it a bit short? Or is it too long?' Mao mused in Zhengzhou. At Beidahe, he had gone so far as to suggest it would take one to three years.[29] Mao also called for the creation of communes in the cities, as a way to push forward the second transition, from exchange of commodities to allocation of products and services, from currency remuneration to a 'supply system'. '[W]hat is of decisive importance here is whether things can be allocated', Mao concluded. 'Things that cannot be allocated by the state cannot count as being owned by the whole people. [Things] allocated under ownership by the whole people are not 'commodities' any more as defined in political economy'.[30] One the goals of the rural communes was precisely to increase the proportion of services and goods that were directly provided by the collective, thus making life in the countryside more similar to that of the SOE urban work-unit. As the theorist Guan Feng highlighted in the *People's Daily*, by the end of September 1958, 70 per cent of the pioneering rural communes in Henan applied a combination of monthly salaries/bonuses and rationing. The latter could include allocation of grain and rice, eating *gratis* at the commune, or even the provision of basic necessities: in addition to food, clothing, lodging, education, birth, medical care, and expenses for marriages and funerals.[31] While everybody, including Mao, agreed that the rural communes remained firmly within the realm of collective ownership, they were supposed to progressively integrate more and more elements of 'ownership by the whole people', thus embodying a clearly transitional form of life.

Urban communes were considered to be at a further stage of evolution than their rural counterparts and the question of what kind of ownership system they belonged to was more uncertain and open to debate. In July 1960, the city of

[27] 'Talk at the First Zhengzhou Conference' (6–10 November 1958), in Roderick MacFarquhar, Timothy Cheek, and Eugene Wu (eds.), *The Secret Speeches of Chairman Mao. From the Hundred Flowers to the Great Leap Forward.* Cambridge, MA: Harvard University Press, 1989, 443–79 at 448.

[28] 'Talks at the Beidahe Conference (Draft Transcript)' (17–30 August 1958), in *The Secret Speeches of Chairman Mao*, 397–441 at 431.

[29] 'Talks at the Beidahe Conference (Draft Transcript)', 431.

[30] 'Talk at the First Zhengzhou Conference', 451.

[31] Guan Feng, 'Xiang gongchanzhuyi guodu de zuihao de fenpei xingshi. Shilun bufen gongji he bufen gongzi xiang jiehe de fenpei zhidu' *Renmin Ribao* (22 October 1958).

Xiamen held a conference over the form of ownership more appropriate to the urban communes, and diverging assessments came with very different policy prescription. Those who argued for 'ownership by the whole people' pointed at the close connection between commune production and state planning. In Xiamen, the commune factories followed a production plan mandated from the top, received materials through the state-owned commercial sector, and they sold finished products through the same path. They were not supposed to have independent access to the market.[32] Other contributors stressed how prices of commune products were regulated and how, unlike rural communes, salaries did not rely on workpoints[33] and were largely independent from profit.[34] In Beijing, a report by the Bureau of Industry (*gongyeju*) reiterated similar points: the urban commune factories had been created with resources provided by the state or by SOEs, they often worked as subsidiaries of the SOEs and, like SOE employees, commune workers were paid salaries. This placed the commune enterprises straight under 'socialist ownership by all the people' (*shehuizhuyi de quanminsuoyouzhi*). However, the author urged, it was advisable not to spread this piece of information to the commune workers, who could then start demanding welfare, salary, services, and conditions comparable to the SOE workers.[35]

This last caveat aside, these descriptions were meant to have a prescriptive power. As mentioned, the experience of urban communes was far from uniform across cities, and highlighting specific cases in which integration with the state and other characteristics made these often makeshift enterprises more similar to SOEs had obvious strategic policy implications for the present and the near future. Precisely because that future was defined by the aspiring normativity of the SOE work-unit employee, with his (that aspiration was largely gendered as male) salaries and his set of benefits. In terms of welfare, while it was recognised that (female) workers in urban commune factories and services did not and should not yet have benefits equal to their (male) SOE counterparts, this was considered only a transitory situation, one to be overcome via economic development and further expansion of rationing.[36] Another implication of these arguments was that urban commune enterprises did not constitute an alternative model of production, but just

[32] Li Chengzong, 'Wo dui chengshi renmin gongshe suoyouzhi xingzhi de kanfa', *Zhongguo jingji wenti* (*Economic Issues in China*), Issue 08 (1960), 12–14.

[33] Chu Hongdao, 'Cong sheban gongye de gongchanxiao qingkuang kanlai, sheban qiye shi shu quanmin suoyouzhi xingzhi de', *Zhongguo jingji wenti* (*Economic Issues in China*), Issue 08 (1960), 14. The assumption that products of urban commune factories were not to be sold on the market was not, however, followed in other localities, such as Beijing, for example.

[34] Su Shucheng, 'Guanyu chengshi renmin gongshe suoyouzhi xingzhi wenti de shangque', *Zhongguo jingji wenti* (Economic Issues in China), Issue 08 (1960), 14–17; Lei Yaoling, 'Chengshi renmin gongshe suoyouzhi jibenshang shi quanmin xingzhide', *Zhongguo jingji wenti* (Economic Issues in China), Issue 08 (1960), 11.

[35] BMA 112-001-00782, 'Guanyu chengshi jiedao renmin gongshe dangqian jige zhengce wenti de yijian (ergao)', (17 March 1961).

[36] Li Chengzong, 'Wo dui chengshi renmin gongshe suoyouzhi xingzhi de kanfa'.

a stage in the process of development leading to the SOE format and to their complete absorption into the planned economy.[37] Stressing that commune enterprises were closer to the SOE model had also less obvious and possibly more ominous consequences for workplace governance and workers' rights: if they were under 'ownership by all the people', commune factories were not supposed to be democratically managed by an assembly of workers – as it was the case for cooperatives and collective enterprise. Potential inscription in the state ownership system meant also the extension of a system of control and labour management, including detailed personnel files that documented the entire life of each employee and tied them to the enterprise.[38]

Officers and cadres who argued that communes instead remained firmly under 'collective ownership' shifted the emphasis to those aspects of collectivisation that were not yet fully realised. First, legally, commune workers were *jumin* (urban residents), they were not state employees.[39] While the state provided help in founding enterprises, capital came from various sources: enterprise below the district level; funds of the handicraft coops; small entrepreneurs (*getihu*) who joined the commune; money raised among residents; donations by overseas Chinese; profit and accumulation by the commune companies.[40] Similarly, while there were city- or district-owned companies belonging to 'ownership by the whole people' that had been placed under the commune administration, only their management rights had been transferred to the commune, not their ownership. They did not 'belong' to the commune, nor did they affect its overall benefit or remuneration structure.[41] So, while it was recognised that, in the urban communes, 'ownership by the whole people' had a dominant position, collective ownership still should and did determine the aspects more directly connected with social reproduction and workers' conditions, i.e. distribution of wages and services. If the communes were still 'collective ownership', workers and members could not even think about asking for the same salaries and welfare of their SOE counterparts, at least not until a distant and always postponed future; forms of more equitable distribution could be indicted as excessive 'egalitarianism' (*pingjunzhuyi*); and commune leadership could deploy management tactics and compensation systems that were usually not possible in

[37] BMA 112-001-00782, 'Guanyu chengshi jiedao renmin gongshe dangqian jige zhengce wenti de yijian (ergao)'.

[38] Li Chengzong, 'Wo dui chengshi renmin gongshe suoyouzhi xingzhi de kanfa'.

[39] BMA 001-028-00031, 'Guanyu Erlonglu gongshe suoyouzhi wenti de baogao'. The term '*jumin*' deserves a closer analysis. While this is not its fixed or original meaning, in the sources the term seems to be applied specifically to urban residents outside the *danwei* system. So urban communes were supposed to mobilise *jumin* and transform them into workers and producers. I am thankful to Lara Kusnetzky for pointing this out.

[40] Wang Congchao, 'Cong guojia caizheng jiaodu lai kan, chengshi renmin gongshe suoyouzhi jiben shang shi jiti xingzhide', *Zhongguo jingji wenti* (Economic Issues in China), Issue 08 (1960), 9–11.

[41] Wang Yongxi, Zhao Guoliang, Wang Fangtian, Guo Shaoxiang, Li Bicheng, Xiao Deyu, 'Guanyu chengshi renmin gongshe suoyouzhi wenti de chubu tantao', *Caijing kexue*, Issue 04 (1960), 21–33.

the SOEs, like piece-rate wage.[42] Under collective ownership, the commune was solely responsible for welfare and salaries, and that responsibility could be pushed down to individual units. Coherently, there were calls to increase the portion of profit kept at the factory level to as high as 80 per cent. That starved the commune administration of resources and made any collective management of welfare impossible.[43] Here we see clearly how the debate on the forms of ownership fit for a 'transitional stage' reflected and refracted the praxis of collectivisation. Asserting that communes were under 'ownerships by the whole people' meant that commune members could possibly claim a series of welfare and service provisions, which pushed them closer to the normative status of a SOE worker and achieved communism. But that also came with costs for the state, which was neither financially nor politically ready for such an egalitarian expansion of benefits.

This debate probed the limits of what inhabiting a temporality of transition allowed and what it did not: it allowed for experimentation in new forms or production, ownership, social relations, and distribution of resources; but it did not allow for any transformation that could place further burdens on the state, as it was programmatically supposed to be only generating benefits without consuming precious resources.

Bourgeois right

The issue of the forms of ownership to which the 'new socialist things' created under the Great Leap belonged had profound and very practical implications, but it nonetheless remained firmly within the relatively narrow theoretical confines of Stalin's argumentation. Yet the Chinese debate emphatically exceeded those confines, starting with Mao himself. Wages and the stubborn permanence of social inequality that salaries embodied, even within the commune structure, constituted the second main trend in the debate over the socialist economy. This trend was related to and overlapped with the discussion about forms of ownership, but it marked a more decisive move away from the parameters set by the Stalinist orthodoxy.

At the beginning of his Beidahe speech, before he even mentioned ownership, Mao addressed the issue: '[We] must eradicate the ideology of bourgeois right. For instance, competing for position, competing for rank demanding overtime pay, high salaries for mental workers and low salaries for physical labourers—these are all vestiges of bourgeois ideology. "To each according to his worth" is stipulated by law; it's also bourgeois stuff'.[44] The term 'bourgeois right' summarised a

[42] BMA 112-001-00783 'Guanyu Erlonglu gongshe suoyouzhi wenti de baogao'. On the reintroduction of piece-rate wages, see BMA 001-006-01864/1, 'Chengshi renmin gongshe diaocha cailiao zhiyi —dui chengshi renmin gongshe xingzhi, renwu, he suoyouzhi wenti de yijian' (24 August 1961). Ironically, Marx had deemed piece-wage to be 'the form of wage most appropriate to the capitalist mode of production'. Karl Marx, *Capital, Volume I*. London: Penguin Classic, 1990, 698.

[43] BMA 112-001-00783 'Guanyu Erlonglu gongshe suoyouzhi wenti de baogao'.

[44] 'Talk at the Beidahe conference', 406.

series of social relationships, legal structures, and economic arrangements typical of capitalism that, according to Marxist theory, had the appearance of producing formal equality while in reality they preserved the capitalist system and the radical inequalities on which it was predicated. The question, for Mao and for others, was how much of this 'bourgeois stuff' still remained and what functions it served under socialism, especially after the massive development and industrialisation spurred by the Five-Year Plan.

At Beidahe, Mao criticised the wage system in the Soviet Union, where wage grades among Soviet cadres were 'too numerous' and 'the gap [between the cadres and] the workers and peasants' too wide.[45] And while the salary system could not 'be abolished immediately' he argued, '[we should] prepare for it in one or two years'.[46] To the salary system adopted in the Soviet Union and (currently) in his own country, Mao counterposed the experience of the anti-Japanese and revolutionary war, 'a twenty-two year history of war communism, with no salaries'[47] when the Red Army, '[t]hrough several decades of battle, … always practiced communism'.[48] The Chinese revolutionaries adopted a 'supply system', in which army and civilians, officers and men were equal. For Mao, the experience of wartime communism proved that you could achieve enthusiastic mass participation, even in what was literally a life-threatening enterprise, without relying on material incentives, and he argued that the reintroduction of salaries and wages was a retreat from what was already a more progressive (revolutionary) form of social relations. Accordingly, he refused the naturalised assumption that salaries were necessary to spur people's activism and initiative: 'Piece-rate wages are not a good system. I don't believe the adoption of the supply system will make people lazy, inventions fewer, or activism lower. Because decades of experience prove otherwise'. The problem, Mao concluded, was not excessive egalitarianism, but the reliance on the Soviet model, with its structural inequalities, and the continuous reproduction of capitalist forms under the influence of bourgeois right: 'The sources [of our problems] are two-fold: one is socialism, which has been borrowed from Elder Brother [the Soviet Union], and the second is capitalism, which is home born and bred.'[49]

Mao's argument and his critique of 'remuneration according to labor' was forcefully restated and pushed to its logical political conclusions in an article by Zhang Chunqiao,[50] first published in *Liberation* (*Jiefang*) and then reprinted in the *People's Daily*, apparently at Mao's behest and with a short unsigned intro allegedly written by Mao himself, who praised the article as 'basically correct, but somewhat one-sided'.[51] Mao's praise is not surprising as Zhang started by re-threading

[45] 'Talk at the Beidahe conference', 425.
[46] 'Talk at the Beidahe conference', 435.
[47] 'Talk at the Beidahe conference', 434.
[48] 'Talk at the Beidahe conference', 435.
[49] 'Talk at the Beidahe conference', 437.
[50] At the time he was a member of the Shanghai CCP City committee and editor at *Liberation*.
[51] Zhang Chunqiao, 'Pochu zichan jieji de faquan sixiang' (Do away with the ideology of bourgeois right), *Renmin Ribao* (13 October 1958). An English translation is available at http://marxistphilosophy.org/BourgeoisRightWeb.pdf.

Mao's analysis of the experience of the revolutionary war. Then the Party instituted a 'wartime communist life' (*junshi gongchanzhuyi shenghuo*) whose main distinguishing and foundational feature was a 'supply system' (*gongjizhi*), according to which soldiers, militiamen, and party members were all provided with the same amounts of (meagre) rations, no matter their rank or position. It was a hard life but one predicated on a principle of radical equality, of which the supply system was both the premise and the realisation. After the 1949 liberation, however, that system came under attack and was progressively abandoned under the influence of the ideology of bourgeois right. The supply system was labelled 'village style', 'a bad guerrilla habit', not suitable for the cities, and that criticism was eventually accepted and absorbed by CCP cadres. The supply system was replaced by remuneration according to labour in the form of wages, the central feature of bourgeois right. The result was that, in place of the radical equality of the revolutionary war, China developed increasing social differences, inscribed in the wage system, between skilled and non-skilled labour, rural and urban labour, mental and physical labour. Zhang recognised how Marx himself had stated that, in the initial stages of socialism, one cannot immediately implement the communist principle of 'to each according to their needs', but one must still stick to 'to each according to their work'. That did not mean, however, that one should let the principles of bourgeois right determine the structure of society; quite the contrary, one should keep striving to destroy them, and that was the essential goal of the socialist period.[52] Marx himself had provided an example with his praise of the Paris Commune's adoption of radical equality in terms of salaries, which was reiterated by Engels and Lenin. To what extent should the socialist state follow the example of the first proletarian commune?

Zhang, like Mao, also rebuffed the main criticism of the supply system, namely that it hampered productivity by offering the same retribution no matter the amount and quality of work performed. 'Whether you work or not, you still eat!' as it was pithily summarised. To this strictly *economic* determination, Zhang responded by recalling once again the *political* experience of the revolutionary war, when people fought and sacrificed themselves without even the faintest notion of material incentives (did anyone on the Long March receive wages?). Zhang also pointed at the distortions that a formal system of 'equal pay for equal work' produced in a society that was supposedly in the process of being restructured under a set of increasingly equalitarian relationships. Wage grades created a separation between cadres and workers, promoting bureaucratism and hierarchies, while the monetisation of work-time through salaries fomented real laziness, because labour became only valuable as a reason for a monetary remuneration. Zhang concluded that, even if some elements of bourgeois right necessarily remained operative under socialism, the goal

[52] It was also difficult to call elements of bourgeois right simply 'remnants', or 'vestiges', of the pre-socialist period when, as Zhang showed, the wage system had been forcefully *reintroduced* after 1949 as a crucial part of the socialist economy.

was not to defend and preserve them but to progressively and resolutely restrict them, reduce their role, and ultimately eliminate them. Reintroducing a form of supply system that would replace salaries with a direct distribution of products and welfare in the communes was a crucial step in this effort.

As Guan Feng had argued in the *People's Daily* piece mentioned above, the partial rationing system introduced by most communes was precisely a way of moving out of 'remuneration according to labour'. By providing those services, communes were socialising birth, rearing, education, and health, but they were also weakening the hold of bourgeois right. While this form of rationing, Guan argued, was different from wartime communism and was not yet 'fully achieved communism', it introduced a form of distribution based on the principle of 'to each according to their needs', and, as such, it belonged to the category of communism.[53] In another essay, published in *Zhexue Yanjiu* (*Philosophical Researches*) in August 1958, Guan stressed how collectivisation was supposed to have not only practical but also massive ideological effects, breaking the shackles of private (bourgeois) ideology. Guan admitted that, under socialism, the continued existence of 'remuneration according to labour' would unavoidably continue to create differences, but the role of social welfare and direct distribution of products and services was precisely to prevent those differences from becoming too big – and eventually to reduce them. He configured collectivisation as part of a progressive expansion of welfare, eventually leading to the transition from one system (socialism) to another (communism). This was prefigured as a gradual transition (*zhujian guodu*), but for both Guan Feng and Zhang Chunqiao, it was clearly a process that, while dependent on the development of the 'forces of production', was far from being an almost natural outcome of economic expansion and industrialisation, as many Chinese economists seemed to imply. On the contrary, this process had to be pushed forward through direct *political* intervention, continuously expanding and elevating those elements and forms that prefigured communist social relationships while actively restricting the power and scope of those that tended to reproduce capitalist relationships, like the Law of Value, commodity exchange, money, and wages.[54]

Zhang Chunqiao's piece elicited a series of responses and reactions, with many rising in defence of salaries and the wage system, positing subtle differences between the functioning of some elements of bourgeois right under capitalism versus under socialism. Space limitations here do not permit a comprehensive summary of the lively and contentious discussion, which eventually merged into the larger debate on the Law of Value and the socialist economy, but I want to highlight a few crucial points in Zhang's (and Guan's and Mao's) approach. The attack on bourgeois right was connected to those changes that the Great Leap Forward was supposed to produce, and that had less to do with industrialisation or economic development, but rather the transformation of social relationships at a granular

[53] Guan Feng, 'Xiang gongchanzhuyi guodu de zuihao de fenpei xingshi'.

[54] Guan Feng, 'Lüeyu renmin gongshe de weida lishi yiyi', *Zhexue yanjiu*, 1958 (05), 2–4.

level. In this perspective, a structure of remuneration and welfare that privileged the direct distribution of products and services in lieu of monetary compensation did not function simply as a facilitator of production allowing for larger participation in labour and increased productivity. It was also truly the motor for a social revolution. In that, Zhang and Mao's interventions were largely outside the sphere of economics, because economics could not prefigure a way of restricting bourgeois right. That could only come through a political decision and political praxis, and in his piece Zhang pointedly mocked the 'economists'. Here the motto 'politics in command' assumes a different meaning, as it postulates politics as 'distinct fields of relations sharply demarcated from the logic of the economic', including Stalin's political economy.[55] Accordingly, Zhang's argument was grounded on the political premise of radical equality as a starting point, rather than as the end result of a long economic process: it was the precondition of a real socialist transition, not the end product of communism. As Benjamin Kindler has observed, Zhang's argument also represented the emergence of a specifically communist concept of labour. If capitalist labour, even under socialism, was still defined by its exchangeability at a supposedly 'equal' value – that is by remuneration based on time or quantity or productivity – communist labour could only be labour 'gratis'. That is, in this system labour was conducted without remuneration, and its duration and effort was determined solely by the voluntary decision of politically conscious members of society.[56] As Kindler makes clear, this is not the humanistic idea of labour captured in Marx's passage about fishing in the morning and philosophising in the afternoon, but labour as necessary collective toil, which constituted the potential foundation of a new set of social relations. Yet, it is one thing to argue that communist labour should be *unpaid*, meaning that it should be *unwaged*, subtracted from a system of commodity exchange and exploitative remuneration, workers being provided with what they need in an egalitarian system of distribution; it is another to reduce human life to labouring bodies, toiling incessantly, without interruption, and seemingly without end. The step from the former to the latter proved to be very small indeed.

Conclusion

In the experience of the urban communes, as I have argued elsewhere, the collective desire to participate in collective labour in order to produce new social relationships ended up being harnessed to the developmentalist push of the state at all levels.

[55] Benjamin Kindler, 'Writing to the Rhythm of Labour: The Politics of Cultural Labour in the Chinese Revolution, 1942–1976', Ph.D. dissertation, Columbia University, 2021, 253. This piece, and in particular the final reflection on bourgeois right, owes a huge debt to Kindler's work, and much of my own research and investigation on this topic has been conducted in close dialogue with him.
[56] Kindler, 'Writing to the Rhythm of Labour', 258.

The logic of the developmental state, which had existed since the Republican era, had disastrous consequences after 1949, especially for how gendered work and social reproduction came to be evaluated, rewarded, and practically organised.[57] While collective welfare under the commune system was progressively restricted or downright eliminated, thus putting social reproduction back into the domestic sphere, gendered work in commune enterprises was reduced to its minimum economic denominator – that of being cheap, unskilled, underpaid, yet useful work. Accordingly, commune enterprises often reverted to the most abject form of monetisation of labour, like piece-rate wage. It is not surprising that, at the same time, the debate on bourgeois right was temporarily halted. In 1959, the Shanghai conference on the Law of Value and commodity production resulted in a consensus on the necessity of capitalist vestigial forms and their positive effects under socialism.[58]

The foreclosure of these experiments is usually viewed through the lens of a reaffirmation of a rational emphasis on productivity and development of the forces of production, but I believe the implications are more profound. The urban commune experiment was short-lived, but, if placed in the context of the coeval debate on bourgeois right and the forms of ownership, it reveals some fundamental contradictions and fissures of the Maoist political economy. The intertwined promises of accelerated development and radical social transformation that the Great Leap embodied proved to be in tension with each other, not simply at a very practical level but, more essentially, in how they were framed within socialist economic theory. It was not just a question of privileging industrial production over social transformation, even if that was indeed the case. More importantly, in an economic discourse and praxis that remained framed by the very categories of 'bourgeois right', an alternative articulation of production, circulation, and distribution could not even be thought, and localised practices that hinted at that articulation looked dangerous and 'extreme'. Outside that articulation, the massive and largely voluntary deployment of labour that collectivisation prefigured and required turned into an unsustainable form of extraction. Yet, those localised practices and political debates that made up the experience of urban collectivisation are worth revisiting not only as the evidence of a historical debacle, but also for what they might tell us about the search for revolutionised social relations, a search that is still very much worth pursuing today.

[57] Fabio Lanza, 'The Search for a Socialist Everyday: The Urban Communes' in ed. Alan Baumler. *The Routledge Handbook of Revolutionary China*. London: Routledge, 2019, 74–88.

[58] Zhongguo kexueyuan jingji yanjiusuo bian, *Guanyu shehuizhuyi zhiduxia shangpin shengchan he jiazhi guilü wenti. 1959 nian 4 yue taoluhui lunwen, ziliao huibian*. Kexue chubanshe, 1959.

4

Seeing Like the Maoist State: Peasant Resistance in Official Court Documents from the 1950s

JING WENYU AND FELIX WEMHEUER

IN THE SPRING of 1960, Y County[1] in Shandong Province suffered from famine like many other places in China. The male villager Hu Jiaxing was 53 years old and had received a good class status of 'poor peasant' during Land Reform in 1946. However, on 22 May 1960, Hu was adjudicated publicly in a market square by legal cadres. The court charged Hu for stealing some fruits, vegetables, a chicken, and a duck. Furthermore, he had secretly eaten some sweet potato seedlings in the field and had illegally killed two piglets. The verdict did not mention the extreme shortage of grain,[2] but instead accused Hu of 'sabotage' and sentenced him to nine years imprisonment.[3]

This chapter will first provide an overview of changing images of peasant[4]-state relations of the Mao era in Chinese and Western research. Scholars have different views on what kind of actions should count as resistance against the state at this time; this chapter will introduce Y County in Shandong Province as a case for studying the debate over resistance against the state. Among 1,533 court files of criminal cases from the county archive, we have decided to focus on three kinds of crimes: 'illegal reseizing', 'speculation', and 'sabotage' as relevant to considering peasant resistance. After the definition of these crimes, an analysis of data in

[1] We decide to anonymise the name of the county to protect providers of sources.

[2] In China, wheat, paddy rice, millet, sorghum, corn, and soybeans count as the six major grains.

[3] *Hu Jiaxing's case: Sabotage of the 15 raises* (1960), Y County Archive, 2-7-2-20.

[4] We decided to use the term 'peasants' in the whole chapter, because after Land Reform all social groups of villagers in the selected county in Shandong would mainly produce for self-sufficiency. Only a smaller part of production was sold on markets or later purchased by the state. In China, this mode of production is called 'small peasant economy' (*xiaonong jingji*). The term 'farmer' is instead used for rural entrepreneurs, who are principally producing for markets to maximise profits.

relation to the verdicts will follow. We further selected several cases for a more detailed qualitative analysis of the files' content. This chapter aims to understand in which ways the Maoist state at the county level identified peasant resistance and how criminal justice was used to punish it. We will argue that harsh legal sentences by courts to long years in prison played a major role to crack down on actions that the local state considered forms of resistance. This finding contradicts earlier Western research on peasant-state relations of the 1950s that saw informal punishment and extrajudicial forms of violence as important features of Maoism. Furthermore, we will argue that legal punishment played a central role to enforce the new order in the countryside from the transition of the 'anti-feudal' Land Reform and Agriculture Cooperatives to the establishment of the People's Communes in 1958. By politicising and criminalising traditional forms of petty trade or survival strategies during famine as 'anti-socialist' resistance, the local courts contributed by harsh punishments to establish a new Maoist set of legal norms that served to protect the state monopoly of grain trade and land.

Changing images of peasant-state relations of Maoist China

The official historiography and cultural productions of the Mao era portrayed the relations between the Chinese Communist Party (CCP) and the peasantry as a natural alliance.[5] Based on a strategy of 'encircling the cities from the countryside', the party led a rural-based revolution to victory in 1949 and 'liberated' ordinary peasants from 'half-feudal' exploitation of landlords. During Land Reform (1946–53), the government labelled the whole rural population in the Han Chinese areas based on class status.[6] Within the logic of class struggle, the CCP considered 'poor peasants' and 'lower middle peasants' as the most reliable allies, 'middle peasants' had to be 'neutralized' and class enemies such as 'landlords' should be expropriated and suppressed. These class labels were heavily influenced by orthodox Marxism-Leninism imported from the Soviet Union. However, in Maoist China the government labelled the whole Han population in a much more systemic way to enforce 'affirmative action' for people from families with good class background and to discriminate against people with bad family background for more than two decades. As Benno Weiner points out in this volume, non-Han ethnic groups in borderlands experienced a very different political process during Land Reform, but, by late 1956, the government had completed the total collectivisation of agriculture in the Han Chinese regions. Several Western scholars have argued that the socialist

[5] For changing narratives see: Susanne Weigelin-Schwiedrzik, 'Re-Imagining the Chinese Peasant: The Historiography on the Great Leap Forward', in Kimberley Manning and Felix Wemheuer (eds), *Eating Bitterness: New Perspectives on China's Great Leap Forward and Famine*. Vancouver: University of British Columbia Press, 2011, pp. 42–6.

[6] For the system of class status see: Felix Wemheuer, *A Social History of Maoist China: Conflict and Change, 1949–1976*. Cambridge: Cambridge University Press, 2019, pp. 29–31.

transformation of the Chinese countryside was achieved with much less violence and had faced little resistance, especially compared to Stalin's 'socialist offensive' between 1929 and 1931 in the Soviet Union.[7]

Since the 1990s, Western and Chinese scholars became inspired by the developing field of Peasant Studies, in which, for example, James C. Scott had shown the importance of 'hidden transcripts' of rural resistance in Southeast Asia. Peasants did not decide often to rise up against the state, but their 'weapons of the weak' were more effective.[8] Field research involving oral history methodologies in Chinese villages brought to light how peasants and local cadres cooperated in the era of collective agriculture (1956–82), working against the state rather than cooperating with it. In fact, the supply system of food, cloth, and other consumer goods had a very strong urban bias in Maoist China. In the 1950s, the state tried to gain control over agricultural 'surplus' to feed the expanding urban population, the army, and to meet quotas for grain exports to the Soviet Union in order to finance the import of industrial technology.[9] Between 1961 and the late 1970s, the government effectively prevented rural-urban migration and forced peasants to stay in their villages by implementing the household registration system.

A key moment that changed the relations between government and peasants was the establishment of state monopsony for purchase and monopoly for sale of grain (*tonggou tongxiao*, Unified Purchase and Sale) in 1953. It was expanded to almost all agricultural products in the following years. The state was empowered to set low prices for agricultural goods, and it would thereby purchase what it deemed the 'surplus' (and sometimes even more). Prices for industrial goods were set high, to the benefit of urban populations where these goods were produced. In contrast to the official logic of 'class status', the whole rural population inside the cooperatives and later the People's Communes had a common interest against the state's logic to extract rural resources, because the state monopsony system threatened their survival. Several studies did show that struggle about grain was a key conflict between peasants and state after 1953, and the 'Great Leap Famine' (1959–61) was seen as a deadly escalation of the conflicts over grain.[10] The rural population had to learn the hard way how to avoid starvation without incurring the wrath of the government.

[7] Yu Liu, 'Why Did It Go So High? Political Mobilization and Agricultural Collectivization in China', *China Quarterly*, 187 (2006), 732–42.

[8] James C. Scott, *Domination and the Arts of Resistance: Hidden Transcripts*. New Haven: Yale University Press, 1992; James C. Scott, *Seeing Like a State: How Certain Schemes to Improve the Human Condition Have Failed*. New Haven: Yale University Press, 1998. For China see: David Zweig, 'Struggling over Land in China: Peasant Resistance after Collectivization, 1966–1986', in *Everyday Forms of Peasant Resistance*, ed. Forrest D. Colburn. New York: M.E. Sharpe, 1989, pp. 151–74; Huaiyin Li, 'The First Encounter: Peasants' Resistance to State Control of Grain in East China in the Mid-1950s', *China Quarterly*, 185 (2006), 145–62.

[9] Thomas Bernstein, 'Stalinism, Famine, and Chinese Peasants', *Theory and Society*, 13 (1984), 339–77; Felix Wemheuer, *Famine Politics in Maoist China and the Soviet Union*. New Haven: Yale University Press, 2014, pp. 43–6.

[10] Kenneth Walker, *Food Grain Procurement and Consumption in China*. Cambridge: Cambridge University Press, 1984; Wemheuer, *Famine Politics*, pp. 128–30.

Based primarily on oral history fieldwork, scholars identified several forms of hiding resources by peasants and local cadres such as 'concealing production and private distribution', building 'black grain stocks', underreporting of land and hiding 'black land', as well as illegal expansions of 'plots for private use'. Individual methods to secure food included outright theft, or 'borrowing' grain from the collective. 'Eating Green' meant to go into the fields to eat unripe crops before the state would get control over the harvest.[11] A common form of reacting to the lack of incentives to increase production was to be idle on the job. Peasants would also try to cheat the government by exaggerating damage from natural disasters in order to demand relief or to falsely report themselves as a 'non-surplus household'. By the mid-1950s, state media would accuse peasants of faking hunger to avoid grain purchases by the government.[12] Common tactics to gain extra income included participation in black markets or engaging in labour migration – with or without permission from the collective. Instead of seeing peasants as loyal supporters of the CCP or victims of a 'totalitarian' regime, several researchers have portrayed the rural population as actors with their own agency, who used many strategies to protect their own interests. The academic narratives since the 1990s have focused on conflicts between 'peasants versus the state' and various forms of 'resistance'.

After developing new information on this period based on the 'hidden transcripts' from the oral histories, all forms of peasant actions mentioned above could also be identified in official documents and internal reports. Nevertheless, scholars could not agree on labelling the behaviour of peasants as 'resistance' (*kangyi* or *fankang*). For example, underreporting or stealing of grain was defined by Gao Wangling as 'counteractions' (*fan xingwei*), that were survival strategies and a response to state policies without rebellious intentions. That being said, such 'counteractions' in the aggregate harmed state policies, once again begging the question what constitutes resistance in this context.[13]

Little is known about how the Maoist state reacted to forms of resistance that were less hidden than early oral history researchers believed. In oral history studies on 'counteractions' of peasants, the role of the law and courts were not mentioned. These case studies argued that local cadres used public humiliations, cut of food rations, beatings, and torture to informally punish theft or

[11] Wangling Gao, *Zhongguo nongmin fan xingwei yanjiu, 1950–1980*. Hong Kong: Zhongwen daxue chubanshe, 2013; Ralph Thaxton, *Catastrophe and Contention in Rural China: Mao's Great Leap Forward Famine and the Origins of Righteous Resistance in Da Fo Village*. Cambridge: Cambridge University Press, 2008, pp. 157–84.

[12] Wemheuer, *Famine Politics*, pp. 108–9.

[13] For the debate see: Wangling Gao, 'A Study of Chinese Peasant "Counter-Action"', in Manning and Wemheuer (eds), *Eating Bitterness*, pp. 273–4; Thaxton, *Catastrophe and Contention in Rural China*, pp. 226–8; Yixin Chen, 'When Food Became Scarce: Life and Death in Chinese Villages during the Great Leap Forward', *Journal of the Historical Society*, 2nd Iss., 10 (2010), 117–65; Frank Dikötter, *Mao's Great Famine: The History of China's Most Devastating Catastrophe, 1958–1962*. London: Bloomsbury, 2010, p. xv.

'underreporting' of grain.[14] Furthermore, peasant memories were focused on the village level, not on state actors from the county level or above. It is to these sources we now turn, to further examine how and why criminal justice at the county level was important to label and punish certain behaviours of peasants as 'resistance' to enforce the new Maoist order in the countryside during the 1950s.

Peasant resistance and punishment by law

In recent years, several scholars have argued that China in the 1950s was not a 'lawless society' as often claimed by scholars during the early 1980s Reform era.[15] The government did not pass a civil or Criminal Law Code in the Mao era, and a constitution was declared only in 1954. Nevertheless, the government announced laws concerning marriage, Land Reform, and labour unions. Furthermore, directives and regulations had a similar function to law in the People's Republic of China (PRC). The CCP considered law as an instrument of class struggle and rejected Western liberal ideas such as the separation of the executive and judiciary powers of government. In the 1950s, some party officials believed that formal court verdicts were in fact needed to punish crimes. In recent years, some case studies on criminal justice at the county level have been published, for example regarding the punishment of 'unlawful landlords'.[16]

For this chapter, we selected Y County in Shandong Province for a case study, because we gained access to a large collection of archival sources, including court records on rural labour. Y County was shaped by agricultural production, but long, pre-revolutionary commercial traditions did exist and several market towns had already developed in the area. The region was part of an old revolutionary base area, in that the CCP had ruled it before the foundation of the PRC in 1949. Its first party branch was established in 1936 and Land Reform was carried out as early as 1946. Our collection of documents from the county archive contains 1,988 volumes with 160,666 pages in total. The types of materials vary from criminal cases to internal organ publications, working reports, meeting minutes, surveys, disciplinary files, and agricultural statistics. Among these documents, there are 1,533 criminal cases from 1949 to the mid-1960s, including counterrevolution, illegal reseizing, sabotage, speculation, fraudulent purchase, murder, rape, and theft.

[14] Thaxton, *Catastrophe and Contention in Rural China*, pp. 204–5; Felix Wemheuer, *Steinnudeln ländliche Erinnerungen und staatliche Vergangenheitsbewältigung der „Großen Sprung"-Hungersnot in der chinesischen Provinz Henan*. Frankfurt (M): Peter Lang, 2007, pp. 191–3, 198–9.

[15] For example see: Daniel Leese, *Maos langer Schatten: Chinas Umgang mit der Vergangenheit*. München: Beck Verlag, 2020, pp. 119–25; Glenn Tiffert, 'Judging Revolution: Beijing and the Birth of the PRC Judicial System (1906–1958)', Ph.D. thesis (UC Berkley, 2015). https://digitalassets.lib.berkeley.edu/etd/ucb/text/Tiffert_berkeley_0028E_15564.pdf

[16] Shigu Liu, '"Shixu" xia de "zhixu": Xin zhongguo chengli chuqi tugai zhong de sifa shijian: Dui Poyang xian bufa dizhu an de jiedu yu fenxi', *Jindaishi yanjiu*, 5 (2015), 91–105.

While oral history studies have shown how peasants remembered their actions in retrospect, we will show ways in which local courts at the county level identified such actions as (illegal) resistance. When analysing the verdicts, we will describe which political narratives formed the basis of adjudication. We will show that the verdicts were related to the class status of the convicts and political campaigns of the time. We will further demonstrate that court verdicts of long-term-imprisonment played an important role in labelling and punishing perceived resistance among the rural population.

In order that the importance of class and political campaigns becomes clear, we will conduct a general examination of all verdicts and qualitative analysis of selected court files from the 1950s to the early 1960s. We decided to choose three kinds of crimes that were related to peasant resistance for this case study: 'illegal reseizing' (*daosuan*), 'speculation' (*touji*), and 'sabotage' (*pohuai*). Searching the digitalised files in our collection, we identified all documents in which at least one of these three crimes is in the title. We will first conduct a general analysis of verdicts in relation to peasant resistance based on the class status of defendants, major content of the crime, and punishment. Next, we select several single files for a qualitative analysis. Little is known about how courts acted and criminal justice worked in the 1950s, so we have used Y County to illuminate this aspect of the Maoist state, even if our example cannot be taken as representative of every context in China.

As noted above, the government did not promulgate a Criminal Law Code in the 1950s. Individual laws and regulations provided often only vague definitions of crimes. Definitions for the three kinds of crimes in this study included various explanations. In the early 1950s, 'illegal reseizing' meant that 'landlords' had seized back land and property from peasants with good class status that had been previously allocated by the revolutionary regime during Land Reform. In the Land Reform Law of 1950, the confiscation of the property of 'landlords' was mentioned, but not the 'rich peasants'.[17] At that point in time, the law still protected the 'rich peasant economy', but, by 1953, the government expanded the target of confiscation to 'rich peasants'. In the following year, the Constitution of 1954 protected the property of peasants in general, but defined the abolishment of the 'rich peasant economy' as an official goal.[18] As a result, 'rich peasants' could be punished for 'reseizing' confiscated property, and the cases not only included conflicts about land, but also houses and trees. Lumber, for example, was an important building material in the countryside. According to the Land Reform Law, 'landlords' and 'rich peasants' could apply to change their class status to ordinary peasants, after they had participated in manual labour for several years

[17] See Articles 2 and 6 of Land Reform Law of 1950, accessed 20 June 2021, www.npc.gov.cn/wxzl/wxzl/2000-12/10/content_4246.htm.

[18] 'Zhonghua renmin gongheguo xianfa (1954 nian)', accessed 6 June 2021, www.npc.gov.cn/wxzl/wxzl/2000-12/26/content_4264.htm.

and had not committed any crimes.[19] However, in some cases from our collection, the class status of people was changed from 'middle peasant' to 'rich peasant' by the government, resulting in accusations that the person had committed the crime of 'rich peasant reseizing'. The reclassification of peasants shows how punishment was determined in relation to the labelling of class status and ever-changing policies. Officially, the punishment of 'illegal reseizing' was justified by the necessity to protect the 'new democratic' order that Land Reform had established.

Similarly, the crime of 'speculation' also threatened the stability of the New Democracy, and was sometimes named a 'disturbance of the market' or a 'fraudulent purchase' (*taogou*). In November of 1950, the government announced a regulation that outlawed 'the speculative businesses disturbing the market',[20] followed quickly by the 'Regulation to Punish Counterrevolutionaries' of 1951, which related 'disturbance of the market' to counterrevolution and, therefore, made it a serious crime.[21] After the establishment of the PRC, one of the most important tasks of the new government was to curb inflation and stabilise consumer markets. It seems that, in the beginning, urban traders and middlemen were the main targets of the crackdown on 'speculation', not ordinary peasants. This changed after the government introduced the Unified Purchase and Sale of grain in 1953 to establish control over the agricultural 'surplus'. The draft version of the Criminal Law Code from 1954 defined speculation in more detail: 'Conducting speculative activities through hoarding, panic or fraudulent purchase, manipulating price, short-selling or all other illegal means to cause local price fluctuations or difficulty in supplying a certain item.'[22] Ordinary crimes could be punished with sentences of two to eight years in prison; courts could, however, announce sentences of more than eight years, life imprisonment, or even the death penalty in serious cases of organised resistance against the leading role of the government in the economy, establishment of groups for speculation, or if the convicts were 'unwilling to reform'.

The crime of 'sabotage' was also defined in the 'Regulation to Punish Counterrevolutionaries', which was aimed at punishing those who incited mass resistance to prevent the implementation of policies by the People's Government, sowed dissent within the people or between the people and government, and those who conducted counterrevolutionary propaganda agitation or had fabricated and spread rumours. People could be punished with more than three years' imprisonment or in serious cases even the death penalty, for example, if they had mobilised

[19] 'Zhengwuyuan guanyu huafen nongcun jieji chengfen de jueding', in *Jianguo yilai zhongyao wenxian xuanbian*, ed. Zhonggong zhongyang wenxian yanjiushi, Vol. 1. Beijing: Zhongyang wenxian chubanshe, 1993, pp. 400–1, 406–7.

[20] 'Guanyu qudi touji shangye de jixiang zhishi', in *Jianguo yilai*, ed. Zhonggong, pp. 466–7.

[21] Article 9 (3) of 'Regulation to Punish Counterrevolutionaries' of 1951, accessed 20 June 2021, www.npc.gov.cn/wxzl/wxzl/2008-12/15/content_1462048.htm.

[22] Mingxuan Gao and Bingzhi Zhao (eds.), *Xinzhongguo xingfa lifa wenxian ziliao zonglan*. Beijing: Zhongguo renmin guo'an daxue chubanshe, 1998, p. 184.

masses to resist or destroy the purchase of grain or tax collection by the government. At the time of announcing the 'Regulation to Punish Counterrevolutionaries' in 1951, the government targeted mainly 'rich peasants' as the most important provider of 'surplus grain' and taxes. Poor and 'non-surplus households' could only contribute a little to the state. However, by 1953, the government had changed its policies from protection to elimination of the 'rich peasant economy'. After the confiscation of land owned by 'rich peasants', the ordinary rural population became the new target of grain and tax collection.

The introduction of Unified Purchase and Sale of grain in 1953 caused serious tensions between the government and peasantry in many areas. For example, peasants complained that the prices for purchase were much too low and the government had set the grain quotas at a starvation level. Furthermore, every improvement in production would result in an increasing of purchase by the state, eliminating any incentive to increase production. In 1955, the central government introduced the so-called 'three fixed' policy (fixed goals for production, purchase, and sale) to curb the tensions in the countryside.[23] Fixed quotas and minimum rations should guarantee that peasants would not stay hungry and an increase of the 'surplus' would not result in higher quotas. However, the policy of the 'three fixed' only lasted until the beginning of the Great Leap Forward (GLF) in 1958. The government decided to increase the percentage of grain purchase to a higher level based on false promises of a bumper harvest. The year of 1959 saw the highest procurement rate of the grain harvest in the history of the PRC, while at the same time production decreased considerably.[24] These twin developments resulted in a famine that peaked in the countryside in 1960. In spite of this devastating and widespread hunger, resistance of ordinary peasants against grain collection could be labelled as 'sabotage', which was a counterrevolutionary crime.

Overview of criminal cases in relation to peasant resistance in Y county

The basic information of class status, content of the charges and sentences contained in the court verdicts is summarised to gain a general understanding of these three crimes – who was targeted and for what causes. Based on the overview analysis, it is clear that the potential targets of the three crimes ranged from old class enemies to the peasants with good class background, and the perceived acts of resistance attached to the state's revolution programme of making rural socialism. In the large majority of cases, these crimes were punished with sentences of several years in prison.

[23] Wemheuer, *Famine Politics*, pp. 96–9.

[24] Zhonghua renmin gongheguo nongye bu jihuasi (ed.), *Zhongguo nongcun jingji tongji daquan 1949–1986*. Beijing: Nongye chubanshe, 1989, pp. 410–1.

Regarding the crime of 'illegal reseizing', convicts had nearly exclusively the bad class status of 'landlords' and 'rich peasants', in addition to being exclusively men. The Land Reform Law of 1950 only announced the confiscation of 'landlord' properties, but later 'rich peasants' were targeted as well and punished after the fact. As previously noted, formal Land Reform had already been completed in Y County by 1946. The peak of 'illegal reseizing' cases occurred during collectivisation of 1955–6 and again in 1958. In the process to establish the first agricultural cooperatives, and later the People's Communes, cadres had to deal with questions of ownership before land, agricultural tools, houses, furniture, or trees could be declared collective property. Furthermore, the CCP saw old class enemies as a potential threat to collectivisation. In contrast to the Soviet Union, 'landlords' and 'rich peasants' were made members of the collectives and therefore the CCP subsequently asserted that they undermined the new institutions from the inside. This might be a reason why many verdicts on 'illegal reseizing' occurred in the context of collectivisation.

The cases of 'speculation' or 'disturbing the market' differed from that of 'illegal reseizing'. Convicts with a great variety of class status were punished for 'speculation', including 'poor peasants' and 'middle peasants'. The first cases occurred in 1954, after Unified Purchase and Sale was established. The main accusations in the verdicts were black market activities and 'fraudulent purchase' in the context of grain purchase and taxes. Almost all crimes were related to grain and other agricultural products, in some cases to trade for (traditional) herbal medicine. Targets of the courts were less commonly the old 'class enemies', and now more likely to be peasants resisting the compulsory Unified Purchase and Sale.

In contrast to 'speculation', the kind of crimes that included 'sabotage' varied greatly. The verdicts punished sabotage of the Land Reform, damage of collective property, illegal slaughtering of cattle, theft, spreading of 'rumours' about the Great Leap, 'rightist critique' of collectivisation, damage of a soldier's marriage, and draft dodging. The convicts of these crimes had various class backgrounds. In 1960, many 'poor peasants' were sentenced to long terms in prison for 'sabotage' of the '15 raisings' (15 *yang*)[25], three of them receiving sentences of up to 10 years. It is very likely that the increase of this crime was related to the famine in 1960, but hunger and starvation were not mentioned in verdicts. The justice system itself masked the motivations of crime: it was very difficult and dangerous for local officials and peasants to address famine, because this could be labelled as 'right-wing opportunism' or 'sabotage of the GLF'. Only in 1961 and 1962, when the central government launched policies to adjust the national economy, were starvation to death mentioned frequently in the files.

[25] The policy of '15 raising', introduced by the state in 1959, meant to raise 15 kinds of farm animals to develop breeding industry, including cattle, horse, mule, donkey, sheep, pig, chicken, goose, duck, rabbit, mink, bee, silkworm, tussah, castor silkworm. Zhaoyuanshi linyeju (ed.), *Zhaoyuan linye zhi*. Beijing: Fangzhi chubanshe, 2007, p. 298.

The harsh punishment of 'illegal reseizing' by courts was an attempt of the government to crack down on class enemies such as 'landlords' and 'rich peasants' in order to consolidate the new order created by Land Reform. Whether or not the convicts were aware of committing illegal acts will be analysed in the next section. The years between 1946 and 1961 saw a permanent change of ownership structures, so it was not always clear which property belonged to private persons or simply to different levels of the collective farms. The class status of those who were punished for 'speculation' and 'sabotage', by contrast, included many labelled as 'poor peasants' and 'middle peasants'. Especially during the famine in 1960, the courts opted for more draconian punishments, even in the cases of the 'poor peasants' whom the state claimed to be defending. While a few numbers of cases were purely political such as 'spreading rumours regarding the GLF', we consider many of the cases that involved all three kinds of crimes to be, in fact, struggles over resources in desperate times. 'Illegal reseizing' was in most of the cases related to land, houses, trees, and the private accumulation of wealth. The punishment of 'speculation' and 'sabotage' was often aimed to enforce state control over the agrarian 'surplus' and farm animals. Therefore, the court verdicts can be seen as part of the tendency in the 1950s to expend state control over all economic resources and activities in the countryside, even during the famine, that were caused by the policies of the GLF.

Qualitative analysis of criminal cases

'Illegal reseizing': Changes of class status and ex-post punishment

Seizing back confiscated properties was regarded by the state as an act that intended to restore the old reactionary order, but not every peasant applauded the expropriation of long-held properties by others during Land Reform.[26] The CCP established Peasant Associations to distribute the main rural properties (e.g., land and houses) during the Land Reform in 1946 in Y County. The Peasant Associations were subsequently dissolved and the village government served as the centre of power in the countryside. Along with later collectivisation, village government was itself replaced by the agricultural cooperatives. In the context of this institutional change, local authority often shifted and the replacement of local authority played a decisive role in the handling of properties. A typical way to take back possessions was getting permission from local cadres.

For instance, 36-year-old Feng Yugeng (male, 'landlord') was arrested for 'illegal reseizing' a variety of properties on 20 December 1955, including 700 *jin* of wheat, four tenths *mu* of a homestead, a cart, two *mu* of land, four beams of

[26] Gao, *Zhongguo nongmin*, p. 5.

lumber and 200 *jin* of straws from a collapsed house, and 800 bricks.[27] His file, a 53-page dossier, contains the interrogation records by the Public Security Bureau and the court, investigation materials by legal cadres, as well as the court's verdict and letters, revealing the way of getting property back and successfully getting himself exonerated at last. He took possession of these with the approval of village leaders on the grounds of 'life difficulties'. The former cadres of the Peasant Associations criticised the present village leaders in their testimonies, stating that 'the village cadres lived close to Feng's family' and 'were irresponsible during work'. Feng was found guilty by the court on 3 February 1956 and sentenced to seven years' imprisonment, but the village cadres submitted two joint letters of support to the court on 25 and 30 August 1956. The first letter claimed that the status of the properties in question was not clear, and the second letter affirmed that these properties had not been confiscated in the first place. As a result, Feng was acquitted by the court in July 1957. For the judicial authorities, permissions of local cadres did not automatically entitle a person to (re-)claim property. However, the crux of Feng's case lay in the joint declaration of village cadres to the court. Feng was one of the few cases in our collection that successfully overturned a verdict, as it was rather infrequent that a 'landlord' obtained full support from local cadres.

Some cases revealed that the accumulation of wealth by richer peasants was condemned as illegal reseizing, including, for instance, land trading and the sale of lumber. 65-year-old Li Jiying, for example, was classified as a 'middle peasant' in the Land Reform and reclassified as a 'rich peasant' during the campaign of 'Transforming Backward Towns' in late 1955. The 'Transforming' campaign was carried out throughout the country, designed to complete the unfinished task of eradicating 'feudal remnants' that had survived Land Reform, and Li was arrested during this campaign, on 10 November 1955, for 'illegal reseizing'.[28] His file contains 57 pages, including a full set of investigation and interrogation materials by police and the court, but the document of his first conviction is missing in this collection. He appealed on 1 August 1956 to the court to maintain his status of 'middle peasant', but, in the retrial, the court rejected his appeal and adjudicated that Li was a 'rich peasant'. The court upheld the original judgement, which sentenced Li to seven years' imprisonment. Li was accused of illegal reseizing three pieces of land, but according to the interrogation records, Li had pawned out the land for the price of several hundred *jin* of wheat for each piece. The alleged acts of 'illegal reseizing' were de facto Li's recovery of the pawned land. Land trading was a basic land right possessed by 'middle peasants' at this time, and was

[27] *Feng Yugeng's case: Illegal reseizing* (1956), Y county Archive, 1-5-7-53-25. The reference of cases was named by the cause of criminal action, the name of defendant and the year of conviction presented on the cover of the case, and the same applies hereinafter.

[28] *Li Jiying's case: Illegal Reseizing* (1955), Y county Archive, 3-4Y-244-73. Detailed analysis of this case see: Wenyu Jing and Shuji Cao, 'Tudi chanquan yu 20 shiji 50 niandai de "funong daosuan": Yi Shandong sheng L xian Li Xuhai daosuan an wei zhongxin', *Zhongguo jingjishi yanjiu*, 4 (2019), 56–68.

supposed to be protected by Article 8 of the 1954 Constitution. However, pursuant to the reclassification of his class status, Li's land market activity was condemned as 'illegal reseizing'.

Similarly, the sale of trees as lumber was another common cause for accusation in reseizing cases. Wood was one of the major properties in rural areas, yet it was not subject to confiscation based on the Land Reform Law of 1950. Timber became a controlled substance by the state in a line with Unified Purchase and Sale in 1956.[29] As a result, tree and timber became one of the valuable and scarce materials in rural areas, which attracted attention, and caused contention as well. In this context, the reseizing cases related to trees increasingly emerged in the court records.

For instance, Wu Haibin (male, 51 years old) was convicted of the 'illegal reseizing' of trees in February 1956.[30] Wu's class status was inconsistent in his files. In the interrogation records, his class status was noted as 'rich peasant', but in his verdict, he was labelled as a 'landlord'. His charges included illegally selling seven willows, one poplar, and one elm for 700 *jin* of grains in 1950, illegally taking four beams of lumber, some firewood and 800 bricks from a collapsed house in the same year, illegally selling an elm for 300 yuan in 1954, and planting several saplings on the confiscated homestead. He was, thus, sentenced to 10 years' imprisonment by the court, which was a punishment meted out well after the activities were considered a crime. According to Wu, no one ever stopped him when he sold the trees in the past few years, and he insisted on his ignorance regarding the confiscation of the trees, so he appealed the ruling. The Intermediate Court at the prefectural level found that both the defendant's class status and the facts of the crime were unclear and returned the case to Y County Court for retrial. The Y County Court made the final judgement that 'the criminal facts were proven, but the sentence was too heavy … the court decided to commute the sentence to six years in prison'. The retrial result suggests that the sentencing was hasty and arbitrary. As revealed in the interrogation records, the alleged 'illegally reseized' trees had grown in Wu's private garden, or next to the house in which he lived. Even for the 'landlords', the 1950 Land Reform Law preserved their houses for personal residence. It means that the trees growing on the personal homestead were also supposed to be his private property, to dispense with as he pleased. However, as several testimonies proved for the judges that he was a 'landlord' – despite the fact that Wu was also engaged in agricultural manual labour himself – and alleged that the trees had been confiscated by another party, Wu was still convicted by the court.

Some other 'reseizing' cases resulted from conflicts between individual peasants and the collective, for instance: Tian Xue (male, 38 years old)'s case of 1958,

[29] Zhongguo nongye quanshu zongbianji weiyuanhui (ed.), *Zhongguo nongye quanshu: Jiangxi juan.* Beijing: Zhongguo nongye chubanshe, 2001), p. 245.
[30] *Wu Haibin's case: Illegal reseizing* (1956), Y county Archive, 1-5-11-17.

which involved direct conflict with CCP members.[31] Because cadres had arbitrarily demolished Tian's house and used the bricks, Tian got into a serious conflict with them. Soon after, Tian was reclassified as a 'rich peasant' and accused on 8 May 1958 of illegal reseizing four one-storey houses and a tree. Tian argued that a district cadre had issued a certificate to him, confirming that he was a 'middle peasant' and therefore the government had not confiscated him anything. Tian's resistant attitude was denounced by the court in the verdict and he was sentenced to seven years in prison. Tian's alleged illegal reseizing fundamentally originated from the peasants' resistance against the ever-expanding seizures of private possessions but, in the context of radicalising campaigns, his resistance was ultimately in vain.

In short, these cases of illegal 'reseizing' reveal a kind of resistance that was both real but also constructed by the state itself. In general, the cases of illegal 'reseizing' in 1950s China increased throughout the decade during the process of collectivisation, to achieve the state/collective's control over land and other major property. With a close look at these criminal files, it can be seen that many cases were not because the 'reactionary classes' proactively and constantly resisted the revolutionary state, as claimed in official documents, but were caused by the state's expanded definition of the crime and extended groups targeted for punishment. Peasants were re-labelled and the retroactive punishments were imposed. The activities that Chinese peasants in the cases pursued, such as land trading, tree sales, and protecting their property from collectivisation, did not meet the original definition of 'illegal reseizing' derived from the 1950 Land Reform Law. The behaviour of the accused peasants in court shows that many were not willing to accept charges and pleaded innocent – indeed, some peasants had already learned to challenge official class designations. However, to defeat the state with its own logic was often not successful in aforementioned cases, because the court insisted on their reclassification.

Reading the court files, it is impossible to know what the accused peasants really were thinking in court. The definition of 'illegal reseizing' violated peasants' traditional concepts of ownership, and perhaps they were fully aware that they were committing a crime – a form of resistance against 'unjust laws'. If so, in order to escape punishment, they may have 'pretended to be naïve' (*zhuang sha*) in court (such as claiming their ignorance of confiscation), which would constitute a form of indirect resistance against state policies. However, it is also possible that, due to the perpetually changing structures of government, ownership, reclassification of class status, and a lack of a Criminal Law Code in the 1950s, many villagers were simply confused regarding which acts could be labelled by the state as crimes, including 'illegal reseizing'. It seems that, in our cases, the accused had not yet internalised a principle of Maoist criminal justice that confession and self-criticism would lead to lighter punishment, while rejection of guilt and 'unwillingness to reform' would result in harsh sentences.

[31] *Tian Xue's case: Illegal reseizing* (1958), Y County Archive, 1-5-8-33.

'Speculation': Enforcing the state monopoly for grain trade

The legal punishment of 'speculation' in rural areas arose during the winter of 1953, when the policy of Unified Purchase and Sale was put into effect. Many agricultural products were both food and raw materials for sideline production (*fuye*) that peasants depended on for survival. In the pre-revolutionary rural economy, most peasants would cultivate approximately around two *mu* of land per capita on average and therefore they could not make a living by merely agricultural income.[32] Whereas, by introducing the crimes of 'speculation', peasants who engaged in sideline production or market activities with state-controlled agricultural products, particularly grain, were punished.

The case of Zhang Dexiang (male, a 45-year-old 'poor peasant'), who was arrested for committing 'fraudulent purchase and collaborating with the enemy' in March 1954, was widely publicised as a typical example by government.[33] In the verdict, Zhang's 'background' (*chushen*) was labelled as 'businessman'. His class status was determined by the variety of businesses he managed before 1949, and he still ran a steamed bun shop when he was arrested. Like other peasants, however, he had always worked on seven *mu* of land passed on to him by his father, although it was just not sufficient to support his family of seven. He made steamed buns with wheat that he had cultivated himself, but also bought additional wheat on the market for his business, to supplement family income. According to the court's verdict, Zhang's financial crimes were especially suspicious:

> … the pseudo-district chief[34] Cui Zhaojun left 300 yuan in Zhang's place in 1947. Zhang sent messages and gave 140 yuan back to Cui in 1948 and then spent the rest 160 yuan. In October 1953, he was fined 190,000 yuan (old currency) for evading taxes while running his steamed bun workshop … meanwhile, he fraudulently purchased 130 *jin* of wheat and 60 *jin* of beans.[35]

In the judgement, the fact that Zhang returned the money to Cui, a cadre of the Kuomintang (KMT), amid the see-saw battles between the KMT and the CCP in Y County during the Civil War became evidence for collaboration with the former regime, and went further to prove his counterrevolutionary background. Running a small sideline business, and his inability to pay a high fine of 190,000 yuan for tax evasion, was regarded by the court as 'refusal to reform'. In accordance with the 1953 'Provisional Resolution on the Grain Market', small amounts of grain transactions were allowed in the rural areas as long as it was not for speculation

[32] Shuji Cao, 'Shandong sheng liangshanxian de liangshi yu nongmin fudan yanjiu (1947–1948)', *Zhonggong dangshi yanjiu*, 6 (2019), pp. 113, 116.

[33] *Zhang Dexiang's case: Fraudulent purchase and collaborating with the enemy* (1954), Y County Archive, 1-9-9-2.

[34] A KMT cadre was called 'pseudo-district chief' (*wei quzhang*) in the verdict, illustrating that the CCP did not recognise the legitimacy of the KMT regime as well as its cadres.

[35] 'Court's verdict', *Zhang Dexiang's case*.

and did not disturb the overall market itself.[36] Zhang purchased grain to operate a sideline business, rather than for speculation, but still was punished for 'fraudulent purchases'.

Before the establishment of the Public Dining Halls in autumn 1958, peasants would receive their individual rations (*kouliang*) of grain after the harvest, store it in their homes and use it for cooking their meals. If the stock was not sufficient to feed the family, peasants could buy grain back from the state. It was not illegal to store grain at home for self-consumption, but peasants could be accused of storing large amounts as a form of 'speculation'. In the case of Jing Wensheng (male, a 40-year-old 'middle peasant'), he was punished for 'fraudulent purchase' on 29 September 1958.[37] The court verdict made clear that the crime in question was profiteering, which, if conducted on a larger scale, would distort the national market for grain that the state was seeking to control:

> Jing stocked 1,000 *jin* of grain in his home, and he bought 515 *jin* of grain from the state in 1958. He sold his hoard of 1,004 *jin* of grain to Chen Village for 0.17 yuan on 4 September 1958 ... the defendant stocked grains, instead of selling it to the state, he bought grains from state and sold it at a higher price in order to gain illegal profits.

Jing bought grains with a ration card from the state, which was legal, but he was still punished for 'fraudulent purchase', because he was accused of stocking grain for the purchase of profitable sale. However, as revealed by a letter from the Provincial Procuratorate to the County Procuratorate after the case review in December 1961, Jing purchased and stored grain from the state during the spring shortage, when prices were high, as opposed to the autumn harvest, when it was sold. Jing sold the grain out of fear that it could be taken by the Public Dining Halls, which collected grain stocks from peasant's homes to use it for their rations for collective eating. Confounding the time of purchase and sale, however, the County Court constructed the charges of 'fraudulent purchase', which itself was dependent on the adjudication that he had 'gained illegal profits'. Jing did not control the prices at which he bought and sold the grain.

In light of personal status, defined by one's current occupation or a long past career, many 'criminals' were labelled as 'traders', suggesting that they used to or still made a living by engaging in private enterprises. Nevertheless, when entering the market, they were basically poor peasants who engaged in sideline production or small businesses for the sake of supporting their families. In the name of 'speculation', the state criminalised peasants' economic activities involving grain and this meant the deprivation of traditional methods of subsistence for many peasants. These crimes, namely to maintain the market order, were de facto a tough approach of forcing peasants to sell agricultural products to the state, which was part of the socialist transition that the PRC government was trying to realise.

[36] Zhengwuyuan, 'Liangshi shichang guanli zanxing banfa' (19 November 1953), in *Jianguo yilai*, ed. Zhonggong, p. 566.

[37] *Jing Wensheng's case: Fraudulent Purchase* (1958), Y County Archive, 2-5-6-5.

'Sabotage': Damaging collective agriculture

According to the early PRC courts, the crime of sabotage can be considered as a pocket crime of the time because of its ill-defined and vague extension. It involved not only some serious acts such as sabotaging policies, but also other issues related to agricultural production, morality, and speech, thus to suppress a series of 'counteractions' of the peasants and to make the new socialist order.

The policy violation in relation to grain, for instance, was one of the core legal issues of the 1950s. In 1958, a leader of a production team, Xian Houde (male, 38-year-old 'middle peasant', who had joined the CCP in 1954), was accused of letting more than 10,000 *jin* of wheat harvest get mouldy on purpose, leading to the failure of the state in its target of acquiring 6,000 *jin* of grain.[38] Xian argued that the mould spoilage was an accident caused by a sudden rain when he was absent from the village, and the damage was only 1,000 *jin*. As this case arose during the GLF, it was very likely that the state's figures were exaggerated. The court, however, determined that Xian had motive to commit this crime, namely that he intended to evade delivering grains to the state, and therefore the court sentenced him to 10 years' imprisonment. It was common at this time that ordinary peasants could also be arrested if they failed to fulfil the quota of selling the 'surplus grains' to the state. The definition of 'surplus' was determined by local cadres, who were under great pressure to please higher authorities and often reported exaggerated production outcomes in 1958.[39]

The accomplishment of policy goals was not the only concern of the state, but also to stabilise the new economic and social order, as well as educating the rural population. Depending on the legal requirements, ordinary disputes and common mistakes could become aggravated and end up as penal sentences. A typical case was Jia Shitu (male, a 53-year-old 'middle peasant'), who was arrested in 1958 for committing 'counterrevolutionary sabotage'.[40] When looking at his long list of accusations, however, most were conflicts with cadres and other peasants that had emerged under the collectivisation of agriculture. For instance, Jia accused a cadre of 'counterrevolution', because the cadre had forced women to hard labour in the fields, and this had caused one woman to miscarry. Furthermore, Jia blamed another cadre for distributing grains to a household with its own surplus. In this case, Jia and the cadre criticised each other, but which accusations prevailed depended on the judiciary's evaluation. In addition to all of this, Jia once struck a woman with his head due to an argument concerning the registration of work points. Jia also suggested male labourers take off their pants when women and men worked together to weed out the crop stalks, which had been deluged by the flood. Unsurprisingly,

[38] *Xian Houde's case: Resisting grain purchase* (1958), Y County Archive, 2-40-6-20.

[39] For the 'wind of exaggeration' see: Ralph Thaxton, *Catastrophe and Contention in Rural China*, pp.146–7.

[40] *Jia Shitu's case: Counterrevolutionary sabotage* (1958), Y County Archive, 1-3-7-1.

many young women were unwilling to work in the fields with them. Jia's indecent acts violated social norms, but they did not quite rise to the level of criminal activity in Chinese society at that time. However, in the eyes of legal authorities, it was 'counterrevolutionary sabotage' because it caused contradictions and instability in agricultural labour and, as alleged, Jia's 'influence was very bad'. The court finally sentenced him to five years in prison.

Some 'counteractions' directly violated policies or regulations, such as 'private distribution' of goods, loafing while at work, failing to take responsibility for work tasks, or disobeying cadres, were technically not illegal actions. Zhang Qingxin (male, a 47-year-old 'rich peasant') was judged by legal cadres publicly in the village for the charge of 'counterrevolutionary sabotage' on 9 June 1960.[41] According to the verdict, his 'criminal acts' were summarised as follows:

> On 10 June 1959, he led a sheep, which was raised by him, to pasture in a sweet potato field, so it would eat seedlings. On 10 August 1959, when ploughing the land, he badly beat a donkey that was raised by the production team. As a result, the donkey was severely hurt and died after two months. Meanwhile, he purposefully sowed the wheat seeds too shallow and some seeds were exposed, and as a consequence five *mu* of land was re-sowed. In December 1957, he told Zhang to not join the agricultural cooperative because it was no good. In 1960, he bought a cabinet for Han's mother and charged her 25 yuan, but the real price is 16 yuan.

Thus, 'for the sake of strengthening national law and safeguarding production for the cause of socialist construction', the court sentenced him to 10 years' imprisonment. These kinds of punishments served as a means to force poor peasants to work during the famine. Zhang's case shows that the peasants' discontent with hunger and collectivisation manifested in their daily lives, in their taking no care for collective property and in their conversations. Such behaviours are more in line with covert 'counteraction', as noted in the first section, rather than direct resistance to state policy. However, as the court verdict demonstrates, in the eyes of the Maoist state, the hidden strategies of peasants turned out to be overt resistance, which were in turn defined as crimes.

Conclusion

Since the 1990s, several oral history studies have shown that Chinese peasants in the era of collective agriculture (1956–82) practiced various forms of resistance against the extraction of resources by the state, including underreporting of production and land, stealing, and organising black markets. However, little is known about how the Maoist state perceived these actions. Based on a large collection of criminal cases from Y County in Shandong, this chapter shows that in the 1950s criminal justice played a key role in the crackdown on peasant resistance. In order

[41] *Zhang Qingxin's case: Counterrevolutionary sabotage* (1960), Y County Archive, 1-3-4-5.

to understand the state's expansion of criminal activity, in service of making the Maoist system, we have analysed court files regarding the crimes of 'illegal reseizing', 'speculation', and 'sabotage'. A general analysis of the data shows that courts punished these crimes with prison sentences of several years. The punishment for these three crimes hit the rural population in general, not only the villagers with bad class status. Many peasants, who were accused of 'illegal reseizing', were originally labelled as 'middle peasants'. 'Illegal reseizing' meant in the beginning only that 'landlords' had seized back property that had been confiscated and distributed to other peasants during Land Reform. The verdicts on 'speculation' and 'sabotage' even included many 'poor peasants'. The government used the system of criminal justice to discipline the rural population for the new state socialist order. Moreover, we consider this legal punishment as part of the struggle over the control of resources between peasants, collectives, and the state, such as land, houses, trees, livestock, and grain.

The qualitative analysis reveals that, with the seemingly limitless expansion of the scope of punishment, many crimes were newly (re-)constructed by the state. The general lack of a Criminal Law Code and constantly changing policies of the government resulted in uncertainty concerning which actions would be considered crimes by courts: first, wealth accumulation from land trading and tree sales could be punished as 'illegal reseizing'; second, grain purchase on the black market to operate side businesses, which was an important financial supplement for many rural families, could be condemned as 'fraudulent purchases'; third, even purchasing grain with ration cards from the state could also be punished, if the judges believed the motive of storage at home was not self-consumption, but 'speculation'; finally, the crime of 'sabotage' involved diverse everyday practices of peasants, including policy violation, regardless of intent, ordinary disputes among villagers, indecent behaviour, or loafing on the job. The courts perceived these actions to be harmful to the social and economic order, and therefore disrupting 'the cause of socialist construction'.

As we have uncovered in many of the cases of 'illegal reseizing', accused peasants did not always accept their punishments as passive victims. Some peasants tried to get support from local cadres for their legal cases, while others 'pretended to be naive' and claimed they were ignorant of confiscation. Some cases showed that the accused peasant insisted on appeals to courts at higher levels. Defendants could also argue within official logic of class status and protest against their reclassification from 'middle peasant' to 'rich peasant' in order to overturn punishment. These forms of 'counteractions' in court, however, were seldom successful and the vast majority of peasants, no matter what their class status, could not avoid being sent to prison for many years. The Maoist state adopted the criminal justice system as a political tool to suppress resistance against its monopoly of rural resources and the agenda of collectivisation.

Part II

The Revolutionary Self

5

The Final Revolution is in Our Hearts: Work and Study in the Personal Diaries of the Early PRC, 1949–1959

AARON WILLIAM MOORE

THE PATH TO communism in China was not only a transformation of material conditions and social relations, but a methodology for enacting revolutionary changes in thought; these changes were directed more by the individual than by a managerial state, but efforts to build a socialist consciousness could produce unacceptable contradictions. This chapter explores that dilemma by examining diarists from former Kuomintang (KMT) strongholds, and in so doing it shows how urban, educated citizens engaged with the new government of the People's Republic (PRC) as an act of self-transformation. While writing a diary under the Maoist state was voluntary, it came with inherent risks: for example, in December 1968, a pile of 1950s personal diaries from the Nanjing area were swept up and investigated by unnamed individuals in a task force (*zhuan'anzu*).[1] Each author nevertheless earnestly used their diary, alongside official documents and discussions with their comrades, to define their personal role in achieving government objectives. Chinese people wrote diaries on a massive scale after 1949, showing that the practice was both widespread and important.

The early-PRC period was a perilous one, especially for those from the former centres of KMT control. In a sense, the successful transformation of educated people in the south was the most crucial for the new regime. By April 1948, the Chinese Communist Party (CCP) directed the People's Liberation Army (PLA)

[1] These 'task forces' often targeted seemingly unimportant intellectuals who kept personal records: Cong Cao, 'Science Imperilled: Intellectuals and the Cultural Revolution', in *Mr. Science and Chairman Mao's Cultural Revolution: Science and Technology*, eds. Chunjuan Nancy Wei and Darryl E. Brock. Plymouth: Lexington Books, 2013, pp. 123–4.

Proceedings of the British Academy, **267**, 95–111, © The British Academy 2025.

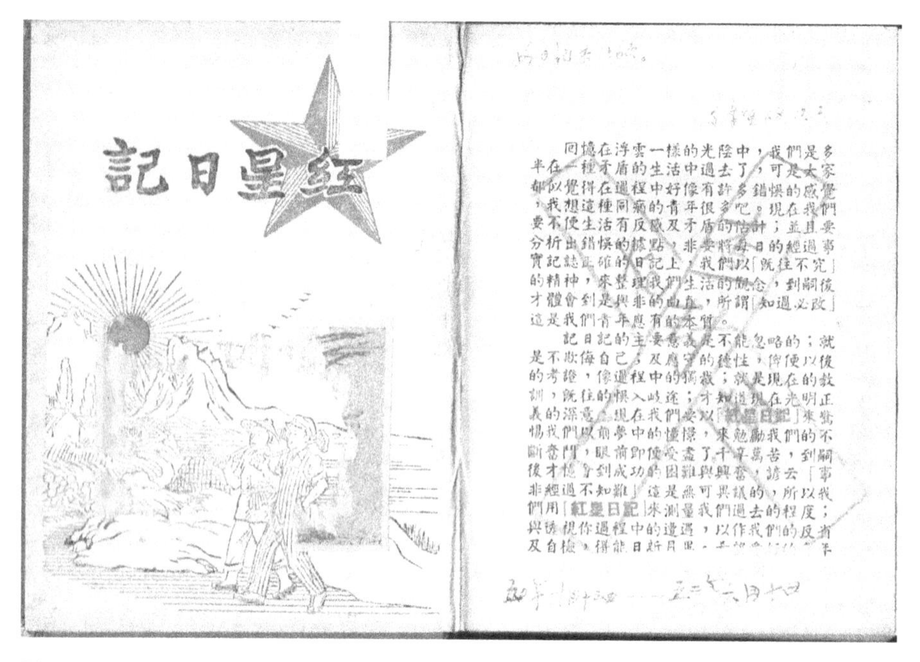

Figure 5.1 Chen Fang's diary, with 1968 inspection marks on inside cover

to keep a careful record of KMT and pro-KMT youth corps' members during their drive south, and local officials staged show trials as a form of legal education and mass discipline. Even in cities like Hangzhou, which offered effectively no resistance to Communist rule, there congregated worrisome numbers of KMT Revolutionary Army officers, party cadres, business leaders, private enterprises in struggles with emboldened labour unions, and their friends and relatives.[2] Perhaps because of the fear of 'corruption by the capitalist class', the reactionary landlords of the south, and the ties that locals had with the KMT, the CCP shipped thousands of newly minted cadres from the north, who arrived like an occupation

[2] Jennifer Altehenger, *Legal Lessons: Popularizing Laws in the People's Republic of China, 1949–1989*. Cambridge, MA: Harvard East Asia Monographs, 2018; James Z. Gao, *The Communist Takeover of Hangzhou: The Transformation of City and Cadre, 1949–1954*. Honolulu, HI: University of Hawai'i Press, 2004, p. 14. Contrast Gao's description of Hangzhou as a non-military city with the fact that elite KMT officers were still collected there – Chiang Kai-shek, after 'retiring' from the presidency in 1948, went directly to Hangzhou to be among his military supporters. Odd Arne Westad, *Decisive Encounters: The Chinese Civil War, 1946–1950*. Stanford: Stanford University Press, 2003, p. 219. On management-labour relations, see Kenneth G. Lieberthal, *Revolution and Tradition in Tientsin, 1949–1952*. Stanford: Stanford University Press, 1980, pp. 81–96. Robert Cliver, '*Minzhu guanli*: the Democratization of Factory Management in the Chinese Revolution', *Labor History*, 50:4 (2009), 409–35.

army with foreign accents and culture.[3] The CCP promoted locals to posts of visible leadership in order to encourage acceptance, but the party leaders considered southern guerrilla and underground operatives to be ill-equipped for governance in major cities.[4] Nevertheless, southerners were not simply running dogs of the *ancien régime*: the KMT had already alienated many members of class groups that were supposed to be their core allies, including industrialists, students who had served in the Revolutionary Army, southern 'bourgeois' intellectuals, and even conscientious or politically leftist KMT officers.[5] Educated Chinese therefore understood critiques of the KMT, but the CCP political orthodoxy that emerged out of Yan'an after 1945 was largely alien to them. Consequently, as in the pre-war Soviet Union, educated Chinese urban citizens used their diaries as a space to carefully work out their personal relationship with the 'new democracy'.[6]

While diary-writing in China was already a long tradition, the use of personal narrative for the political discipline of national citizens and political party members was very recent. As I have explored elsewhere, the KMT was inspired by Japanese and Soviet applications for personal and group diaries, which coincided with pedagogical uses of vernacular diaries to learn basic composition.[7] These techniques were not innovated by the CCP, but applied vigorously by the party as a tool for building a new society in which citizens shared a common set of values.[8] The scale of diary production was vastly expanded after 1949, and there were several kinds of mass-produced blank diaries during the 1950s.[9] Paratextual inscriptions

[3] Frederick C. Tiewes, *Provincial Party Personnel in Mainland China 1956–1966*. New York: Occasional Papers of the East Asian Institute, Columbia University, 1967. The proportion of cadres from outside the province who took on leadership roles was most pronounced in areas where the KMT was strongest, as well as some borderland areas. On theatre, see Chang-tai Hung, *Mao's New World: Political Culture in the Early People's Republic*. Ithaca: Cornell University Press, 2010, p. 88.

[4] Gao, *Communist Takeover of Hangzhou*. Tiewes, *Provincial Party Personnel*, pp. 15–20.

[5] Sherman Cochran, 'Capitalists Choosing Communist China: The Liu Family of Shanghai, 1948–56', in *Dilemmas of Victory: The Early Years of the People's Republic of China*, eds. Jeremy Brown and Paul G. Pickowicz. Cambridge, MA: Harvard University Press, 2010, pp. 359–85. Wen-hsin Yeh, *The Alienated Academy: Culture and Politics in Republican China, 1919-1937*. Cambridge, MA: Harvard East Asian Monographs, 2000. On disillusionment in the armed forces, see Kevin Landdeck, 'Under the Gun: Nationalist Military Service and Society in Wartime Sichuan, 1938–1945' (Ph.D dissertation, University of California, Berkeley, 2011). Odorick Y.K. Wou, *Mobilizing the Masses: Building Revolution in Henan*. Stanford: Stanford University Press, 1994, pp. 336–42.

[6] On Soviet life writing, see Igal Halfin, *Terror in My Soul: Communist Autobiographies on Trial*. Cambridge, MA: Harvard University Press, 2003; Jochen Hellbeck, *Revolution on My Mind: Writing a Diary under Stalin*. Cambridge, MA: Harvard University Press, 2009.

[7] Joan Judge and Hu Ying (eds.), *Beyond Exemplar Tales: Women's Biography in Chinese History*. Berkeley: University of California Press, 2011; Aaron William Moore, *Writing War: Soldiers Record the Japanese Empire*. Cambridge, MA: Harvard University Press, 2013, chapter 1; Marjorie Dryburgh and Sarah Dauncey (eds.), *Writing Lives in China, 1600–2010: Histories of the Elusive Self*. London: Palgrave Macmillan, 2013.

[8] Jan Kiely, *The Compelling Ideal: Thought Reform and the Prison in China, 1901–1956*. New Haven: Yale University Press, 2014.

[9] See my contribution, 'Personal Diaries', to the website curated by Jennifer Altehenger, 'The Mao Era in Objects', accessed 25 November 2023, https://maoeraobjects.ac.uk/object-biographies/personal-diaries/.

suggest that many were gifts, such as those given to 'delinquent youth' in urban 'work-study' rehabilitation centres as described by Aminda M. Smith, or handed out among members of collective farms and urban work groups.[10] Despite popular views of Maoist totalitarianism, most PRC diaries show no evidence of having been regularly reviewed by others. Stephen A. Smith wrote that to 'address the theme of individuality in the context of working-class formation and Communist revolution may seem perverse' but, like Smith, I also see the self-discipline of individuals as crucial to the development and maintenance of any political and social order; conversely, it is unproductive to overlook the importance of group dynamics in the development of an individual's revolutionary consciousness.[11] By the 1950s, the Chinese Communist movement was no longer, as David Apter and Tony Saich described, the intentional community of Yan'an, where willing adherents entered trembling into a 'sacred space', but the whole nation was now subject to the same 'logocentric ... narrative reconstruction of reality'.[12] Life-writing became even more important for learning one's place.

By analysing six diarists in former KMT areas – Chen Fang (Shanghai-Wuxi), Jiang Jianmin (Nanjing), Lu Yuren (Nanjing), Hu Chenming (Kunming), Xu Wenrui (Kunming), and Dai Jihui (Chongqing) – I hope to show how they used their personal accounts as a workspace to master socialist discourse; they did so by digesting party documents (*wenjian*), actively participating in small and large group debates, and conducting self-examinations. I show the centrality of 'study' (*xuexi*) as a method for subjectification, which was inextricably linked to the transformation of working relationships (*gongzuo*). 'Work and Study' (*Gongzuo yu xuexi*), which was the title of an argosy of blank diaries used by PRC citizens, were the twin engines of CCP hegemony in the socialist aircraft that was to transport China to its communist future.

Essential reading: Central documents and the making of Maoist discourse

When understanding the role of the diary in socialist revolutionary practice, knowing what authors read is as important as what they wrote. Most important, the diarists made constant reference to 'documents' (*wenjian*), reflecting the extent of the CCP's programme to disseminate the language, philosophy, and political values of the new regime. None of the diarists had experienced the CCP's first systematic

[10] Aminda M. Smith, *Thought Reform and China's Dangerous Classes: Reeducation, Resistance, and the People*. Lanham: Rowman & Littlefield, 2012.

[11] S.A. Smith, *Revolution and the People in Russia and China: A Comparative History*. Cambridge: Cambridge University Press, 2008, p. 69.

[12] David E. Apter and Tony Saich, *Revolutionary Discourse in Mao's Republic*. Cambridge, MA: Harvard University Press, pp. 12–15.

ideological training of cadres during the Yan'an Rectification Campaign (1942–5), and persistent confusion about communism, Mao Zedong Thought, and socialist transformation opened up a window for idiosyncratic interpretations. Three other factors made matters difficult: first, after 1945, the CCP and PLA absorbed and hurriedly trained thousands of new cadres and officers, with inconsistent results; second, older, local revolutionaries harboured, and openly expressed, heterodox views, including pre-revolutionary ideas about 'nation-building' (*jianguo*); third, as Nicolai Volland shows in this volume, the PRC was still importing Soviet discourse, which often departed from or outright contradicted Mao Zedong Thought, in the form of translated publications, films, and advisers.[13] It is not a common feature of industrial societies that workers and managers should study philosophy together, but the CCP devoted considerable resources to this activity and consequently early-PRC diaries reflect the systemic importance of documents.

The theoretical approaches to problems faced by the regime were made and remade throughout the 1950s, and they had to share space with the discourses and ideologies of the pre-1949 society that came before it. Especially for people like Hu Chenming, who lived in a Yunnan borderland area where campaigns often came late and with many local problems, the 'vigorous establishment of socialism' required not so much ideological training, but 'modernisation, regularisation, and the establishment [of infrastructure]' – a common KMT era refrain. Invoking pre-1949 tropes such as 'awakening' (*juewu*) or 'spirit' (*jingshen*), Hu exhorted his comrades to embrace the 'reconstruction of the borderlands of our fatherland' (*zhuanjian zuguo bianjing*), which overall required thorough study (Hu, late 1954 or early 1955).[14] In Chongqing, which was the wartime capital of the KMT, workers heard many old, familiar phrases from their cadres alongside CCP and Soviet concepts. While promoting exemplary progressive producers (*xianjin shengchanzhe*), cadres also encouraged Stakhanovite-style competitions between both workers and managers, emphasised technical training, and tied work into nation-building concepts.

> [We must] understand how completing the national [five-year] plan and the individual is connected, how it's clearly a contribution to national construction (*guojia jianshe*) ... Through Mao Zedong Thought we understand that all professions are part of the connection between national interest and personal interest (*guojia liyi he siren liyi de guanxi he yizhixing*), and we can [thereby] change the backwards attitudes we currently have (11 May 1956).

The revolutionary discourse that southerners encountered was therefore a mix of Mao Zedong Thought crafted in Yan'an and KMT developmental discourses of 'nation-building' that were often inconsistent with the CCP worldview; the northern

[13] On Soviet film in the 1950s, for example, see Tina Mai Chen, 'Internationalism and Cultural Experience: Soviet Films and Popular Chinese Understandings of the Future in the 1950s', *Cultural Critique* 58 (2004), 82–114.

[14] This entry was likely written during the First Taiwan Straits Crisis (3 September 1954 to 1 May 1955), following his last dated entry of 14 October 1954. Chenming connects the movement to liberate Taiwan with the reconstruction of mainland China's borderlands.

cadres who were meant to teach the party's vision and overcome pre-revolutionary thinking were not always clear on the difference between them.

Until the late 1950s, diarists also contended with the suffusion of Soviet discourse alongside Mao Zedong Thought, although not every writer embraced 'Elder Brother'. The Chinese government did not aim to reproduce the Soviet system, and most educated southerners, like Chen Fang, prioritised the CCP over the foreign whenever the two conflicted: he also criticised students who only read 'Russian war stories', which he considered a waste of time (7 July 1949). Combining data, philosophy, mathematics, and notes from political meetings, urban construction engineer Jiang Jianmin was enthusiastic about Soviet advisers, which reflected the importance of Soviet developmental economics on PRC reinventions of pre-revolutionary 'nation-building'.[15] Jiang's sporadic attempts to learn Russian dapple his diaries, right up to 1958, when he bought a blank diary with 'The World' (*mir*) written in Cyrillic on the cover, and then abandoned Russian entirely during the Sino-Soviet split. Meanwhile, Hu Chenming admitted Soviet culture 'had a large educational function' in his understanding communism. Hu had seen as many as 153 Russian films, which provided the 'teachings of Marx and Lenin on communism' and from which he could 'recite the lines by heart' (early 1955). Hu weaved the titles of Soviet cinema into his own life story:

> When I 'Started My Life' (*Zou xiang shenghuo*) 'Among the People' (*Zai renjian*), I wanted to become 'A Brave Man' (*Yonggan de ren*), and consequently joined 'Young People Entering the Army' (*Qingnian jinxingjun*), then carried out the struggle against our enemies, just like 'Marx in His Youth' (*Makesi qingnian shi*). (Undated, early 1955)

Still, the Soviet Union was a distant mystery for the vast majority of Chinese people, and educated southerners were understandably more interested in comprehending the new government closer to home.

Despite the confusing ideological landscape of the 1950s, the party leaders in Beijing were constantly talking to their citizens through document dissemination but also organised group study. During the Eighth National Congress, Dai Jihui recorded in his diary all of the significant CCP figures participating (Mao, Liu Shaoqi, Deng Xiaoping, and Zhou Enlai), but also the means through which understanding about communism should be effected:

> Zhou Enlai has explained the changes to the 2nd Five Year Plan, and how to study this. Set up a work group (*danwei*). Collect and review the newspaper reporting, and then carry out a debate (*taolun*). This includes the intermediate study groups (*zhongjizu*) and cultural study groups (*wenhuaban*). In order to facilitate understanding of the bigger issues, each work group's important, leading cadres can study on their own and offer their opinions in order that everyone can understand them.

[15] Charles Musgrove, *China's Contested Capital: Architecture, Ritual, and Response in Nanjing*. Honolulu: University of Hawai'i Press, 2013. Chang-tai Hung, *Mao's New World*, pp. 36–41, Chapter 2. Soviet writers also mixed science and philosophy in their diaries: Thomas Lahusen, *How Life Writes the Book*, p. 85.

The documents and philosophy generated by the CCP were not just for individual study, because changes in thought were only achievable through collective activities in the workplace. Collective study took place among 'small groups' (*xiaozu*) or 'culture groups' (*wenhuaban*) at the lowest level (*chuji*), but could also occur among cadres in provincial offices (*shengchu*) who were at the intermediate level (*zhongji*) in their study of political philosophy and theory.[16] Errors in thought (*sixiang cuowu*) should be confronted through criticism by fellow cadres, and a focus on material concerns, such as how to increase production, gave progressive elements on the factory floor the encouragement they needed to unleash their potential for change.

> Work units set the agenda for study, beginning with carrying out readings of the broadsheets [like the *Zhongguo qingnianbao*] and then conducting debates (this includes the intermediate groups and the cultural study groups). In order to help everyone understand some of the larger issues, each work unit's leading cadres could conduct individual study (*ziwo xuexi*) and then offer their views in order to help everyone else (or have comrades from the intermediate study group do this).

The correct interpretation of central documents was crucial and controlled by cadres operating at the provincial level, but equally important was moving from individual comprehension to a shared understanding among comrades. Dai estimated it would take six to eight hours per week until the job was done, and could interrupt other forms of study in the small groups (24 September 1956); in some cases, study led to halts in production, underlining just how important understanding documents was for the early PRC system.

The PRC government thus disseminated documents for 'study' through several media outlets, and everyone had to examine them closely from provincial leaders to factory floor workers. The diarists were all consumers of government discourse, which they immediately applied in their work lives, using the diaries as a work*space*, in the work*place*, to master the use of language through the process of 'work and study'. The only way to ensure that individuals fully comprehended the meaning of these documents, however, was to conduct their studies on the job, and furthermore to enforce changes to consciousness through democratic means involving workers and farmers.

Democracy in action: Contributions and contradictions

For each writer, the diary was a useful tool for integrating official documents into their own language, but it also functioned as a space of inter-subjectification, where the views of others in 'debate' or 'study' shaped the language community. Chen epitomised the sort of essential grassroots cadres and student activists who emerged

[16] For an early explanation on the use of 'small groups' to educate workers, and the history of this practice, see Martin King Whyte, *Small Groups and Political Ritual in China*. Berkeley: University of California Press, 1974.

at the end of Nationalist rule, for whom self-criticism in the diary was 'a weapon of the proletariat's revolution (*wuzhan jieji geming*), because through it one is able to understand one's daily progress' (8 July 1949).[17] One could not do it quietly and alone, however: the diarists understood that the revolution affected all aspects of life, and repeatedly discussed 'reforming the system' (*gaishan zhidu / xitong*). In Kunming, Xu Wenrui wrote that 'the system is not sound (*jianquan*)', that corruption and waste were still rife across China. She believed that democratising the workplace and critically examining production failures were part of the process of the CCP's continual party (re-)formation (*jiandang*) in areas that were formerly controlled by the KMT (14 August 1952). As the diarists worked harder to contribute to the democratic building of socialism, however, they unwittingly created new contradictions within the project.

As Joel Andreas showed at Tsinghua University, the educated elites began to embrace their status as 'red and expert' quickly in the 1950s, but this inevitably created hierarchies and power contests.[18] Experts worked hard to achieve production quotas, which paradoxically involved reinforcing their technical authority in the workplace in spite of the new democracy. Following Liu Shaoqi's speeches critical of rural cadres, the CCP expected educated people from the cities to 'lead' the countryside in their social and economic organisation, for which Lu Yuren was directly responsible.[19] At first, he worked as an accountant who collected on bank loans against the communes (*shoukuan gongzuo*), which was the third piece of the (often dysfunctional) puzzle that constituted the communal production system.[20] CCP leaders asked educated urbanites like Lu to oversee the establishment of people's communes, which involved the extraction of grain in rural areas to support higher per capita levels of consumption in the cities – which in turn undermined local farming practices.[21] At the beginning of the Five-Anti Campaign, Jiang Jianmin

[17] Yung-fa Chen, *Making Revolution: The Communist Movement in Eastern and Central China, 1937–1945*. Berkeley: University of California Press, 1992. On self-criticism, see Joseph Yick, *Making Urban Revolution in China*, pp. 123–4. On the importance of anarchism in the pre-revolutionary history of the CCP, see Dirlik, *Anarchism in the Chinese Revolution* and Apter and Saich, pp. 16–17.

[18] Joel Andreas, *Rise of the Red Engineers: The Cultural Revolution and the Origins of China's New Class*. Stanford: Stanford University Press, 2009, chapter 3.

[19] For analysis of Liu and the dominance of urban over rural revolution after 1949, see Kenneth Lieberthal, *Revolution and Tradition in Tientsin, 1949–1952*. Stanford: Stanford University Press, 1980; and Jeremy Brown, *City versus Countryside in Mao's China: Negotiating the Divide*. Cambridge: Cambridge University Press, 2012, pp. 20–2.

[20] Mao famously criticised shortcomings in the communes by adding 'loan collection' (*san-shoukuan*) to free canteen meals (*yi-ping*) and gratis products (*er-diao*) as a general descriptor of a crisis in the Chinese rural economy. There was little coherency across China in the commune system: Xin Yi, 'On the Distribution System of Large-Scale People's Communes', translated by Jiagu Richter and Robert Mackie, in Felix Wemheuer and Kimberley Ens Manning (eds.), *Eating Bitterness: New Perspectives on China's Great Leap Forward and Famine*. Vancouver: University of British Columbia Press, 2012, pp. 130–47.

[21] Kenneth R. Walker, *Food Grain Procurement and Consumption in China*. Cambridge: Cambridge University Press, 1984.

wrote on the importance of the 'democratisation' of the workplace, which meant not merely relying on the educated managers and technical specialists but also the consciousness of the labouring masses:

> We must eliminate all poisons from the old system and way of doing things (*zuofeng fangfa*) (including the leadership position of the capitalist class). In order to have the consciousness (*sixiang*) of the workers lead us, we must correct our thoughts (*sixiang gaizao*) and establish a working class mindset (*sixiang*)—this will complete the democratic revolution. Through the Five Antis, we will establish a democratic, new workplace. At the level of consciousness (*sixiang*), we will start anew and draw a clear line (*huaqing jiexian*) [with the past].

Nevertheless, while Jiang wrote of the value of applying the mass line to his technical field, he was consistently committed to protecting the status of intellectuals, engineers, and scientists in the new regime. Debates in the workplace at this time could be vituperative, impassioned, and mean-spirited – no one knew precisely how to interpret political theory on the factory floor – which led managers like Dai Jihui to reflect on problems of temperament: 'Looking back over the last few years, my thoughts and feelings never took into account of how my words might cause suffering for other people. This is not just impolite, it's also a form of brutality. It can cause opposition and resistance' (12 October 1956). As a rural production manager, Lu Yuren also emphasised the importance of the 'masses' (*qunzhong*) in his diary, even when the opinions of others turned against him during the crisis of his corruption charges. For him, this involved listening to the opinions of others and selecting the one that matched the evidence, or was argued the most persuasively – although effectively he valued the (privately shared) views of technical experts, especially personal friends, most highly. Educated southerners' inability to be convinced by the views of workers was a failure of their understanding of the new democracy, while ironically endeavouring to overcome any obstacles in order to meet the production objectives of China's socialist system.

Indeed, making socialism work required careful management of working people, a task to which educated southerners quickly applied their talents. In her Kunming factory, Xu Wenrui noted how 'we have all had an awakening and established a revolutionary person's perspective', and that this was a consequence of the 'Great Three-Antis Campaign'. While fighting waste to increase productivity, however, resistance rose against grain requisitions, which in turn formed a perilous environment wherein recently demobilised KMT forces allied with outraged farmers.[22] To raise living standards and hit higher quotas, Xu and her colleagues in Kunming tried to increase sales of their goods to the Yunnan countryside, chasing 'profit in sales' as they had pre-1949, but the prices of rural inputs and industrial outputs could not generate too much disparity. Profits were certainly necessary

[22] Brian DeMare, *Land Wars: The Story of China's Agrarian Revolution*. Stanford: Stanford University Press, 2019, pp. 21–3.

for contributions to state coffers, even as the threat of overproduction in the early 1950s rapidly became a concern, and this led Mao Zedong to call for a greater integration of rural and urban economies. Xu saw the problem of urban-rural disparity as inherently inseparable from the quandaries around political consciousness, but also understood it as an issue of resource management; as she put it: 'profitability influences Chairman Mao's call for accelerating the flow of capital and goods between town and country' (*chengxiang wuzi jiaoliu*) and that 'the alliance of rural and urban production' represents both 'economic and political significance' for the success of socialism. Linking together political education and economic production was a project ideally suited for the pre-revolutionary managerial class. If they understood anything in this new world, it was systems and processes. 'If the system is chaotic, thought is chaotic,' Xu observed, where 'using one's brain ... to improve the methods of administration' would have immediate positive impact on sales and profit margins, bringing productivity to her sector and support to the accumulation of wealth in rural areas – exactly as Mao Zedong had recommended (14 August 1952). In her revolutionary fervour, then, Xu had arrived at an argument for the value of an over-arching, ever-growing bureaucracy within the socialist system, which was one of the new contradictions in the PRC system that she was meant to overcome.

Educated individuals in former KMT urban areas constructed an idea of 'the masses', proactively engaged with those people, defined their own positions and roles, and then discovered that their requirements for increasing production somehow put them back in control just as before. Superficially, those involved in the process of industrial production embraced the creation of a collective subjectivity in order to increase effectiveness on the factory floor. For example, Jiang Jianmin admitted that the first Five-Year Plan had been very problematic. In his diary, he bemoaned the fact that many of the country's projects achieved 30 per cent of their targets, but he concluded that the best response was to 'listen to the workers' and 'strengthen the work of leading the rectification of thought'. Thinking on the past failures, he wrote, 'What was the origin of these issues? Basically, especially recently, it is a lack of understanding (*buhao renshi*). The important issue is that the cadres do not rely on the masses. The right level of awakening only occurs among the masses' (31 October 1955). Xu too repeatedly blamed 'attitude toward work' (*gongzuo taidu*) as a reason for failing to meet targets (14 August 1952). Nevertheless, criticism and debate, which was supposed to increase production, often thwarted it, because of the importance of expertise. As Jiang observed, factory chiefs cautioned workers and cadres against attacking technicians, engineers, and scientists, many of whom had received their educations under the old regime:

> Cadres from the executive government leadership have put forward opinions regarding technical and scientific experts that are irrational. We are one unified group (*women shi tuanjie yizhi de*). In the past six years we have achieved much, but we may need another fifty or sixty to truly come together ... We must seek the views

> of the technical experts, accept one another in good faith, and be able to discuss our shortcomings. Reforming the intellectuals—we've seen significant improvement in the last five or six years, even if there are some lingering faults. There is progress. After five years, the capitalist class has also progressed greatly. … Because we must take a proactive and positive attitude, we must help them, take care of them, and not adopt an adversarial stance.

Technical experts continually and carefully considered their role: 'How do we carry on with our research?' asked Jiang, and then supplied his own answer: 'It is the most important work of the Five Year Plan. Knowledge is reason enough.' The view that scientists and engineers should be free to study what they deemed important was a crucial point for Jiang, but this could hardly be considered an orthodox view within the CCP. Still, Jiang insisted on describing the intellectual class's research for knowledge's sake as a form of collective subjectivity – only one that departed from the mass line: 'Scientists all feel the same, and this will be in a unified mode of expression' (26 December 1955). Within the collective subjectivity that was emerging within the workplace, the diarists attempted to carve out a separate language community for bureaucrats, scientists, and technicians.

Even though Jiang upheld the importance of technical and scientific expertise, like the other diarists he also maintained views of industrial development that were consistent with CCP central documents: for example, Jiang regularly asserted that 'right now our problem isn't technical or economic, it's mental' (*sixiangshang de wenti,* 8 March 1956). He went so far as to rely on the mass line for responding to a serious bodily accident on the factory floor (14 November 1955). Nevertheless, despite their insistence on the transformative power of a revolutionary collective self and the importance of the mass line, educated citizens on the factory floor still desired to carve out an autonomous, authoritative world for themselves within the new socialist society.

Further stoppages on the mass line: Gender, ownership, and performativity

Socialism produced contradictions to be overcome, and those contradictions could exacerbate pre-existing social conflicts. Dai's diary reveals that there were problems with workers in 1950s' Chongqing factories that persisted from the Republican era, including youth rebellions, exposure of sensitive industrial information, and worker absenteeism (4 May 1956). Some of the conflicts, however, were newly introduced by the PRC political system. The establishment of a collective subjectivity at a time of national socioeconomic reorganisation inevitably exposed people who did not learn the language correctly, or refused to use it because it did not reflect the 'truth' of their lived experience. This included the most vulnerable members of Chinese society who were to be served by the party, including women and subsistence farmers, but also the cadres themselves. While the diarists, as educated people, did not become members of the CCP's *nomenklatura* – what Mikhail Voslensky called the 'Soviet ruling class' – they gradually embraced authoritative roles.[23]

As Delia Davin and Christina Gilmartin noted regarding the gender dynamics of the CCP leadership, the fact that the targets of male diarists' frustration were often women is reflective of the persistence of male dominance within leftist movements.[24] As early as December 1949, Chen recorded an interrogation of a woman who was born into a landlord family, but not 'earnestly engaging in landlordism herself'. Nevertheless, she was charged with composing loan contracts (*baodanshu*) on behalf of landlords 'and being unable to solve faults in her thinking'. She claimed that she 'sought the truth with the best intentions' (*dui de qi ziji liangxin jiu shi*), and was deemed by Chen to have committed a grievous error: confusing a 'good heart' (*ziji de liangxin*) with authentic revolutionary morality (*geming de liangxin*, 25 December 1949). Chen also recorded his ups and downs with another female comrade, Ms Xie, with whom it seemed he may have had occasional relations.

> She told me, 'I think, ah, I think sometimes I like you,' but she also said before that she 'thinks romance (*lian'ai*) is foolish.' ... She thinks that going about life observing things and people only captures superficialities, but one has to see the real quality of things [*shizhi*]. This is true, and her thoughts make for good study (29 October 1950).

Less than a week later, however, Chen was irritated with her. They had endured a 'very unpleasant argument' regarding government policy, after which they turned to other comrades for their input. Cadres often accused women of putting emotions above party work, echoing the discourse of pre-revolutionary patriarchy.[25] 'Xie was full of resentment (*yan'e*), and was unwilling to accept it', Chen complained, 'she cannot separate her emotions from the task'. In this case, the politics of the new regime and personal relationships were not overcome by a new language, as Chen himself pointed out: 'We discussed things of a political nature, and she told us her thoughts and feelings. She said our solutions were unacceptable. She really is the sort of person who refuses to listen to the opinions of others' (2 November 1950). As men like Chen exercised power over these women, and emotional ties developed simultaneously, such disagreements seemed less theoretical and more concerned with the use of newly granted political powers to resolve personal disputes.[26]

Poor and middling farmers also challenged Chen directly during his time conducting Land Reform in Jiangsu, and he sometimes referred to the 'obstacles in

[23] Michael Voslensky, *Nomenklatura: The Soviet Ruling Class*. Trans. Eric Mosbacher. Garden City, NY: Doubleday & Company, Inc., 1984.

[24] Delia Davin, *Woman-Work: Women and the Party in Revolutionary China*. Oxford: Oxford University Press, 1976. Christina Gilmartin, *Engendering the Chinese Revolution: Radical Women, Communist Politics, and Mass Movements in the 1920s*. Berkeley: University of California Press, 1995, Chapter 4.

[25] Martin King Whyte, *Small Groups and Political Rituals in China*. Berkeley: University of California Press, 1974, p. 67.

[26] Neil Diamant, *Revolutionizing the Family: Politics, Love, and Divorce in Urban and Rural China, 1949-1968*. University of California Press, 2000.

[their] thinking needing resolution' (*jiejue sixiang-shang de zhang'ai*, 27 July 1950). Nevertheless, as Chen discovered, the countryside did not exist simply as an abstraction in the debates of study groups, but was a highly complex ecosystem whose evolution occurred throughout centuries of human management and investment. Farmers refused to move their homes and pointed out 'inequities', which they perceived as 'punishments', in the state's land distribution system, such as apportioning less fertile land to the poorest farmers and the apparently autocratic decision process (Chen, 9 July 1950). In addition to the proprietary views farmers had over their land, including the cultivation of tea bushes and plum trees over many years, they also resented the imperious attitudes of some neighbours and cadres – particularly those who felt free to enter their property. Chen, an urban educated propagandist at heart, had to tackle thorny questions: when a farmer was granted land abutting a waterway, did he own right of access to the waterway, including the capture, consumption, and sale of the fish within it (Chen, 29 July 1950)? Despite hours spent in debates and group criticisms with women and farmers, Chen spent little time reflecting on his own shortcomings, and often became a commanding presence during Land Reform.

Chen's issues paled in comparison to Lu Yuren's, who used his diary not to arrive at truth or adopt an authoritative persona, but to embrace a falsehood in order to support the integrity of the socialist system. In August 1958, Lu was sent to help a commune with its accounting work, mainly because they were short-staffed and over-worked. 'As usual,' Lu wrote, 'I was tasked to collect [state bank] loans [from the commune].' He had a bad premonition about this, especially after he was given multiple duties, writing: 'I feel things are very difficult. Basically, I think my abilities are inadequate for the job' (*yewu nengli tai di*, 18 August 1958). He was quickly caught up in a torrent of accusations concerning corruption in the commune system. Nearly half a year later, in February 1959, Lu recalled this terrible ordeal, writing 'it was a day I won't easily forget, a day in which I was insulted and humiliated'. His accusers were in a local (town or *zhen*) Office of Grain Provision (*Liangshi bangongchu*). According to Lu, on 1 December 1958, a minor error ruined his life, and thereafter he was known as a corrupt official (*tanwu fenzi*) to be placed in confinement.

> This is what really happened: on 1 December 1958, I was assigned to the No. 2 work group for collecting bank notes [on state loans]. That day, we were very busy at work, from morning until night, while the farm workers laboured without stopping. Apart from the times we left our workplace to eat, there was no rest. That day there was a lot of cash around, such that by the afternoon when we issued bank notes, we hadn't taken account of all the money, so one of my comrades even quit [normal duties] early just to count the notes. The farm workers were in a rush, wouldn't wait, and begged us to speed things up. I was a member of the Communist Youth League, so I couldn't watch things pile up or ignore the suffering of the masses. At the same time, I trusted (*yilai*) the party and its leadership, so I did not stop working. I asked the station chief to look over the money, so Comrade Wang took over recording the bank notes. Then, while things were moving quickly, I discovered from the ledger that we were over 58

> yuan short. Comrade Wang said there was no way the collection could be that short. The records of our reserves were poor and there weren't any other people watching the money, so we asked the station chief what to do. He said we'd speak again tomorrow.

Wang and the others would not accept that there were problems with the accounting of bank notes to and from the collective farms. Consequently, all work came to a grinding halt in order for them to sort out what had happened. The second time they counted all the money and found 60 yuan missing. The third time they analysed the books more closely and found it was actually 59 yuan. At a contentious meeting of those responsible for the farm's accounts, it was decided that Lu had embezzled the difference (whatever it was), and that he should publicly confess his shortcomings (*tanbai*). This provoked a crisis for Lu who, claiming innocence, naturally could not have had the bank notes to return:

> What could I say after this? Acknowledging my corruption would be easy, but the money wasn't just sitting in my pocket, was it? [...] In reality, I hadn't taken it, so how could I admit to being crooked? If I don't accept this blame, then their retribution would be to lock me up at once.

Engaging in the speech act of declaring oneself to be a 'corrupt official' was certainly easier than accounting for these discrepancies, or admitting incompetence among the leadership. He maintained his faith in the party, however, claiming that the leadership 'cannot forget a good man. I won't be lost like a stone thrown into the water' (8 February 1959). Lu left his old post, and took up custodial work at a different commune. Two days later, Lu's boss at his new job was notified of his past, and Lu realised that he had to face the old accusations in order to continue working: 'Under this sort of punishment, I will make a temporary confession, and see if I can clear my name' (10 February 1959). The very next morning, he wrote his confession, 'but my purpose was to get a temporary acknowledgement of fault and await [the situation] to be dealt with'. Another colleague, Liu Xueshi, encouraged him to 'proactively resolve the problem', but promptly warned him that 'a fake confession is wrong. The truth is truth, and deceit is deceit', adding that, 'The leadership still hasn't determined your punishment' (11 February 1959). Other members of his work team professed ignorance, or warily asked him if he was 'still angry'.

In the end, Lu was only able to resolve the situation by accepting responsibility and declaring himself a 'corrupt official'. Friends took up a collection and helped him repay the funds he had not stolen. Lu suggested that the leadership knew he was not a thief, but his role in this narrative showed that he was cooperative and helped them to iron out the rough edges of the commune economy. He later went on to be a moderately successful demographer and housing census factotum for Nanjing in the 1960s – that is, until his diary was taken in 1968 and he disappeared from the record. China's women, farmers, and even the cadres had to change themselves in order to find their place in the process of socialist making, and sometimes the fall from grace came very quickly. The diary was useful for working out what one's role in the new society should be, even if it was ultimately a lie.

Conclusion

It is unclear what happened to the Nanjing diarists like Lu whose personal records were investigated in 1968. Some may have walked away completely clean, others may have endured various forms of struggle as the political landscape, and discourse, shifted in the late 1960s. During the 1950s, however, they all managed to find individual understandings of what revolutionary life entailed, and where they fit into it. Chen was quick off the mark, but his idealistic student days had to give way to the regimented, and frequently rigidly enforced, methods of Land Reform. Jiang Jianmin, Hu Chenming, Xu Wenrui, and Dai Jihui embraced the inseparability of democratisation, self-discipline, and systemic change as a way to transform the old ways of production into socialism, but they never fully relinquished their special claims of expertise. Lu Yuren discovered the importance of narratives in the maintenance of order, donning a title for himself that was inaccurate but useful to his superiors. In each of these cases, the diaries helped the authors acquire a new language and think through the contradictions that the new order presented them, so that their revolutionary self could become a resonant, indispensable part of the community that would build a new world.

It is extraordinary that we now see socialist regimes of the twentieth century as inimical to individual development and self-actualisation, even if such claims were once understandable given the historical record of those who suffered under Maoism. Mao Zedong once proclaimed self-realisation as the paramount concern of all human beings, and that whenever individuals were repressed, 'there can be no greater crime'.[27] As Julia Strauss and many others have observed, the party and the state, from Mao down to the lowest level of cadres, took seriously the views and concerns of locals because they firmly believed that popular support delivered victories throughout modern history.[28] Organising philosophical study groups including managers, political leaders, workers, and farmers was not a common characteristic of capitalist industrial modernisation, but the CCP considered it an essential part of the socialist project; the continual act of reinvention and reconstruction, in theory, opened the path to a communist utopia. Dai wrote:

> All life is in a state of perpetual development. There are no absolutely pure objects, as they are all in motion and developing as living things. The newly born can perish, and we humans are not recent developments. From birth to youth, middle age, and elder years, we inevitably die. The socialist economy continually expands productivity, whereas the capitalist economy is in a terminal state. What is something new, and what is something dead? Marxist philosophy teaches us the difference. A new thing is growing, developing, and has a future path ahead of it (8 August 1956).

[27] Quoted in Smith, *Revolution and the People*, p. 80.

[28] Strauss, Julia (2002) 'Paternalist Terror: The Campaign to Suppress Counterrevolutionaries and Regime Consolidation in the People's Republic of China, 1950-1953'. *Comparative Studies in Society and History*, 44 (1). pp. 80–105.

Despite the prevailing view of Maoist totalitarianism, the party made no effort to systematically suppress or monitor life-writing, but instead encouraged its use as a method of self-transformation during the critical phase of consolidating political power. Chinese diarists' reconstruction of the self was the lynchpin to democratising the workplace, which would inevitably produce a new social and economic system across the PRC. For such critical work on the self, the diary was still the most accessible, accepted, and effective technology in the 1950s. What emerges from a close reading of the diaries was how the authors' extraordinary efforts to make Maoism work created new, unexpected problems, which were often the direct consequences of those efforts; just as no transformation of labour, social relations, and production was immediate or perfect, so too overcoming the contradictions between the lives they tried to create and the problems they were meant to solve required constant work and study. As the blank diary title 'Work and Study' suggested, these two activities could never be pulled apart in the making of the socialist system, and the collective labour of individual subjects was essential.

Diarists

N.B.: All diarists' identities have been protected by pseudonyms.

Chen Fang was a political writer, activist, and schoolteacher in the Shanghai-Nanjing-Kunshan area. He was born in 1926 in Zhejiang, and was a graduate of the Zhejiang Province Jintang Normal College at Guanhaiwei, between Ningbo and Shaoxing. From August 1945, he accepted a series of teaching posts until he became an unemployed drifter, eventually finding work after liberation at age 23. In 1957 he co-published a book about a propagandist that, judging from its current distribution in library collections, enjoyed some modest popularity. Chen's diaries begin in 1949, which included an old 'Record of Achievements' (*Baicheng riji*) from the end of the Nationalist period and a post-revolution 'Red Star Diary' (*Hongxing riji*). These five diary notebooks continue on to 1960, detailing his experience of Land Reform, worker organisation, and political mobilisation.

Jiang Jianmin was a construction manager for projects in and around Nanjing, whose seven diary notebooks begin in 1952 and end in 1962; the titles of the blank diaries are varied, from 'Age of Greatness' (*Weida shidai*) to 'Work and Study' (*Gongzuo yu xuexi*). The building projects enjoyed ties with Russian advisers up through the Great Leap Forward (GLF), reflecting both the importance of infrastructure projects in the early PRC and the power of Soviet influence.

Lu Yuren was dispatched from Nanjing in the late 1950s to a collection of villages in rural Jiangsu to oversee the operation of a large collective farm during the time of the GLF. Lu had previously worked in Changzhou, Jiangsu Province, in a down feather processing factory, and appears to have been a local. He also worked in 'sales' (*fahuo*), which presumably meant the nationalised distribution system of consumer goods. Among other things, he helped transport fertiliser to agricultural workers, carrying out several runs into the countryside from a dispatch centre by boat from about 7:30am into the evening (5–8 March 1959).

Hu Chenming purchased his 'Work and Study Diary' (*Gongzuo yu xuexi*) by June 1952. Hu was from a poor farming family in Qiubei County, equidistant from Kunming and the Vietnamese border. When he was a baby, his mother briefly gave him to a landlord family due to the family's poverty, but he eventually returned home when he was three. Hu was educated from the age of 7 to 13, when his father died, and thereafter began collecting coal for the local train station while continuing his schooling. Despite the fact that Hu claimed that he was 'very poor' and had a 'low cultural level', he briefly worked as a teacher prior to the revolution coming to Yunnan, which he described as a 'good way to make a living'. Hu participated in the revolution in 1949 by joining a PLA guerrilla unit (*youjidui*).

Xu Wenrui was a mid-tier manager in a medium-sized factory, which produced tools for rural communities in Yunnan. She is the only female author in this study. She purchased her blank diary, published as 'The Everyman's Diary' (*Dajia riji*), in Kunming on 28 July 1952, and began recording in it that August. Her last recorded entry was on 14 November, but undated content continues thereafter. Much of her diary describes the content of her discussions with the factory head manager about meeting production targets.

Dai Jihui was a Party member, influential cadre, and bureaucrat in the Chongqing City Commission's Ministry of Industry (*Chongqing-shi weiyuan gongye-bu*). In particular, Dai oversaw steel production in foundries that were originally established in the Republican era, and was preoccupied with meeting national industrialisation targets leading up to the GLF. Dai purchased the 'Red Flag' (*Hongqi*) diary and began recording from the first week of May 1956, concluding on 12 October 1956. A note at the front page of the diary states that it was 'collected' on 27 August 1956.

6

Brainwashing and World Revolution

AMINDA SMITH

ONE OCTOBER DAY in 1951, a small group of men and women sat around a table inside an old poorhouse in the heart of Beijing. The Chinese Communist authorities had arrested each of the individuals for petty offences, such as picking pockets, prostitution, or panhandling, and incarcerated them for reeducation. On this particular day, the internees were learning about American soldiers who were also undergoing Chinese Communist reeducation, as captives in Korean War prison camps. A political instructor led a discussion in which one former beggar, erstwhile sex worker, and accused 'labor shirker' reportedly said: 'After I read about these soldiers, their great sacrifice made me see that I should devote myself to increasing production so that China can beat back the American imperialists'.[1]

This quotation was taken from a work report, written by the political instructor, who was a member of the Chinese Communist Party (CCP). Thus, it is difficult to know if a real reeducatee actually made these statements, or made them sincerely, but the fact that these words were included in an official report does demonstrate the close connection the CCP drew between domestic reeducation centres and Korean prisoner of war (POW) camps. Despite those clear links, however, the two have not been treated as elements of the same thought reform project since the 1960s, when the psychiatrist and public intellectual, Dr Robert Lifton, published the last of his many works on the subject.[2] Much recent work tackles the topic of POWs and

[1] Beijing Bureau of Civil Affairs, 'Work Report on Educating Internees about the campaign to Resist American and Aid Korea', October 1951, Beijing Municipal Archives, 196-2-20.

[2] Robert Lifton's many publications on Chinese Communist thought reform included Robert J. Lifton, 'Home by Ship: Reaction Patterns of American Prisoners of War Repatriated from North Korea', *American Journal of Psychiatry,* 110 (1954), 732–9; 'Thought Reform of Chinese Intellectuals: A Psychiatric Evaluation', *Journal of Asian Studies* 16, no. 1 (Nov 1956), 76–7; *Thought Reform and the Psychology of Totalism: A Psychiatric Study of 'Brainwashing' in China.* New York: W.W. Norton, 1961; 'Peking's 'Thought Reform' – Group Psychotherapy to Save Your Soul', in *The China Reader: Communist China Revolutionary Reconstruction and Internal Confrontation 1949 to the Present*, ed. Franz Schurmann and Orville Schell. New York: Random House, 1967.

Proceedings of the British Academy, **267**, 112–125, © The British Academy 2025.

their Chinese captors, and there is no shortage of research on reeducation in Mao's China, but these are treated as two very separate fields of inquiry.[3] This chapter argues that domestic Chinese thought reform efforts should be read together with the Korean War 'brainwashing' scare as two facets of a global revolutionary praxis.

Brainwashing and thought reform: Entangled histories

After the official change of state power in 1949, the newly victorious CCP intensified their efforts to win the hearts and minds of the Chinese people with massive propaganda and political education campaigns, carried out in schools, offices, workplaces, and neighbourhoods. As a part of that national thought reform campaign, the Communist state also orchestrated large-scale internments of everyone thought to pose a potential threat to the success of the new People's Republic of China (PRC). By 1952, there were 1,800 incarceration-based reeducation centres in China.[4] That number did not include the similar number of more traditional prisons, in which political educators also attempted to remould individuals convicted of crimes like larceny, murder, treason, war crimes, and 'severe acts of counterrevolution'. The 1,800 new centres were for individuals whose actions were categorised as falling somewhere 'between crime and error', such as sex workers, petty criminals and hooligans, itinerants, monks and nuns from various sects and religious orders, fortune tellers, grifters, secret society members, drug addicts, minor counterrevolutionaries, and low-level spies. Not all of these internees were Chinese. Shanghai's Public Security Bureau interned resident foreign sex workers, for example, and many foreign scholars and missionaries living in China were also incarcerated. Some Chinese intellectuals were interned as well in the early 1950s.

[3] Some recent work on the brainwashing of UN POWs includes Charles S. Young, *Name, Rank, Serial Number: Exploiting Korean War POWs At Home and Abroad*. Oxford: Oxford University Press, 2014; Matthew W. Dunne, *A Cold War State of Mind: Brainwashing and Postwar American Society*. Amherst: University of Massachusetts Press, 2013; Susan L Carruthers, *Cold War Captives: Imprisonment, Escape, and Brainwashing*. California: University of California Press, 2009; David Seed, *Brainwashing: The Fictions of Mind Control, A Study of Novels and Films since World War II*. Kent: Kent State University Press, 2004.

Some of the recent work on thought reform includes Jan Kiely, *The Compelling Ideal: Thought Reform and the Prison in China, 1901–1956*. New Haven: Yale University Press, 2014; Aminda M. Smith, *Thought Reform and China's Dangerous Classes: Reeducation, Resistance, and the People*. Lanham: Rowman and Littlefield, 2013; Klaus Mühlhahn, *Criminal Justice in China: A History*. Cambridge, MA: Harvard University Press, 2009; Børge Baaken, *Crime, Punishment, and Policing in China*. Lanham: Rowman and Littlefield, 2005; Fu Hualing, 'Punishing for Profit: Profitability and Rehabilitation in a laojiao Institution', pp 213–30 in *Engaging the Law in China: State, Society, and Possibilities for Justice*, eds. Neil Diamant, Stanley B. Lubman, and Kevin J.O'Brien. Stanford: Stanford Univ. Press, 2005; James D. Seymour and Richard Anderson, *New Ghosts Old Ghosts: Prisons and Labor Reform Camps in China*. Armonk: M.E. Sharpe, 1998.

[4] *Zhongguo da baike quanshu*, 'Shehui xue' (Beijing: Zhongguo da baike quanshu chubanshe, 1001), p. 422.

Importantly, however, while much of the scholarly (and journalistic) literature portrays Chinese intellectuals as the primary targets of Maoist thought reform, intellectuals never comprised the largest portion of reeducation-centre internees, and large-scale internments of scholars, doctors, researchers, writers, and artists did not occur until 1957, in the aftermath of the Anti-Rightist campaign.[5]

Reeducation was not solely a domestic project; it had an important international dimension, beyond the incarceration of foreigners living within China's borders. In the early 1950s, education aimed at non-Chinese people consisted primarily of multi-language publications circulated abroad, as the Chinese Communists rarely had the opportunity to do more intensive thought reform outside of China. One exception, however, was the effort to reeducate captured soldiers during the Korean War, and thus in late 1951, the Chinese established roughly 15 reeducation camps in the north of Korea, just near the Yalu River. Per an agreement with their allies, the Chinese mainly interned United Nations troops; South Korean soldiers were generally held in separate camps run by North Koreans.

What happened in the Chinese-run camps became front-page news around the world. Some of the most shocking of those news reports came in 1953 and 1954, as POWs made radio broadcasts decrying US involvement in the Korean War, a small but resolute number of those POWs eventually refused repatriation to their home countries and accepted the official invitation to immigrate to China, where they received a heroes' welcome.[6] The 'brainwashing' story was also big news in China. Tales about soldiers undergoing mental transformations began circulating before the Chinese officially took over POW work in Korea. As early as July 1950, the Party's official newspaper, *The People's Daily*, ran a story in which a 'special correspondent on the ground in Korea' gave a first-hand account of what he had seen in the POW camps:

> The American Army's 24th division, 52nd battle regiment's Captain Nugent was originally the manager of a small theatre in the US. In this small group of captives, he was the highest ranking official and the most highly educated. Before he had become conscious (after which he said, "We in this group of captives believe that all foreigners should immediately leave Korea and let Korean people solve their own domestic problems"), he "only knew that Korean apples were delicious". Now he was the first to sign the joint declaration resisting American military interference in Korea.[7]

Similar stories filled newspapers, magazines, radio broadcasts, and films throughout the early 1950s.

Strikingly, given the importance of the topic at the time, there has been almost no discussion in the historiography of the PRC, either in Chinese or English, about

[5] Smith, *Thought Reform and China's Dangerous Classes*, p. 2.

[6] *They Chose China*, dir. Wang Shuibo, 2006; Susan L Carruthers, *Cold War Captives*; Adam Zweiback, 'The 21 "Turncoat GIs": Non-repatriations and the political culture of the Korean War', *The Historian* 60:2 (Winter 1998).

[7] *The People's Daily* 20 July 1950

Chinese brainwashing in Korea. There is plenty of excellent English-language scholarship on the brainwashing of POWs, but that work is primarily authored by historians of the US and the UK and scholars of American cultural studies.[8] Scholars with expertise in Chinese history, language, and culture, on the other hand, have not written widely on brainwashing. In Chinese language, there are a number of popular treatments, journalistic accounts, and memoirs by former POW-camp workers – but little scholarship. Historians of China write about Chinese POWs in UN camps but not about UN POWs in Chinese camps.[9]

The lack of scholarship written by China experts can be explained, in part, by the way the field has dismissed the term 'brainwashing' itself. Robert Lifton noted early on that the Chinese Communists did not seem to use the term to describe their own work.[10] When the CCP talked about what Americans were calling 'brainwashing', they generally called it 'thought reform' (*sixiang gaizao*), or one of that phrase's many variants. In fact, Edward Hunter claimed that he was the one who coined the English-language term 'brainwashing', and the Oxford English Dictionary still credits him with the invention of the word. Hunter asserted that 'brainwashing' was his direct translation of the Chinese word, '*xinao*' (in which the first character, *xi*, can mean 'wash' and the second, *nao*, can mean 'brain'), but he also claimed that *xinao* was only used in spoken Chinese and that he was 'the first person to write it down in any language'.[11] China scholars have long doubted Hunter's claim. Until very recently, received wisdom was that the Chinese word *xinao*, which is now used in Chinese, first entered the language as a translation of the English word 'brainwashing'. That English word, most scholars claim, was fabricated by Edward Hunter. The Chinese Communists themselves make a similar claim. It turns out, however, that Hunter was probably telling the truth.

As Ryan Mitchell has shown, the metaphor of 'brainwashing', and sometimes the Chinese word '*xinao*', featured regularly in the writings and speeches of reform-minded Chinese from at least the turn of the twentieth century. Yan Fu, for example, in his translation of Herbert Spencer's *The Study of Sociology*, inserted his own line insisting that 'those who engage in the study of sociology wash our brains (*xinao*) and purify our hearts'.[12] The CCP continued to use very similar metaphors of washing and purifying during their thought reform campaigns. In 1952, for example, *The People's Daily* published a letter ostensibly written by a reeducated

[8] See footnote 3.

[9] In English, see David Cheng Chang, *The Hijacked War: The Story of Chinese POWs in the Korean War*. Stanford: Stanford University Press, 2020. Another important work on the topic of the US treatment of POWs (not written by a China specialist) is Monica Kim, 'Empire's Babel: US Military Interrogation Rooms of the Korean War', *History of the Present*, Vol. 1: no. 3 (2013), 1–28.

[10] Robert J. Lifton, 'Peking's 'Thought Reform', p. 135.

[11] United States Congress Senate Committee on the Judiciary, *The Effect of Red China Communes on the United States: Hearing before the Subcommittee to Investigate the Administration of the Internal Security Act and other Internal Security Laws of the Committee on the Judiciary, United States Senate Eighty-sixth Congress First session*. Washington, DC: US Government Printing Office, 1959.

[12] Ryan Mitchell, 'China and the Political Myth of "Brainwashing,"' *Made in China Journal*, 8 Oct 2019.

sex worker who had become a factory worker. She said, 'like dumping foul water out of a bottle and refilling it with fresh water, thought reform leaves the mind clean and clear'.[13] Later, during the 1960s 'Four Cleanups', a movement aimed at purifying the thoughts of Communist Party cadres (rank-and-file members and agents), a popular slogan urged cadres to 'go upstairs, wash your hands, and bathe'. In the same decade, during the Socialist Education Movement, cadres accused of corruption were told to 'take a warm bath', which meant conscientiously admitting to their mistakes and making financial restitution. Another example comes from 1971, when a group of American scholars visited China and met with Zhou Enlai. Speaking in Chinese, Zhou used the word *xinao* (brainwash) to reference the 1950s accusations about POWs and others, but then he said, 'I'd like to wash my own brain because I still have old ways of thinking in my mind'.[14] Finally, in the late 1980s, the author of a famous Chinese novel on thought reform wrote in her introduction that 'brainwashing' was a Western word, but she named her novel *Xizao*, or *Bathing*, gesturing to the CCP's use of the metaphor in their campaigns.[15]

The Chinese democracy advocate, Hu Ping, has pointed to similarities between the term *xinao* (brainwash) and *xizao*, which is generally translated as bathing or showering and which can be found throughout the CCP's documents as a synonym for thought reform. The fact that *xizao* (bathe) and *xinao* (brainwash) are rhyming words with a similar grammatical structure further suggests a Chinese origin, as such word play is common in China's political vernaculars. *Xinao* (brainwashing) also alludes to another Chinese word with a long cultural history – *xixin*, or 'heart-washing', appears in common Chinese idioms, such as *xixin gemian*, which is akin to, 'make oneself new' or 'turn over a new leaf'. The word 'heartwashing' comes from ancient Daoist texts, where it refers to purifying oneself by washing away the pollutions and (false) attachments of the heart.

If Edward Hunter had excellent Chinese-language skills, he certainly could have invented this wonderfully evocative word, with its close historical and etymological connections to ideas of reform and regeneration, but there is no reason to think Hunter did so. As Hu Ping writes, 'One can infer that *xinao* or 'brainwashing' originated [in China]' and that 'while the term did not get written down in formal documents, it nevertheless was a popular oral usage'.[16] The word was clearly consistent with the officially sanctioned meanings, language, and metaphors associated

[13] *The People's Daily* 19 March 1952.

[14] 'Zhou Enlai tong Meiguo qingnian xuezhe de tanhua', from Aidejia Sinuo [Edgar Snow], *Zhou Enlai jinian ji* (Xianggang tiandi tushu youxian gongsi, 1977), pp. 310–45. For an English-language transcript from this meeting, see *Bulletin of Concerned Asian Scholars*, Vol. 3, 3/4 (Summer/Autumn 1971).

[15] Yang Jiang, *Xizao* (Beijing: Renmin wenxue chubanshe, 1988). In the English translation, the title is rendered as *Baptism*, which elicited some controversy but is interesting in the context of this discussion. Translation citation: Yang Jiang, *Baptism*, trans., Judith M. Amroy and Yaohua Shi. Hong Kong: Hong Kong University Press, 2007.

[16] Hu Ping, *The Thought Remoulding Campaign of the Chinese Communist Party-state*. Amsterdam: University of Amsterdam Press, 2012, p. 13.

with thought reform. It is actually surprising that it was not incorporated into the Communist Party *tifa*, or the formal, official rules about which words can and must be used for certain subjects. As Hu Ping points out, however, there is probably an obvious reason this did not happen: once Edward Hunter picked up the word and used it as the centrepiece of his anti-Communist agitation tour, the CCP opted to abandon it. Once Mao himself began claiming that the word was a Western smear tactic, this became the party line.[17]

This party line was well articulated in late twentieth-century Chinese dictionaries where *xinao* (brainwashing) is defined as 'a word originally used by antagonistic elements to slander our country's early 1950s campaign of thought reform (*sixiang gaizao*). Later it entered into common usage'.[18] That definition is likely inaccurate, but it does correspond to and highlight a key divide we see in the scholarship: thought reform is a Chinese concept, and a Chinese subject of study, whereas brainwashing is a non-Chinese concept, and a foreign, and especially US/UK, subject of study.

Thus, historians who work on China have written a lot about thought reform in China, but those studies do not offer much analysis of the holistic project, of the fact that at the same time the new state was incarcerating millions of people in domestic reeducation centres, that state was also becoming infamous around the world for its supposedly sinister mind-control technologies. That accusation of sinister mind-control technologies is taken up primarily by scholars of the US and the UK. While that literature is excellent, most of it doesn't use Chinese sources, and the very few studies that list Chinese texts in their bibliographies do not engage with scholarship on Chinese thought reform or even Chinese communism most broadly. Robert Lifton was the last person to genuinely and fully contextualise brainwashing in the context of Chinese thought reform and the history of the PRC. Yet, from the disputed etymology of the word 'brainwashing' alone, we can see how entangled these histories are. The very words that the CCP used to talk about their domestic thought reform effort were directly shaped by the international dimensions of that project and the way their Cold War enemies responded to it.

Thought work as global praxis

One of the most striking things about the records from the CCP's thought reform project during the early 1950s is how similar the various reports were, whether they were produced in domestic or international reeducation centres. For example, when officials from the Beijing Bureau of Civil Affairs met with reeducators in training in 1949, they warned their subordinates to remember that sex workers, vagrants, petty

[17] Hu, pp. 13–14.

[18] See, for example, *Xiandai hanyu xin ci yu cidian*. Beijing: Zhongguo qingnian chubanshe, 1994, p. 975.

thieves, and other reeducatees were not criminals but 'unemployed members of the labouring masses who have been driven by oppression to take up improper occupations [...] They are members of the Chinese people who have suffered'. Officials went on to instruct reeducators to treat such oppressed people leniently, if firmly. They were not to beat, threaten, or intimidate their internees but were to rely instead on persuasion and education.[19] The records from Beijing reeducation centres suggest that reeducators did not always follow these instructions perfectly, but non-criminal internees did indeed receive more lenient treatment than those prisoners accused as enemies of the people.[20]

A similar situation was to be found in Korea, where reeducators also emphasised the way their targets had been 'oppressed' and advocated more 'lenient' treatment (at least for rank-and-file soldiers) as a result. While treatment was certainly not 'lenient' by many standards, a number of POWs did later claim that the Chinese did not treat them violently.[21] Thought reformers regularly claimed they advocated the 'Lenient Policy' because most of the UN POWs had not freely chosen to fight. Jiang Kai, who worked with POWs in Korea, illustrated this point with the specific example of a POW surnamed Jones (*Qiongsi*): 'He himself was not willing to fight in the war. He was forced to join the Army in order to support his family'.[22] Another reeducator, who was sent to Pyoktong POW camp in 1950, later wrote, 'During our interrogations we listened to the autobiographical narratives of the prisoners to determine their pre-war jobs, lifestyle, family situation, and we got a look at the vast gulf between the upper strata and the ordinary people in the US'. With the 'ordinary people', she argued, they used 'persuasion and education' to bring soldiers to oppose the war.[23] One POW remembered a Chinese reeducator explaining things in similar terms: 'We have the right to liquidate you,' his captor told him, 'but we know that you are just little pawns in the great capitalist game. You are tools in the hands of your capitalist masters, and therefore, we are going to give you the chance of being reeducated'.[24]

This rhetoric was common and memorable, and it went beyond the POW camps. In 2014, a veteran Volunteer Army soldier (who fought in Korea but did not regularly work with POWs) remembered the time he captured an Australian soldier in 1951. He recalled that as he prepared to escort the captive to the POW transport station, his superior said to him, 'Treat the prisoner according to the Lenient Policy. Most of these foreign soldiers are not our enemies; they are ordinary working people who enlisted in the army because they had no other way to make a living. It

[19] Beijing Bureau of Civil Affairs, *Beijing shi Minzhengju guanyu chuli qigai wenti yu jinü wenti de gongzuo buzhi*, May 18, 1949, Beijing Municipal Archives, 196-2-191.

[20] Smith, *Thought Reform and China's Dangerous Classes*, p. 74.

[21] See interviews in Young and in *Korea: The Unknown War*.

[22] Radio interview with former POW reeducator, Jiang Kai, transcript available here: www.am828.com.cn/zxz8d/201011/t20101118_629304.htm.

[23] Zhao Da, 'Zai Bitong he 'Lianheguo jun zhanfu' da jiao dao', *Renmin zhengxie bao* (December, 2011).

[24] *Korea: the Unknown War* (miniseries), dir. Mike Dorner, Max Whitby, and Phillip Whitehead, 1988.

is not their war'.[25] Whether reeducators were talking about their work at home or abroad, if they were reforming individuals they classed as 'ordinary people', they described the nature of their work in the same way, as illustrated in the following two quotations:

> From a report on the reeducation of prostitutes in Beijing:
>
> 'We used persuasion and education to reveal the truth to them. Eventually, the vast majority of them came to understand and agree'.[26]
>
> From a report on the reeducation of POWs in Korea:
>
> 'Our tactic was to tell them the truth. And after repeated persuasion and education, the vast majority of them came to understand that they had been duped'.[27]

It might appear that the similarities in these early thought reform narratives are easy to explain – whether reeducators were speaking of Catholic nuns in Tianjin or US soldiers in Korea, French professors in Shanghai or Chinese beggars in Beijing. Observers and analysts have long argued that this sort of rhetoric should be taken with much more than a grain of salt. As Seymour and Anderson summarise: 'Government reports [about prisons and thought reform institutions], even when of an 'internal' nature, can be deeply flawed. Prison personnel produce documents more to please superiors than to reflect reality… Everyone – prisoners, guards, and administrators – participates in a conspiracy to 'prettify' the picture, for all stand to benefit if a prison is declared an exemplary unit'.[28] If reeducators, then, were not describing their genuine thoughts and experiences and were instead expected to repeat banal platitudes about oppression and liberation, this would certainly explain why reports from thought reform institutions read similarly. Some scholars largely dismiss the official record on these grounds.

There is indeed an element of 'whitewashing' in these sources, and report writers often tried to construct an ideal vision of their practice. Yet I caution against dismissing altogether the ways that communist thought reformers described their own practices.[29] I have worked with these thought reform documents for many years, including handwritten drafts and notes; classified sources not kept in state archives but circulating in quasi-legal document markets; descriptions written by low-ranking cadres who had not mastered the official language and who thus violated its rules – while lower level agents may have tried to 'please their superiors', responses from supervisors, and repeated rectification campaigns, suggest that they often failed to do so. These documents are idealised, but they are not *merely* idealised.

[25] Interview with Mr Wang (surname changed), Volunteer Army soldier, June 2014, Shanghai.

[26] Beijing Public Security Bureau, 'Shourong jinu gongzuo baogao', (December 1950), Beijing Muncipal Archives, file 14-1-51.

[27] Zhao Da, 'Zai Bitong he "Lianheguo jun zhanfu" da jiao dao'.

[28] Seymour and Anderson, p. xiv.

[29] Aminda Smith, 'Long Live the Mass Line: Errant Cadres and Post-Disillusionment PRC History', *positions: asia critique* 29:4 (November 2021, 786–9.

Furthermore, even if we restrict our analysis to the most idealised form of this rhetoric, a question remains: why did a Communist Party that took itself very seriously construct a formal narrative that seems to be a caricature of itself? For all of these reasons, I argue that these documents deserve to be taken seriously, if critically.

A critical reading might first ask whether the similarities in the narratives about thought reform practice reflect a tension in the way the Chinese Communists understood the mind. According to the theorists who shaped Chinese communism in those early years (notably Mao Zedong, Liu Shaoqi, Peng Zhen, and Ai Siqi, among others), the human mind is almost entirely a post-birth construction, shaped by a person's material and social realities. The way any given individual perceives the world is, thus, dependent on where they stand relative to the mode and the means of production and the various social and political hierarchies associated with those things.[30] A 'one-size-fits-all' reeducation method does not appear to be at all consistent with that theory. If one's worldview (or 'perspective', in Maoist terms), is determined by one's Marxian class position, as well as one's position in a colonial or post-colonial system (following Lenin), how could the same techniques that turned Chinese sex workers into factory workers also turn US soldiers into Communist defectors?

In trying to answer that question, I have spent a great deal of time looking for ruptures in these narratives, analysing the differences that race, gender, sexuality, nation, and class made to the way thought reform was carried out and experienced in different contexts. In the end, however, I decided that it was my understanding of thought reform, and not thought reformers' understanding of the mind, that needed to be clarified. While I had tended to see brainwashing as a process of indoctrination with new ideas, that was not the way reeducators understood their work, at least not in its earliest stages. In the early 1950s, thought reformers saw themselves as de-programmers. The first, and by far the most important step in the process, was stripping away the indoctrination the Chinese Communists believed people had already received.

Like many of their contemporaries, the Chinese Communists understood capitalists and imperialists to be reliant on forms of mass mind control to keep workers working, consumers consuming, and cannon fodder available for foddering. Taylorism, for example, sought to make capitalist workforces more efficient by instilling new ways of thinking that naturalised a certain mode of production. Thinkers like the architect and urban planner, Le Corbusier, asserted that Taylorist 'states of mind' were the only thing that could stave off otherwise certain revolution. In that context, it would make sense if would-be revolutionaries thought they could launch revolution simply by un-creating that state of mind.[31] Similarly, Mao Zedong argued that the US and other 'imperialist' militaries instilled fear of

[30] Aminda Smith, *Thought Reform and China's Dangerous Classes*, pp. 25–6; Donald Munro, *The Concept of Man in Contemporary China* (Ann Arbor: University of Michigan Press, 1977), pp. 13–20.

[31] Tessa Morrison, *Unbuilt Utopian Cities 1460–1900: Reconstructing Their Architecture and Political Philosophy*. London: Routledge, 2015, p. 170.

communism to motivate individuals to fight off imminent revolution. If that were true, then would-be revolutionaries might foment revolution merely by ridding people of their anti-communist fears. And indeed, early Maoist thought reformers described their work in precisely those terms, as did a number of their targets. Clarence Adams, one of the POWs from the US who chose to move to China, aptly summarised this vision of practice: 'The Chinese didn't brainwash me. They unbrainwashed me'.[32]

Of course, Adams could have argued that the Chinese did in fact brainwash him, were it not for the entangled histories of the terms '*xinao*' and 'brainwashing', because 'washing the brain' was indeed a good metaphor for what Chinese reeducators claimed to be doing: they imagined themselves scrubbing and rinsing away the manufactured mindsets that stood between capitalism/imperialism and world revolution. Thus, when thought reformers narrated the reeducation process, in China or Korea, they almost invariably began by demonstrating that their targets had indeed been indoctrinated against revolution. Interestingly, they often did this by describing how fiercely detainees fought the Communists when they were captured: POWs, their jailers said, battered the soldiers who attempted to imprison them. Interned pickpockets reportedly rioted and beat up cadres, and groups of incarcerated sex workers apparently pummelled the Chinese soldiers who guarded them. Thought reformers used this resistance to prove their captives were very much in need of brainwashing. These individuals resisted the Communists, according to the Communists, not because they were true supporters of capitalist or imperialist regimes, but because they had been swindled into acting against their own interests.

POW reeducator, Zhao Da, claimed that when she interrogated the soldiers, she discovered 'they had been duped' in that 'They all said they'd been told that after they finished fighting, they'd be able to get rich. They also said that senior officials told them this war wouldn't last long and that by the end of 1950, they'd be back with their families to celebrate Christmas'.[33] Similarly reeducators claimed that sex workers, beggars, and pickpockets were led to believe that they were fated to such livelihoods.[34]

Added to that, according to reeducators, exploiters such as high-level military officials and crime bosses had further hurt ordinary people by spreading vicious rumours about how the Communists were the ones who mistreated people. Thought reformer Huang Jiyang, reported that, before the POWs were captured, 'The U.S. military constantly disseminated deceitful propaganda to them, saying things such as 'If you're captured by the Volunteer Army, you'll be tortured, you'll be beheaded. And so when they were first captured, they were all terrified'.[35] Thus, according to thought reformers, POWs fought savagely against the Communists,

[32] Zweiback, p. 43.
[33] Zhao Da, 'Zai Bitong he 'Lianheguo jun zhanfu' da jiao dao'.
[34] Smith, *Thought Reform and China's Dangerous Classes*, pp. 80–1.
[35] Huang Jiyang, 'Zhiyuan jun fu guan chu chengli zhichu', *Renmin zhengxie bao 26 March 2014*, www.rmzxb.com/sjzz/xsg/2014/03/26/309897.shtml.

not because of any antipathy toward China or communism, but because they had been tricked with false information.

Reeducators claimed that these mistaken ideas ran so deep that even once they subdued initial resistance, most internees were still unwilling to fight against their 'real' enemies. In 1949, for example, a cadre in Beijing explained that when reeducators asked newly incarcerated sex workers if they'd like the opportunity to fight back and punish the brothel keepers who had abused them, internees mostly declined. When reeducators continued to push the issue, one internee reportedly sighed and said, 'Oh, don't you get it? That's just the way things are. How else could it possibly be? Those of us in this line of work – we can't blame other people for what's happened to us; it's our fate, that's all there is to it'.[36] Likewise, a reeducator in Korea reported that he asked one soldier, 'Didn't you think it was wrong that the senior officials started this war, but they sent you to fight it?' The POW replied: 'That's just the way things are'.[37]

The task then, as reeducators imagined it, was to show that POW that things did not need to be as they were, or to show that sex worker that she could, indeed, blame other people for what had happened to her. Yet thought reformers did not propose to accomplish this task through indoctrination; they did not imagine themselves to be implanting manufactured mindsets as they accused capitalist and imperialist powers of having done. Instead, they invoked Mao Zedong's mass line, which stipulated that all truth came 'from the masses' themselves and that the Party's only role was as that of 'a processing plant' – it gathered the masses' truths, interpreted them within a Marxian framework, and helped the masses to see how their own experiences actually proved the truth of Marxian claims.[38] Thought reformers argued that oppressed and exploited individuals already knew the truth of their situation, but their prior capitalist and imperialist indoctrination had hidden that truth from them. All reeducators claimed they did was teach reeducatees to reinterpret the facts they already knew in a different, and more convincing way.

As the first step in this programme, thought reformers encouraged internees simply to tell their stories, to recount the life events that had shaped their paths and led them to the situation in which they currently found themselves. This was not a technique reserved for incarcerated reeducatees or for the thought reform process alone; it was a revolutionary mobilisation tactic the CCP had used since its founding in the 1920s. William Hinton's Maoist classic, *Fanshen* (a locus classicus for many non-Chinese leftist intellectuals and activists in the 1960s), described a practice, which the Chinese called 'speaking bitterness': One day, in a village called Long Bow, two CCP cadres called together 30 of the poorest locals, most of whom 'owned nothing but the clothes on their backs' and encouraged each of them 'to tell his or her life story and figure out for himself the root of the problem'. The

[36] Beijing Public Security Bureau, *Beijing fengbi jiyuan*, pp. 72–3.
[37] Interview with Mr Liu (surname changed). Shanghai 2014.
[38] Smith, *Thought Reform and China's Dangerous Classes*, pp. 97–101.

cadres launched the meeting with a question: 'Who lives off whom?' One cadre, formerly a hired labourer, told his own story first. He had been living with his uncle, who had borrowed 20 silver dollars 'in order to get married'. Before the year was up, the man recalled, 'the interest plus the principle amounted to more than 300 dollars. We could not possibly repay this. The landlord seized all our lands and our house and I became a migrant wandering through the province looking for work'.[39] Hearing this example, the assembled villagers began to tell their own stories.

> Many wept … as they remembered the sale of children, the death of family members, the loss of property. All the while the village cadres kept asking 'What is the reason for this? Why did we all suffer so? Was it the "eight ideographs" that determined our fate or was it the land system and the rents we had to pay? Why shouldn't we now take on the landlords and right the wrongs of the past?'[40]

In the same way, Communist thought reformers attempted to 'raise consciousness' by eliciting the life stories of reeducatees. Those reformers insisted that the truth about the world became obvious once reeducatees simply looked at the facts of their own lives. The tales people told did seem to confirm those claims. One POW, for example, reportedly told the following story:

> In the US, it was very hard to find an occupation, so if I had the skills to do something, I did it. This year, I'm 23. I've been a truck driver, a boxing coach, a wrestler, a circus motorcycle performer. I joined the army so I could save a little money so that in the future I could go to school and raise a family. Before we set out, our superiors told us that we weren't going to fight but that we were going to do something more like policing. But we did end up in battle. […] That's how I was captured'.[41]

In a reeducation centre in Beijing, an interned beggar and pickpocket told his story: 'When I was young, my family was very poor. My father and I cultivated land for other people to make our living. Every year, the grain we got wasn't enough to pay the rent. When the landlord asked for the rent, my father didn't have it. The landlord killed my father, leaving me all alone, roaming the streets, begging in order to survive'.[42]

As they recorded these quotations, thought reformers editorialised, insisting that when reeducatees narrated the specific details of their own lives within the analytical framework the CCP provided, the truth of oppression became obvious. After discussions of bitterness-speaking sessions, report writers invariably added some version of the following statement, this one written by a cadre working with sex workers: 'As internees told their stories,' they 'recognized who *our* true Enemies were, and their minds gradually became clear'.[43] In addition to using the words

[39] William Hinton, *Fanshen: A Documentary of Revolution in a Chinese Village*. New York: Vintage Books, 1966, p. 132.

[40] Hinton, pp. 132–3.

[41] *The People's Daily*, 31 July 1950.

[42] Beijing Municipal Bureau of Civil Affairs, *Shourong qigai gongzuo zongjie*.

[43] Beijing Bureau of Civil Affairs, *Beijing shi chuli jinü gongzuo zongjie*.

'our' and 'we', reeducators used other methods to link themselves to their reeducatees, to suggest they all belonged to a single community – the pre-indoctrinated members of the global masses, in need of consciousness-raising and brainwashing.

Reeducators certainly argued that soldiers from an enemy army and people who broke the law within China were suffering from more severe ideological damage than many other members of the oppressed masses and thus needed intensive, incarceration-based reform. But this constituted a difference of degree, according to thought reformers, not type. In type, they were all ordinary people for whom interconnected capitalist-imperialist systems of domination had created a shared history of suffering. If POWs or petty thieves failed to see themselves as part of a community that included their captors, thought reformers were not surprised, as one of the key things about the global exploited majority, in the CCP formulation, was that very few of its members recognised themselves as oppressed, and almost none recognised themselves as part of an international brother- and sisterhood that included everyone from Chinese sex workers to UN soldiers. In the reeducators' Marxian understanding, the primary reason that people could not see the 'truths' of their own oppression and the way that oppression linked them to others, was because they had been duped with religious, philosophical, and sociological concepts that hid the mechanisms of domination and sought to divide the oppressed and turn them against each other. Many CCP consciousness-raisers argued that people would turn against the 'American imperialist government' and its 'Wall Street backers' as soon as they understood the way they had been used and betrayed by those institutions.

Conclusion

Because thought reform was intended to teach individuals to critically reinterpret their own lives, thought reformers saw their methods as translatable, as modifiable to almost any context, precisely because theirs was a praxis rooted in the investigation of specific, situated events and understandings. In reeducation, individuals focused on their own most personal histories, endlessly narrated and analysed the events of their own lives, and (ideally) critically reflected on the relationship between their situation and global political/economic systems.

Reeducators were as aware as we are that people's experiences are shaped by categories such as gender, race, and class, and they highlighted those differences during their bitterness-speaking sessions. The early years of the 1950s, however, were a hopeful period in the history of the global left, when people believed that the world could be other than it was and that oppressed peoples could unite to successfully fight for equality and justice. It was in that spirit that the discourses and the practices of reeducation focused far more on the similarities between individuals than on the differences that divided them. For example, reeducators often focused on sexualised personal stories from female sex workers, arguing that critically examining these accounts was crucial to successful reeducation for women who had suffered under a manifestly patriarchal system. Thought reformers nevertheless did not portray these feminised experiences as something that divided women from men or female sex workers from male soldiers. In these discussions, even rape was often rendered as one of the many miseries that united oppressed people through that shared history of victimisation, which feminised everyone.[44]

Their understanding of thought reform as a practice that revealed truths about the nature of oppression allowed reeducators to detail the particularity (gendered and otherwise) of individual stories, while simultaneously synthesising the meanings of those experiences. Thus, the narrative form of the bitterness accounts they elicited was remarkably similar, even as the details diverged, whether the alleged speakers were beggars, farmers, soldiers, or factory workers. That was precisely the effect consciousness-raisers were trying to achieve as they attempted to bring all of the masses to a powerful realisation, namely: there were countless variations on key political lessons, the most important of which was that capitalism and imperialism had created two opposing groups: the oppressed masses and the enemies who perpetrated and profited from that oppression.

Of course, in practice there were a number of intermediate categories, but the binary of 'the oppressed people' versus 'their oppressors' was politically and rhetorically powerful. Within that binary, there could still be a great deal of diversity on either side, but ultimately thought reformers were trying to forge broad solidarities by crossing and overcoming divisions such as sex, race, and even class. Reeducators argued that in the specificity of individual experiences, people could see, not only the way they were oppressed but also the way their oppression was connected to the oppression of others. Chinese Communist thought reformers believed that this was a powerful political lesson that could galvanise ordinary people everywhere to forge broad alliances. During the middle of the twentieth century, people around the world shared that vision, and it led them to believe in the possibility of world revolution. One might read the historiography of the late twentieth and early twenty-first centuries as evidence of the decline of such views. In writing about reeducation within China as entirely separate from international projects such as Korean War 'brainwashing', we erase the global scope of the communist project.

[44] Smith, *Thought Reform and China's Dangerous Classes*, pp. 55–6. For more on the discourse of feminization in the Chinese revolution, see Gail Hershatter, *Dangerous Pleasures: Prostitution and Modernity in Twentieth-Century Shanghai*. Berkeley: University of California Press, 1997, pp. 239–45.

7

Constructing the Maoist Sexual Subject: 1950s Hygiene Guides and the Production of Sexual Knowledge

SARAH MELLORS RODRIGUEZ

ACCORDING TO THE official rhetoric of the Chinese Communist Party (CCP), the period before the 1949 revolution – the Republican period (1911–49) – was not only ideologically backward but also rife with 'feudal' sexual practices, such as prostitution, concubinage, and child marriage. When the Communists founded the People's Republic of China (PRC), they boldly claimed that a new socialist society would be born out of the dregs of the previous regime. Not only would China undergo radical economic and political reforms, including land redistribution and mass political mobilisation, but social life would also change substantially in the 'new society'. Such changes were not simply limited to bans on the formerly widespread practices of polygamy and arranged marriages, nor did they stop at establishing women's rights to freely marry and divorce.[1] At the most personal level, creating a new society involved the establishment of a new, socialist sexual morality.[2]

Scholars have analysed many aspects of the 1949 power transition, revealing important political, economic, and social ruptures, as well as continuities.[3] In terms of family organisation and gender relations, Neil Diamant has shown how the implementation of the historic 1950 Marriage Law, which promoted free-choice, monogamous marriages and eased access to divorce, empowered rural women

[1] Women already had the right to divorce in the Republican period, but the 1950 Marriage Law further eased restrictions on divorce.

[2] Wang Wenbin, Zhao Zhiyi, and Tan Mingxun, *Xing de zhishi*. Beijing: Renmin weisheng chubanshe, 1956, p. 2.

[3] Jeremy Brown and Paul Pickowicz (eds.), *Dilemmas of Victory: The Early Years of the People's Republic of China*. Cambridge, MA: Harvard University Press, 2010; Frank Dikötter, *The Tragedy of Liberation: A History of the Chinese Revolution, 1949–1957*. New York: Bloomsbury Press, 2013.

and helped reshape the family structure.[4] Despite contemporaneous campaigns to improve the status of women and expand their workforce participation, Harriet Evans also reveals that official discourses in the 1950s used the rhetoric of science to frame gender difference as biologically determined, which in turn legitimised the male-dominated social order.[5] A related topic that has been given less scholarly consideration is the role that narratives of normative sexuality played in socialist state-building and the relationship between Mao-era sexual norms and those from the Republican period. Indeed, in the early years of the PRC, producing a new sexual morality and normative sexual subjectivity were seen as critical for throwing off the yolk of the 'Old Society'.

Drawing on a collection of sexual hygiene guides published by state and commercial presses in the 1950s, this chapter probes the construction of the ideal socialist sexual subject: what social and political currents shaped the production of Maoist sexual knowledge? How did these texts distinguish between normative and transgressive sexual behaviours?[6] These guides, which employed a combination of biomedical language, moralising narratives, and ideological rhetoric, sought to position so-called 'socialist sexual norms' and the 'new society' as antidotes for the ills of the Republican period. Yet, much of the information the texts provided merely repackaged Republican discourses under the guise of scientific authority and Communist ideology. In fact, these sexual hygiene guides drew heavily on globally circulating discourses about modernity and selectively appropriated elements from eugenics, sexology, and Traditional Chinese Medicine (TCM).

This chapter highlights the processes through which a Maoist sexual morality was theorised and concretised. The first section introduces the historical background in which early Mao-era sex guides were produced, situating the production of sexual knowledge within broader social and political trends. The second and third sections of this chapter examine the ideas presented in these texts in greater detail, revealing how official discourses fused global and indigenous ideas about sexual desire, perversion, and disease to fashion a model of sexual normativity. On the one hand, these guides sought to reconcile individual sexual agency with the

[4] Neil Diamant, *Revolutionizing the Family: Politics, Love, and Divorce in Urban and Rural China, 1949–1968*. Berkeley: University of California, Press, 2000, p. 2.

[5] Harriet Evans, *Women and Sexuality in China: Dominant Discourses of Female Sexuality and Gender Since 1949*. Cambridge: Polity Press, 1997, p. 34; Jiping Zuo, 'Women's Liberation and Gender Obligation Equality in Urban China: Work/Family Experiences of Married Individuals in the 1950s', *Science & Society* 77, no. 1 (2013): 109–10.

[6] At this time, there was no clear distinction between 'sexual hygiene' (*xing weisheng*) and 'sexual health' (*xing jiankang*). However, the dominant expression found in these guides was the more clinical and prescriptive term, *weisheng*. Ruth Rogaski translates *weisheng* as 'hygienic modernity' because of the perception in the early twentieth century that China was inherently deficient and could only achieve modernity through scientific rationality and improved public health. Thus, 'hygiene' (*weisheng*) is imbued with a much broader range of meanings than 'health' (*jiankang*). Ruth Rogaski, *Hygienic Modernity: Meanings of Health and Disease in Treaty-Port China*. Berkeley: University of California Press, 2004, pp. 2–3.

collective goals of labour productivity and social stability, and at times to foster a more progressive sexual culture. On the other hand, they also drew a sharp demarcation between 'normal' and 'abnormal' sexual behaviour and ultimately legitimised both the Maoist social order and expert intervention into private life.[7]

The historical context

When the Communists came to power, they inherited a vibrant commercial publishing industry and the remains of what had once been a robust public sphere. During much of the Republican period, numerous private publishing houses, aided by advances in mass production, competed to supply reading material for general audiences as well as niche markets.[8] One flourishing area of popular consumption was reading material – books, news articles, magazines, and tabloids – about love and courtship, sex, and eugenics.

More than simply a form of cheap entertainment, during the May Fourth/ New Culture Movement (1910s–20s), Chinese reformers seeking to modernise and 'save' the nation also turned to publishing articles on these topics as a way to express their political concerns.[9] As part of their critique of Confucianism, iconoclastic intellectuals promoted the Western ideal of the nuclear family in place of the traditional extended family.[10] In conversation with other reformers around the world who endorsed the notions of individual choice and sexual liberation during the interwar period, some Chinese reformers also espoused marriage based on 'free love' in lieu of arranged marriages.[11] Although, like other regimes in the post-World War II (WWII) era, the CCP would later abandon some of its more radical family and gender reforms, in the 1950s debates about companionate marriage and intimacy remained at the forefront of discussions about sexual normativity.[12]

[7] Theorist Michel Foucault coined the term 'biopower' to explain how state disciplinary power is used to control and surveil certain facets of life, including birth, sexuality, illness, and health. Hygiene guides in the early PRC reflect this modernist impulse to regulate and intervene in sexual life; Michel Foucault, *The History of Sexuality, Vol. 1: An Introduction*. New York: Pantheon Books, 1978, p. 140.

[8] Frank Dikötter, *Sex, Culture and Modernity Medical Science and the Construction of Sexual Identities in the Early Republican Period*. Honolulu: University of Hawaii Press, 1995, p. 4.

[9] Wenqing Kang, *Obsession: Male Same-Sex Relations in China, 1900–1950*. Hong Kong: University of Hong Kong Press, 2009, p. 7.

[10] Susan Glosser, *Chinese Visions of Family and State, 1915–1953*. Berkeley: University of California Press, 2003, p. 3.

[11] Mirela David, 'Bertrand Russell and Ellen Key in China: Individualism, Free Love, and Eugenics in the May Fourth Era', in *Sexuality in China: Histories of Power and Pleasure*, ed. Howard Chiang. Seattle: University of Washington Press, 2018, p. 78.

[12] The US, the Weimar Republic, and the early Soviet Union similarly experienced an era of social and intellectual openness in the 1920s and early 1930s in which gender roles became less rigid. Subsequent regimes, however, would renege on earlier promises to advocate gender equality. Barbara Einhorn, 'Mass Dictatorship and Gender Politics: Is the Outcome Predictable?' in *Gender Politics and Mass Dictatorship*, ed. Jie-Hyun Lim and Karen Petrone. London: Palgrave Macmillan, 2010, pp. 43–9.

Between the fall of the Qing dynasty in 1911 and the start of the Second Sino-Japanese War in 1937, the preoccupation with modernising China also meant that dozens of texts on biology and sexology were translated into Chinese from Russian, Japanese, French, English, and German.[13] Sex first emerged as a site of empirical inquiry in nineteenth-century Europe, and sexology – the 'scientific study of sex' – was subsequently established as a field of specialisation. Claiming that sex could only be understood through modern science, sexologists used their new-found authority to identify normative sexual behaviours as well as sexual pathologies.[14] In the 1920s and 1930s, Chinese sexologists similarly sought to categorise sexual tendencies as either normal or deviant and, as in Europe and the US, sexual behaviour was also increasingly linked to psychological health.[15] In this context, foreign ideas about sexology, psychology, and medical modernity melded with indigenous conceptions of health and normative sexual behaviour.[16]

Eugenic ideas also drew a large following among members of the Chinese intelligentsia concerned about the decline of Chinese civilisation and the Chinese 'race' in the face of foreign imperialism. The eugenics movement was indeed a global one with adherents hailing from every continent. In the interwar period, in places as diverse as Brazil, India, Japan, and Sweden, a growing number of public health reformers embraced eugenics as a 'scientific' way to combat disease and racial degeneration.[17] For some prominent Chinese intellectuals, eugenics promised to foster a healthy, modern society while balancing the needs of the individual with those of the collective.[18] As products of translingual exchanges, Western eugenic texts translated into Japanese at the turn of the century and then translated again into Chinese played a particularly important role in shaping the language and norms associated with modern sexuality in China.[19] At various

[13] To decentre the Eurocentric periodisation of World War II in Chinese history, here I employ the term 'Second Sino-Japanese War' instead.

[14] Howard Chiang, *After Eunuchs: Science, Medicine, and the Transformation of Modern China*. New York: Columbia University Press, 2018, p. 4.

[15] Emily Baum, 'Healthy Minds, Compliant Citizens: The Politics of 'Mental Hygiene' in Republican China, 1928–1937', *Twentieth-Century China* 42, no. 3 (2017): 222.

[16] Dikötter, *Sex, Culture and Modernity*, p. 12.

[17] Matthew Connelly, *Fatal Misconception: The Struggle to Control World Population*. Cambridge, MA: Harvard University Press, 2008; Cassia Roth, *A Miscarriage of Justice: Women's Reproductive Lives and the Law in Early Twentieth-Century Brazil*. Stanford: Stanford University Press, 2020, pp. 62–4.

[18] In the 1920s and 1930s, for example, sex reformers in Germany and Sweden similarly looked to eugenics as a new sexual morality that could provide individual fulfilment while strengthening the race and society. These ideas appealed to liberal Chinese social reformers, such as Zhang Xichen and Zhou Jianren. Atina Grossman, *Reforming Sex: The German Movement for Birth Control and Abortion Reform, 1920–1950*. New York: Oxford University Press, 1995, pp. 15–16; Mirela David, 'Bertrand Russell and Ellen Key in China: Individualism, Free Love, and Eugenics in the May Fourth Era', in *Sexuality in China: Histories of Power and Pleasure*, ed. Howard Chiang. Seattle: University of Washington Press, 2018, p. 78.

[19] Lydia H. Liu, *Translingual Practice: Literature, National Culture, and Translated Modernity—China, 1900–1937*. Stanford: Stanford University Press, 1995, pp. 32–4; Yuehtsen Juliette Chung, *Struggle for National Survival: Eugenics in Sino-Japanese Contexts, 1896–1945*. London: Routledge, 2002, pp. 13–14.

points, the Nationalist authorities sought to regulate the dissemination of this material, which could be construed as 'obscene' (*yin*) and therefore illicit, yet it continued to circulate.[20] The legacy of these debates, ideas, and texts would endure long after the Communist victory, despite state efforts to bring publishing into the Communist fold.

In the early 1950s, China modelled itself after its closest ally, the Soviet Union, and adopted a largely natalist approach to population. Soviet leaders argued that China should initially follow the path taken by the Union in its early years of economic development, the period from 1918 to the late 1920s, before socialism was achieved and Joseph Stalin came to power. Yet, it was Stalin's 1936 decision to criminalise abortion, which was legalised by the Bolshevik government in 1920 and would subsequently be decriminalised again in 1955, that inspired the CCP's decision to take a harsher stance on birth control and abortion.[21] The intellectual fathers of socialism, Karl Marx and Friedrich Engels, had claimed that overpopulation was an issue unique to the capitalist mode of production with its weak social welfare system.[22] Therefore, they argued, birth control was bourgeois. Stalin further asserted that encouraging large families would augment the workforce and yield a stronger economy. As in the Soviet Union, Chinese women who produced many children were lauded as 'heroine mothers' and given rewards for their reproductive labours.[23] In this way, childbearing became a form of gendered labour and an essential component of socialist production. The combination of limited access to contraceptives and inducements for having large families would lead to unprecedented population growth with the population expanding by more than 45 million people between 1949 and 1953.[24]

In light of the shift to natalism, fragmented efforts were made to constrain access to information about sex and birth control. Yet, only a few of the most popular books, such as *A Guide to Married Life*, *Common Knowledge about Female Physiology*, and *Arts of the Bedchamber*, were banned, and other similar books continued to be published.[25] These books were often reprints of Republican-era

[20] Y. Yvon Wang, 'Whorish Representation: Pornography, Media, and Modernity in Fin-de-siecle Beijing', *Modern China* 40, no. 4 (2014): 381.

[21] The 1955 decriminalization of abortion in the Soviet Union similarly led to a loosening of abortion restrictions in China in the mid-1950s. Mie Nakachi, *Replacing the Dead: The Politics of Reproduction in the Postwar Soviet Union*. Oxford: Oxford University Press, 2021, p. 9.

[22] By this logic, since Communism offered universal access to employment and social services, poverty, scarcity, and other issues associated with a large population could never arise. Nakachi, *Replacing the Dead*, p. 8.

[23] Wang Chongyi *et al.*, *Dangdai zhongguo de weisheng shiye*, Vol. 2. Beijing: Zhongguo shehui kexue chuban she, 1986, p. 230; Hua-Yu Li, 'Instilling Stalinism in Chinese Party Members: Absorbing Stalin's *Short Course* in the 1950s', in *China Learns from the Soviet Union, 1949–Present*, eds. Thomas Bernstein and Hua-Yu Li. Lanham: Lexington Books, 2010, p. 113.

[24] Guojia jihua shengyu weiyuanhui zonghe jihua si, *Quanguo jihua shengyu tongji ziliao huibian* (N.p.: n.p., 1983), p. 1.

[25] Shanghai Municipal Archive, B1–2–3622–152.

guides to hygienic sex and eugenic births. In spite of growing government surveillance of printed media, Republican-era texts related to sex could also be purchased on the black market throughout much of the early Mao period.[26] In addition, new guides to sexual hygiene continued to be produced in limited quantities. The fact that many of the individuals involved in publishing before 1949 retained their positions after the Communists consolidated power over the industry in the mid-1950s helps explain why many of the health guides from the early Mao era bear a striking resemblance both in content and organisation to those from the Republican period.[27]

Guides to sexual hygiene were particularly abundant between 1956 and 1958. This period aligns approximately with the Hundred Flowers Movement (1956–7) and the Anti-Rightist Movement (1957–9), when intellectuals were encouraged to air their political grievances only to meet with a violent backlash. Shao Lizi, a member of the National People's Congress and a long-time advocate of birth control, and Ma Yinchu, a prominent economist and president of Peking University, were two of the most vocal proponents of family planning. After Shao and Ma voiced their concerns about China's rapidly increasing population in 1957, both men were purged.[28] Yet, combined with ongoing debates about the most efficient ways to improve public health and mitigate grain shortages, these calls for change produced the first official attempt to promote family planning in the PRC.[29] The Ministry of Health, as well as provincial bureaus of health and state-run publishers, issued a series of sexual hygiene guides that aimed to promote healthier births and foster a stronger, more productive population. Totalling more than two dozen, these texts each had several editions and print runs from tens of thousands to several million. At 0.10 yuan, the least expensive guide would have been affordable to members of the literate public (in 1957 the average urban couple in China had an annual expenditure of 220 yuan).[30] The most expensive guide was 0.18 yuan, and it is probable that doctors, midwives, and other medical practitioners purchased these guides for use in their clinics. Although the hygiene guides all sought to combat illness in the service of the nation, the books' content varied considerably with some focusing solely on sexual hygiene and others covering health and wellness in general with a subsection about sex.

[26] Y. Yvon Wang, 'Yellow Books in Red China: A Preliminary Examination of Sex in Print in the Early People's Republic', *Twentieth-Century China* 44, no. 1 (2019): 82.

[27] Robert Culp, *The Power of Print in Modern China: Intellectuals and Industrial Publishing from the End of Empire to Maoist State Socialism*. New York: Columbia University Press, 2019, p. 8. For more on publishing and knowledge production in the PRC, see Robert Culp's chapter in this volume.

[28] Judith Shapiro, *Mao's War Against Nature: Politics and the Environment in Revolutionary China*. Cambridge: Cambridge University Press, 1999, p. 22.

[29] In 1953, the central government issued a preliminary statement ordering local governments to loosen restrictions on abortions and sterilizations, a policy that would be made permanent in 1955.

[30] 'Per Capita Annual Income and Expenditure Urban and Rural Households', All China Data Center, chinadataonline.org.

The most widely disseminated guide, *Knowledge about Sex*, was first published in 1956 to spread awareness about sexual hygiene.[31] The book was loosely based on an eponymous text published first in 1926 and then again in March 1949 as part of a series on 'women's problems'. Although the Republican and Communist-era editions of *Knowledge about Sex* contained similar information, because the original book was a translation of an American book, the PRC edition did not acknowledge its Republican antecedent or its foreign origins.[32] In choosing what material to publish for the PRC edition, readers submitted questions to the authors (two gynaecologists and a neurologist) and the authors responded in subsequent editions of the book. With its detailed illustrations and frank discussions of menstruation, childbirth, and birth control, *Knowledge About Sex* was so popular that it was reprinted in 1957 and again in 1958.[33] Altogether, more than 3.6 million copies of this book were sold, making *Knowledge about Sex* by far the most well-known of the 1950s sex guides.[34]

Sexual hygiene guides like *Knowledge about Sex* largely addressed a female audience. Because women were viewed as responsible for policing the moral bounds of the family while safeguarding familial health, the target readership was adolescent women of marrying age who were on the cusp of having sex and producing children for the first time.[35] The publications' purpose was likely two-fold: to prepare young women for marriage and to instill in them a sense of right and wrong regarding sex. Viewed in this light, sexual hygiene guides were meant to empower, and even protect, women while indoctrinating them with certain ideas about what constituted normal sexual desires and behaviours. It is the tension between the paternalistic desire to help young women overcome structural barriers to knowledge and the broader goal of controlling sexuality that comes to the fore in these texts. Such hygiene guides, even those that appear to be pushing for greater gender equality, ultimately worked to preserve the gender hierarchy while validating the new political order.[36]

Many of the ideas these guides espoused overlapped with or even repeated verbatim those presented in the main journals associated with private life in the 1950s, *Women of China* and *China Youth*.[37] Published by the All-China Women's Federation, an organisation tasked with representing women's interests in the CCP,

[31] Wang, Zhao, and Tan, *Xing de zhishi* (1956), introduction.

[32] W.J. Robinson, *Xing de zhishi*, trans. Fang Ke. Shanghai: Kaiming shudian, 1949.

[33] *Knowledge About Sex* was even translated into Korean and Vietnamese.

[34] In 1981, when China's social and economic policies were relaxed, a new edition of *Knowledge about Sex* was printed. Wang Wenbin, Zhao Zhiyi, and Tan Mingxun, *Xing de zhishi*. Beijing: Renmin weisheng chubanshe, 1981.

[35] Evans, *Women and Sexuality*, pp. 125, 198.

[36] This finding affirms Harriet Evans' claim that 1950s narratives on sexuality perpetuated gender inequality and made women the object of disproportionate surveillance.

[37] Evans, *Women and Sexuality*, p. 35.

and the Communist Youth League, respectively, the articles in these two magazines were also written for a literate, female, and primarily urban audience.[38] Between 1956 and 1957 the two magazines boasted a combined readership of nearly three million.[39] Yet, the relatively open public dialogue on sex apparent in the mid-1950s ended in late 1958 with changing political winds and the onset of the Great Leap Forward (GLF).

In sum, the Communist Party benefitted greatly from the Republican-era publishing apparatus as well as the intellectual legacy of the May Fourth/New Culture Movement debates about sexuality. The intertwined goals of socially engineering the population and legitimising state authority converged in the production of numerous hygiene guides in the 1950s.

Sex by the book

Like many aspects of post-1949 life, Mao-era sexuality – at least rhetorically – was defined in complete opposition to Republican sexual practices, and many sex guides from the 1950s begin with editorial notes decrying the backwardness of Republican society. These texts universally condemned the arranged marriage system, concubinage, and the widespread moral laxity of the Republican era.[40] Yet, the attitudes toward marriage reflected in 1950s sexual hygiene guides were still very much a product of the May Fourth/New Culture Movement and global interwar debates about individual desire versus state imperatives. Some texts from the 1950s specifically promoted the ideas that friendship is the basis for love and marriage and that companionate marriage benefits both the individual and society. Meanwhile, other texts denounced Republican society for commodifying everything, including love, and criticised the individualistic emphasis on 'free love' that emerged during the May Fourth/New Culture Movement.[41] One guide, *Youth Marital Hygiene*, claimed that the idea of 'free love' was inherently capitalist and therefore posed an obstacle to the building of Communist youth morality.[42] Indeed, free-choice marriages were popularly believed to encourage promiscuity and loose behaviour, accusations that could ruin a young woman's reputation.[43] Moreover, cavorting with the

[38] Zhonghua quanguo funü lianhe hui, 'Zhongguo funü' zazhi she', accessed 18 May 2021, www.women.org.cn/zhuanti/, www.women.org.cn/quanguofulian/zhishudanwei/funvzazhishe.htm.
[39] Evans, *Women and Sexuality*, p. 17.
[40] Gao Yunhan, *Xing jiaoyu yu xing weisheng*. Shanghai: Zhongguo tushu chubanshe, 1954, p. 3.
[41] Shanghai di er yixue yuan fu chan ke jiaoyanzu, *Qingnian hunyin weisheng*. Shanghai: Shanghai kexue jishu chubanshe, 1958, p. 1.
[42] Shanghai di er yixue yuan, *Qingnian hunyin weisheng*, pp. 1–3.
[43] Harriet Evans points out that premarital courtship (euphemistically referred to as 'making friends') was only practiced among educated urbanites in the Republican era, and that this practice was not widespread in the early Mao era; Evans, *Women and Sexuality*, pp. 84–5.

opposite sex and focusing too much on romance – even if it led to marriage – was believed to negatively influence productivity at work and school.[44] The tension between love and sex for personal gratification and marriage for the purpose of maintaining social stability undergirds all the sex-related advice of the 1950s. Each of the sexual hygiene guides acknowledged both the biological validity of sexual desire within marriage and the centrality of stable, productive households to meeting state goals.[45]

In all of the discussion about the right and wrong way to approach marriage and sex in the 1950s, romantic sentiment remained a marginal consideration. Rather than taking intangible passions with their capitalist and imperialist influences as their focus, most texts purported to advocate a modern and 'scientific' understanding of bodies and sexuality, thus positioning the Maoist state as a progressive guiding force. In reality though, science was often simply the handmaiden of politics, wherein biomedical language was used to justify socially constructed claims about sexuality.[46] For example, Maoist sexual hygiene guides contended that all people regardless of gender desired sex because it was biologically 'natural'. Yet, these same books also claimed that women were sexually passive (*beidong*) and that even their orgasms were less active than men's.[47] Such claims justified male dominance in the bedroom specifically and in marriage more generally.

The hygiene guides also mobilised the language of science to claim that sex was only permissible between a man and a woman.[48] Scholars have debated at length the extent to which homosexuality was tolerated in late imperial China. Some argue that it was widely accepted while others maintain that it was viewed as a threat to the social hierarchy.[49] During the debates of the May Fourth/New Culture Movement, same-sex relations became intertwined with China's political crisis. Some conservative thinkers blamed same-sex relationships for China's weakness and semi-colonisation, whereas other intellectuals idealised same-sex relations as a higher form of love and the foundation for a human utopia.[50] By the 1950s, however, the topic of homosexuality had practically been eliminated from

[44] Yi Weizhi, 'Xianzhi lian'ai de "jueding"' ,*Renmin ribao*, 9 August 1956; Gu Yunfeng, 'Xianzhi lian'ai de fu jiaodao zhuren', *Renmin ribao*, 2 September 1956.

[45] Wang, Zhao, and Tan, *Xing de zhishi* (1956), p. 2; Shanghai di er yixue yuan, *Qingnian hunyin weisheng*, p. 6.

[46] Evans, *Women and Sexuality*, p. 3.

[47] Shanghai di er yixue yuan, *Qingnian hunyin weisheng*, p. 32; Wang, Zhao, and Tan, *Xing de zhishi* (1956), p. 26.

[48] Wang, Zhao, and Tan, *Xing de zhishi* (1956), p. 21; Frank Dikötter explains how, in the first half of the nineteenth century, European professionals posing as scientists came to view sexual desire as a natural, biological impulse. This notion certainly influenced Republican and Mao-era ideas about sex; Dikötter, *Sex, Culture and Modernity*, p. 63.

[49] Bret Hinsch, *Passions of the Cut Sleeve*. Berkeley: University of California Press, 1990; Matthew Sommer, *Sex, Law, and Society in Late Imperial China*. Stanford: Stanford University Press.

[50] Kang, *Obsession*, p. 1; Tze-lan Deborah Sang, *The Emerging Lesbian: Female Same-Sex Desire in Modern China*. Chicago: University of Chicago Press, 2003, p. 16.

public discourse. A foundational principle of Mao-era sex guides was the gender binary between men and women, a distinction based on anatomy and reproductive functions rather than gender performance.[51] This construction of 'essential difference' between the genders foreclosed possibilities for discussing alternative gender identities.[52] Moreover, all sex was assumed to be procreative or at least to have the potential to result in procreation, ideas that dovetailed with the notion that women had a 'natural duty' to be both wives and mothers.[53] Homosexuality, then, by virtue of its erasure from public discourse, was rendered aberrant.

In addition to being heterosexual, sex was also supposed to be confined solely to marriage. Therefore, hygiene guides regularly conflated sex and marriage, with marriage being the inevitable outcome of courtship and sex the outcome of marriage.[54] Every sex guide reiterated this point in one way or another: 'sex is natural and important for building a family' and 'having children after marriage is a natural desire and requirement'.[55] Such discourses positioned sexual desire as a service to both the individual and to preserving the social fabric of Chinese society. To lack sexual desire or the desire to procreate, then, was considered a biological deficiency. For example, several hygiene guides endorsed Sigmund Freud's argument that certain women – those experiencing 'frigidity' – lacked sexual desire altogether or desired clitoral stimulation in place of vaginal intercourse due to a psycho-sexual affliction linked to lesbianism.[56] Insufficient sexual desire, particularly for women, could also be construed as intentionally denying one's reproductive potential to the household or the collective.

Beyond possessing the appropriate amount of sexual desire, understanding the mechanics of sex and procreation was essential to being a modern socialist subject. Indeed, one of the complaints repeated in many different hygiene guides was that sex was taboo during the Republican period.[57] The books alleged that, because discussing sex was considered indecent in the 'Old Society', young people lacked a basic understanding of how bodies worked with some even wrongly believing that sex could cause illness.[58] Such concerns ostensibly made young people afraid to marry and caused them needless worry.[59] For those reasons, the authors of the

[51] Judith Butler, *Gender Trouble: Feminism and the Subversion of Identity*. New York: Routledge, 1990, p. 1.
[52] Evans, *Women and Sexuality*, p. 27.
[53] Evans, *Women and Sexuality*, p. 121.
[54] Evans, *Women and Sexuality*, p. 113.
[55] Wang, Zhao, and Tan, *Xing de zhishi* (1957), p. 63; Wang, Zhao, and Tan, *Xing de zhishi* (1956), p. 2.
[56] Although women were generally perceived as sexually passive in comparison to men, not having any sexual desire whatsoever was considered to be a biological abnormality; *Xing jiaoyu zhinan* (Shanghai: Zhonghua shuju, 1949), p. 3; Wang, Zhao, and Tan, *Xing de zhishi* (1956), pp. 58–9; Katherine Angel, 'The History of 'Female Sexual 'Dysfunction as a Mental Disorder in the 20th century', *Current Opinion in Psychiatry* 23, no. 6 (2010): 536.
[57] Chen Xiyi, *Funü weisheng wenda*. Shanghai: Shanghai weisheng chubanshe, 1957, pp.3–4.
[58] Shanghai di er yixue yuan, *Qingnian hunyin weisheng*, p. 29; Gao, *Xing jiaoyu yu xing weisheng*, p. 1.
[59] Shanghai di er yixue yuan, *Qingnian hunyin weisheng*, p. 29.

hygiene guides had set out to liberate young people from unwarranted fears and fortify them with Soviet-style 'scientific' knowledge about sex.[60]

The result was a series of sexual hygiene guides that offered a primer on a range of topics, including male and female anatomy and physiology, the physical changes that occur during puberty, the biology behind conception, and to a degree the mechanics of sex. According to *Knowledge about Sex*, intercourse was divided into the 'three stages of sex life': the preparation stage, the intercourse stage, and the finishing stage.[61] The preparation stage involved arousal, the intercourse stage focused on achieving orgasm, and the finishing stage required confirming that both partners had in fact orgasmed (rather than solely the man). Aside from vaginal intercourse, other types of sexual acts, even those that could be said to fit within the bounds of heteronormativity such as oral and anal sex, were completely elided. This was likely because such acts were viewed as gratuitous, unproductive, and even obscene.[62] *Knowledge about Sex* aside, the focus of most hygiene guides was not pleasure itself but intercourse that would leave both partners fulfilled enough to preserve the marriage while producing heirs.[63] Even though their intended audience was virginal newlyweds, almost none of the guides mentioned the lead-up to sex; kissing, caressing, and other forms of foreplay were completely omitted.[64]

That is not to say that pleasure was never mentioned. Some texts warned about lack of sexual fulfilment (that is, the woman failing to orgasm during sex due to premature insertion of the penis and therefore premature ejaculation) as a common source of marital discord.[65] In fact, women's sexual satisfaction, particularly early in the marriage, was considered important for conjugal and familial stability. One guide alleged that women whose husbands could not satisfy them were prone to hating their partners and rejecting their sexual advances.[66] Therefore, some guides featured graphs charting the ideal timing of female orgasm and male ejaculation for producing sexual (and conjugal) fulfilment (Figures 7.1a and 7.1b). Although *Knowledge about Sex* generally took a more thoughtful approach to sex than comparable books, most hygiene guides – particularly those that were not simply reprints of Republican-era texts – walked the reader through the sex act in a methodical and detached fashion.[67]

[60] Gao, *Xing jiaoyu yu xing weisheng*, p. 1.

[61] Wang, Zhao, and Tan, *Xing de zhishi* (1956), pp. 35–8

[62] Although other scholars have contested this point, Frank Dikötter argues that authorities had long viewed sexual acts that could not lead to conception, such as anal sex, as illegitimate and therefore dangerous. Dikötter, *Sex, Culture, and Modernity*, p. 63.

[63] Despite contemporaneous efforts to promote the use of contraception, rather than being viewed as a bonding activity between partners, sex was often depicted as a way to meet biological needs while enabling the creation of offspring.

[64] Francis Lee Bernstein also notes that 'sex play' was discouraged and deemed wasteful in Soviet discourses on sexuality from the 1920s; Frances Lee Bernstein, *The Dictatorship of Sex: Lifestyle Advice for the Soviet Masses*. DeKalb: Northern Illinois University Press, 2007, p. 13.

[65] Wang, Zhao, and Tan, *Xing de zhishi* (1956), p. 39.

[66] Wang, Zhao, and Tan, *Xing de zhishi* (1956), p. 39.

[67] Song Hongjian and Zhao Zhiyi, *Biyun changshi*. Shanghai: Shanghai weisheng chubanshe, 1957, p. 7.

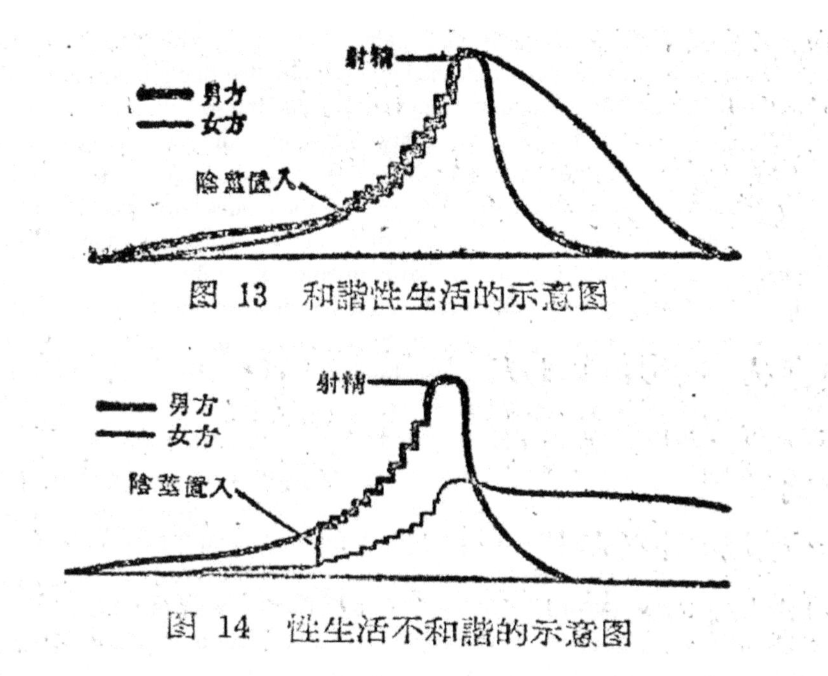

Figures 7.1a and 7.1b Charts demonstrating the correct (7.1a) and incorrect (7.1b) timing for penal insertion, male orgasm, and female orgasm to ensure a 'harmonious' sex life

In short, 1950s sexual hygiene guides recognised sexual desire as a biological necessity and one that was critical to companionate marriage, the foundation of Maoist society.[68] However, these texts also coopted the rhetoric of science to justify a particular vision of sexual and conjugal normativity.

Defining deviance

The guides also used prescriptive language to establish the boundaries of normative sexual tendencies, a tactic likely adopted from early Soviet tracts on sex and reproduction.[69] As with too little sexual desire, excessive sexual desire was viewed as abnormal and unhealthy.[70] According to *Youth Marital Hygiene*, too much sexual desire among young men could cause exhaustion, reduced sexual functioning, and impotence. Excessive sexual desire could also manifest through sleeplessness,

[68] Shanghai di er yixue yuan, *Qingnian hunyin*, p. 28.
[69] Bernstein, *The Dictatorship of Sex*, p. 7; As Jennifer Altehenger demonstrates in her chapter in this volume, standardisation – whether in medical discourses or guidelines for furniture manufacturing – played a critical role in Maoist nation-building.
[70] Evans, *Women and Sexuality*, p. 44.

fatigue, memory decline, dizziness, back pain, and poor appetite. Not only would the body become weak, but it would negatively influence one's work and study, both of which were critical for supporting the state and the social order.[71]

Like excessive sexual desire, (excessive) masturbation was also viewed as unhealthy. Some of the potential side effects of excessive masturbation among males were anxiety, regret, and guilt, and therefore the guides laid out steps to limit the temptation to masturbate. In the case of young women, masturbation could supposedly lead to health conditions like emmeniopathy (abnormal or irregular menstruation). One hygiene guide even advised young women to avoid reading stimulating short stories or wearing tight pants that rub the body since both of those things could lead to arousal.[72] Thus, although sexual desire was the biological norm, it needed to be closely regulated.[73]

The preoccupation with masturbation as a precursor to disease was partly rooted in global discourses on the subject. The increasing medicalisation and pathologising of sex during the late nineteenth and early twentieth centuries further legitimised state intrusions into private life. Indeed, European and American, and later Japanese, sexologists wrote at length about the dangers of excessive sexual desire and masturbation. Rather than claiming that masturbation was sinful as had long been argued in the West, some began contending that it led to nervous exhaustion and even mental illness, both of which warranted professional medical intervention.[74] In that sense, early Maoist texts warning against the dangers of excessive sexual desire and masturbation were very much in line with prevailing global discourses on these topics.

Yet, these ideas were also in dialogue with indigenous medical discourses. According to TCM, the body is made up of *qi* (the essence of matter), and the balance of *yin* and *yang* – the complementary feminine and masculine aspects of *qi* – governs bodily health. For women, regular menses indicated balanced levels of *yin* and *yang*, which in turn meant a woman had good health. In this way, masturbation was dangerous because it destabilised the equilibrium of *yin* and *yang* and might potentially impact a woman's reproductive capacity. In the case of men, Daoist teachings argued that men needed a certain amount of female *yin* to balance out their abundant *yang*. Excessive sex could drain a man of his *yang* and lead to illness, but an appropriate amount of sex would benefit him by replenishing his *yin*.[75] According to the authors of *Knowledge about Sex*, masturbation was not bad

[71] Shanghai di er yixue yuan, *Qingnian hunyin weisheng*, p. 30; Ma Yuru, *Jiating weisheng*. Beijing: Renmin weisheng chubanshe, 1956, p. 18; Gao, *Xing jiaoyu yu xing weisheng*, pp. 8–9.

[72] Chen, *Funü weisheng wenda*, p. 16.

[73] Gao, *Xing jiaoyu yu xing weisheng*, p. 5.

[74] Sabine Frühstück, *Colonizing Sex: Sexology and Social Control in Modern Japan*. Berkeley: University of California Press, 2003, pp. 7–8.

[75] Hugh Shapiro, 'The Puzzle of Spermatorrhea in Republican China', *Positions* 6 (1998): 553–4; Everett Yuehong Zhang, *The Impotence Epidemic: Men's Medicine and Sexual Desire in Contemporary China*. Durham: Duke University Press, 2015, pp. 137–8.

because it reduced a man's 'vitality' (*yuanqi*), as was wrongly believed in the 'Old Society'. Rather, masturbation was harmful because it could negatively influence one's health and work, and therefore needed to be overcome through 'correct thinking' (*duanzheng sixiang*) and new habits.[76] For the authors, the vague term 'correct thinking' referenced pairing modern and scientific behaviour (getting enough sleep, washing regularly, and exercising to reduce sexual desire) with political correctness (acknowledging the scientific basis for sexual desire but channelling that impulse toward collective goals). In reality, the authors simply reframed the line of thinking espoused in TCM – that sexual moderation is essential for good health – as an ideological issue. In either case, though, excessive sex and sexual desire were viewed as dangerous, and even deviant and counterproductive. Both Republican-era discussions of masturbation and those from the 1950s emphasised the importance of self-discipline and sexual self-restraint.[77] In this way, curbing excessive sexual desire became a political act and one that, like so many other behaviours in Maoist China, was conflated with commitment to the Communist cause.[78]

In another important carryover from the past, premarital chastity – particularly for women – was viewed as critical for conforming to a 'correct' sexual trajectory. As many other scholars have demonstrated, in the late imperial and Republican periods, female chastity was not only a prized virtue but often a requirement for marriage, and for that reason it is not surprising that nearly every 1950s sex guide featured a section devoted entirely to the hymen (*funü mo*).[79] These guides mentioned that unmarried women with broken hymens often faced extreme prejudice and might even be divorced by their husbands upon marriage.[80] Thus, while the authors walked the reader through the process of breaking a woman's hymen on her wedding night, they actually sought to counter the stigma associated with unmarried women already having a punctured hymen. *Women's Premarital Hygiene*, for example, counselled that a woman's hymen could be broken before marriage due to reasons other than intercourse, such as strenuous exercise or labour.[81] In spite of their progressive intentions, by providing alternative explanations for why a woman might already have her hymen broken the authors implicitly affirmed the idea that a woman must remain 'pure and clean' (*qingbai*), and therefore virginal, until marriage. In other instances, rather than referencing morality, some writings simply justified female chastity in terms of science. They claimed that women who have sex too young will not develop correctly and will suffer physiological damage.[82] In

[76] Wang, Zhao, and Tan, *Xing de zhishi* (1956), p. 30; Ma, *Jiating weisheng*, p. 18.

[77] Frank Dikötter, *Imperfect Conceptions: Medical Knowledge, Birth Defects, and Eugenics in China*. New York: Columbia University Press, 1998, p. 69.

[78] Wang, Zhao, and Tan, *Xing de zhishi* (1956), p. 32.

[79] Matthew H. Sommer, *Sex, Law and Society in Late Imperial China*. Stanford: Stanford University Press, 2000, pp. 7–8; Janet Theiss, *Disgraceful Matters: The Politics of Chastity in Eighteenth-Century* China. Berkeley: University of California Press, 2004, p. 211.

[80] Lu, *Nüzi hunqian weisheng*. Shandong: Shandong renmin chubanshe, 1955, p. 9.

[81] Lu, *Nüzi hunqian weisheng*, p. 9.

[82] Ma, *Jiating weisheng*, p. 18.

this way, the focus on female chastity inherited from previous eras endured but was now explained and justified using the language of science.

As with excessive masturbation and premarital sex, the guides also painted infertility and lack of children in marriage as 'abnormal'. According to *Knowledge about Sex*, there is a small number of couples – 5–10 per cent of people according to Soviet statistics – who desire children but cannot have them.[83] In an attempt to rectify the tendency to blame women for infertility, the authors of sexual hygiene guides repeatedly stressed that infertility is an affliction that affects both men and women.[84] *Knowledge about Sex* also attributed infertility to the poor quality of life and worker oppression under the Nationalists but claimed that improved material conditions since 1949 had helped solve fertility issues.[85] Such an explanation presumes that the standard of living in the 1950s increased uniformly, thus entirely eradicating the underlying causes of infertility. Moreover, the authors did not explicitly address the effects of chronic malnutrition and poverty on fertility, despite the fact that both were widespread in the Republican period and remained so into the 1950s.[86] By contrast, another hygiene guide, *Common Knowledge about Women's Hygiene*, discussed at length how gonorrhoea and tuberculosis, excessive drinking, smoking opium, and anxiety can negatively impact fertility.[87] Echoing the idea that excessive sexual desire is unhealthy, *Common Knowledge about Women's Hygiene*, counterintuitively, also attributed infertility to excessive sex.[88]

These explanations for infertility all served to justify medical surveillance. The book *Fertility and Infertility* stressed the fact that being unsound mentally may impede pregnancy and therefore people struggling to conceive require mental health examinations.[89] *Knowledge about Sex* similarly provided extensive guidance for curing infertility through testing, lifestyle changes, and treatment. This text also encouraged prospective parents to remain optimistic about having children since the PRC was making rapid scientific advances.[90] In every case, though, not having children, especially within the first few years of marriage, was viewed as a deficiency that needed to be remedied through expert investigation and intervention.[91]

As for other sexual behaviours or characteristics that were characterised as aberrant, although sexually transmitted infections (STIs) certainly fell under this rubric, coverage of the topic in 1950s health guides was highly uneven. In the Republican

[83] Wang, Zhao, and Tan, *Xing de zhishi* (1956), p. 63.
[84] Xin zhongguo funü zazhi she, *Funü weisheng changshi*. Beijing: Xin Zhongguo funü zazhi she, 1956, p. 20; Wang, Zhao, and Tan, *Xing de zhishi* (1956), p. 64; Liu Benli, *Shengyu yu buyu*. Shanghai: Jia chubanshe, 1953, p. 45; Gao, *Xing jiaoyu yu xing weisheng*, p. 59.
[85] Wang, Zhao, and Tan, *Xing de zhishi* (1956), p. 64.
[86] Evans, *Women and Sexuality*, p. 145.
[87] Xin Zhongguo funü zazhi she, *Funü weisheng changshi* (1956), p. 20.
[88] Xin Zhongguo funü zazhi she, *Funü weisheng changshi* (1956), p. 20.
[89] Liu, *Shengyu yu buyu*, p. 55.
[90] Wang, Zhao, and Tan, *Xing de zhishi* (1956), p. 69.
[91] Liu, *Shengyu yu buyu*, p. 54.

period, news columns, magazine articles, tabloids, and books addressing STIs were quite numerous in part due to the intelligentsia's association of venereal disease with racial decline.[92] Records from the Shanghai Health Bureau suggest that STIs were endemic in cities, particularly Shanghai, in the late Republican period and that the commercial sex industry played a significant role in the transmission of venereal disease.[93] The CCP maintained that STIs were a by-product of prostitution, which in turn was the result of Republican-era capitalist exploitation and foreign imperialism.[94] Two hygiene guides even claimed that the influx of American films, stories, and songs into China under Chiang Kai-shek was responsible for 'eroto-mania' (*seqingkuang*), rampant prostitution, and widespread venereal disease.[95] To address these issues, in the early 1950s the new government closed thousands of brothels and sought to treat and retrain former prostitutes.[96] Perhaps because the CCP claimed to have eradicated most STIs by the early 1960s, quite a few books about sexual hygiene (or hygiene in general) published in the late 1950s completely ignored this topic.[97]

However, a few guides still addressed venereal disease in detail and in the same style as books from the Republican period. *Women's Premarital Hygiene*, for one, featured graphic images not only of the bacteria that cause syphilis and gonorrhoea under the microscope but also of wrinkled and blistered babies born with congenital gonorrhoea (Figure 7.2).[98] A major concern was the transmission of STIs to offspring.[99] The guides that did address STIs, like those that addressed infertility, generally acknowledged that these diseases could afflict both genders. Yet, by featuring these topics in books designed for women, the books implicitly affirmed the idea that STIs are a female problem.

Discussions of STIs went beyond recommending simple preventative measures to include more invasive procedures. As part of courtship, *Women's Premarital Hygiene* advised readers to not only consider things like political character, personality, appearance, age, and hobbies when choosing a spouse but also whether that

[92] Dikötter, *Imperfect Conceptions*, p. 106.

[93] Gail Hershatter, *Dangerous Pleasures: Prostitution and Modernity in Twentieth-Century Shanghai*. Berkeley: University of California Press, 1997, p. 296.

[94] Xin Zhongguo funü zazhi she, *Funü weisheng changshi* (1956), p. 23; Wang, Zhao, and Tan, *Xing de zhishi* (1957), p. 77.

[95] Lu, *Nüzi hunqian weisheng*, p. 36; Gao, *Xing jiaoyu yu xing weisheng*, p. 5.

[96] Xin Zhongguo funü zazhi she, *Funü weisheng changshi*, p. 20; Gail Hershatter, 'Regulating Sex in Shanghai: The Reform of Prostitution in 1920 and 1951', in *Shanghai Sojourners*, eds. Frederic Wakeman Jr, and Yeh Wen-hsin. Berkeley: University of California Press, 1992, p. 179. Although legal prostitution persisted, imperial Japanese authorities were similarly concerned with managing sex workers and the spread of STIs; Sheldon Garon, 'The World's Oldest Debate? Prostitution and the State in Imperial Japan, 1900–1945', *The American Historical Review* 98, no. 3 (1993): 712.

[97] Evans, *Women and Sexuality*, p. 160; *Jiating weisheng, Funü weisheng wenda, and qingnian hunyin weisheng*, for example, largely ignore the topic of venereal disease. However, *Xing de zhishi* claimed that venereal disease had been 'thoroughly eradicated'; Wang, Zhao, and Tan, *Xing de zhishi* (1957), p. 76.

[98] Lu, *Nüzi hunqian weisheng*, pp. 42–3.

[99] Wang, Zhao, and Tan, *Xing de zhishi* (1957), p. 78.

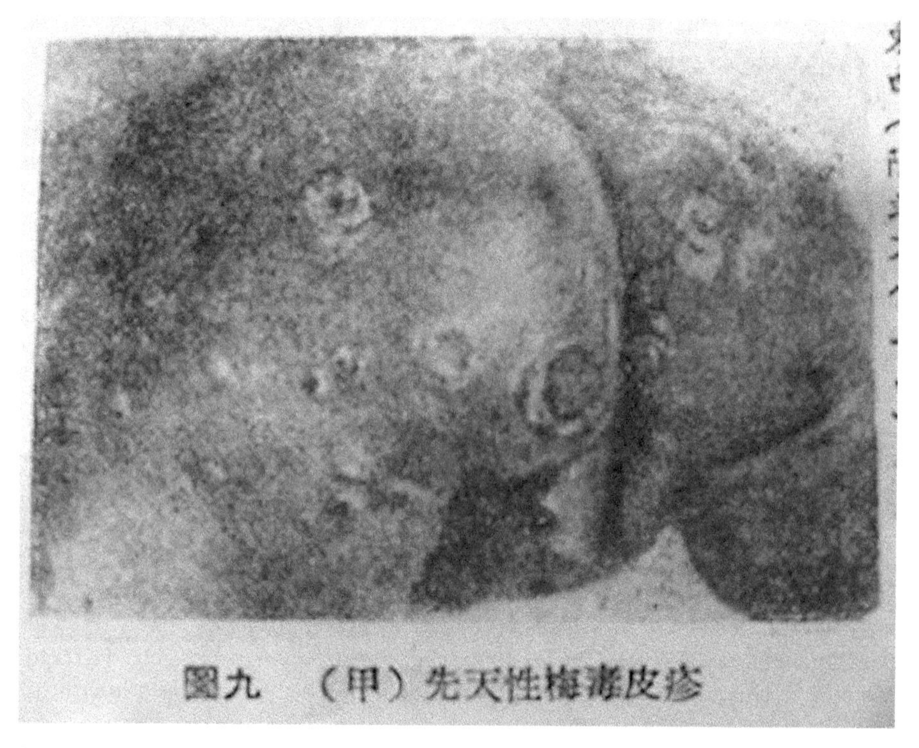

Figure 7.2 An image of a baby with blisters from congenital syphilis

person had a venereal disease.[100] Responsibility for determining a partner's biological viability, though, was not simply left up to individuals or their families. Rather, premarital health exams were stipulated as a way to avert marital strife, fetal illness, and national decline.[101] Premarital health examinations were first introduced under the Nationalists in the 1940s when eugenic marriage laws were passed prohibiting people with STIs, leprosy, or tuberculosis from marrying.[102] Like its forerunners, the 1950 Marriage Law also forbid marriage (and sex) between people with mental illness, physical deformities, or STIs and relied on premarital testing to identify individuals ineligible for marriage.[103]

[100] Lu, *Nüzi hunqian weisheng*, pp. 36–7.

[101] Lu, *Nüzi hunqian weisheng*, p. 38; Shanghai di er yixue yuan, *Qingnian hunyin weisheng*, p. 29.

[102] Chung, *Struggle for National Survival*, p. 161.

[103] Because marriage was assumed to be synonymous with sex, forbidding marriage among the 'unfit' was equated with banning sex and reproduction among the affected parties. The 1950 Marriage Law also prohibited marriage among men under 20 and women under 18, raising the legal marriage age from the Republican period by two years for both sexes. Part of the rationale for this change was the belief that physically and emotionally immature youth would produce 'weak' offspring, who in turn would limit their parents' ability to contribute to society. Huo Xuanji, 'Dayuejin zhihou de jihua shengyu, 1962–1966' (Family Planning Policy after the Great Leap Forward, 1962–1966) (Master's thesis, Nanjing University, 2015), p. 54; Wang, Zhao, and Tan, *Xing de zhishi* (1957), p. 21.

As a product of global anxieties about racial fitness and national prowess, these policies were consistent with eugenic policies that already existed in other parts of the world. In fact, laws prohibiting 'degenerates' from marrying had existed in parts of the US, Europe, and Japan since the early twentieth century.[104] *Women's Premarital Hygiene* even acknowledged that the US policy of forcing the 'unfit' to undergo sterilisation had inspired China's marriage ban. However, the book's author was also critical of the American policy. He argued that forced sterilisation was too extreme given the prevalence of STIs and that this policy reflected the flaws within capitalism and Western pseudoscience. This ambivalence toward eugenics was likely a product of China's close relationship with the Soviet Union. Eugenics had enjoyed some support in the Soviet medical community in the 1920s. Yet, in 1930 Joseph Stalin disbanded the Soviet Eugenics Society and abandoned eugenics altogether on the grounds that it was 'bourgeois' and 'fascist'.[105] Following suit, in the early 1950s Western biology and genetics, and by extension eugenics, became politically suspect in China.[106] Political ideology aside, government efforts to police sex and reproduction in China reflected the global zeitgeist of the era.

Indeed, Mao-era sexual knowledge owed much to global and domestic discourses about health and medicine. 1950s hygiene guides positioned premarital chastity and invasive health exams as the norm while constructing masturbation, infertility, and venereal disease as deviant. In this way, individual needs and desires were often subordinated to the political project of facilitating social stability, and women's bodies in particular continued to be the object of state surveillance.

Conclusion

With the establishment of the PRC in 1949, CCP leaders vowed to reimagine and rebuild Chinese society. In order to extend China's political transformation to the bedroom, in the 1950s commercial and state presses – in consultation with doctors and political activists – published sexual hygiene guides that combined the authority of modern science with socialist rhetoric to fashion a model of Maoist sexual normativity. Although they ostensibly rejected the moral logic of the Republican period, in reality early Mao-era hygiene guides selectively drew on ideas that had been circulating since at least the early twentieth century. At the intersection of local and global conversations about medical modernity, these texts also synthesised concepts from TCM, sexology, and eugenics to forge a sexual model that was both ideologically correct and pragmatic.

[104] Frank Dikötter, 'Race Culture: Recent Perspectives on the History of Eugenics', *American Historical Review* 103, no. 2 (1998): 471; Chung, *Struggle for National Survival*, p. 16.
[105] Bernstein, *The Dictatorship of Sex*, pp. 172–3.
[106] Laurence Schneider, 'Lysenkoism and the Suppression of Genetics in the PRC, 1949–1956', in *China Learns from the Soviet Union, 1949-Present*, eds. Thomas Bernstein and Hua-Yu Li. Lanham: Lexington Books, 2010, pp. 337–8.

Examining constructions of normative sexuality in the 1950s reveals that Chinese authorities shared many of the same anxieties and assumptions about public health, gender roles, and the relationship between state and society as their counterparts elsewhere. Like medical practitioners and policymakers in other countries across the political spectrum, the authors of Chinese sexual guides sought to alleviate the uneasy tension between individual interests and collectivist goals. Although some efforts were made to foster a more liberatory sexual culture, ultimately social stability and productivity were privileged over other considerations in the making of early Maoist sexual norms.

Part III

Material Culture and Everyday Life

8

How to Standardise Life in 'New China': The Case of Furniture

JENNIFER ALTEHENGER

IN OCTOBER 1976, China's Technical Standards Press officially published *Basic Measurements of Commonly Used Furniture* (SG 98–75, *Basic Measurements* hereafter), a leaflet written by the National Furniture Technical Group (*Quanguo jiaju jishu zu*). It had been commissioned by the Ministry of Light Industry and, on its first page, the authors explained that its purpose was to set out national standards for 'wood and metal furniture used for work, study and life', including furniture made of 'bamboo, rattan, plastic, and other kinds of materials'. Over 20 pages it specified measurements for chairs with and without armrests, a folding chair, stool, bench, square and rectangular table, three types of desk (with drawers on either, both, or neither side), single and bunk beds, double bed, as well as a wardrobe, chest of drawers, and bookcase.[1]

People in the world of furniture took note of this thin, unremarkable leaflet that cost only 0.08 yuan. The National Furniture Technical Group was not the first to write guidelines; most provinces, municipalities, and several leading research institutes had worked on furniture standards in earlier years. The national Ministry of Light Industry, too, had repeatedly tried to unify measurements. In 1965 and 1966, it convened major conferences of furniture producers, which resulted in specifications for six types of timber furniture: an office desk, office chair, cabinets, wardrobe, and bed.[2] What made *Basic Measurements* significant was that it focused on a much larger range of everyday furniture and was among the first to

[1] Zhonghua renmin gongheguo qinggongyebu, *Bubiaozhun – Changyong jiaju jiben chichun*. Beijing: Jishu biaozhun chubanshe, 1976. The leaflet's publisher, the Technical Standards Press, was founded in 1963 to organise and disseminate designs and measurements. During the early 1980s, it was renamed the China Standards Press.

[2] Fujian Provincial Archives 186-003-1059, 044, 'Guanyu mu jiaju dangqian shengchan de jidian yijian', 4 November 1966.

Proceedings of the British Academy, **267**, 147–166, © The British Academy 2025.

be published for public circulation in the name of the Ministry of Light Industry. It was the product of several years of work, starting with investigations into industrial practices and conferences in Shanghai in 1974, and ending with months of drafting in 1975. As the editorial team went about its work, they could not have known that their measurements would circulate publicly in the wake of Mao Zedong's death. This coincidence might have seemed of little consequence to them at first, but, with hindsight, it made it a small harbinger of an impending momentous transformation. Over the next two decades, everyday furniture was to become one of China's light industrial model goods, a central focus of party-state industrial policy, and for many people material proof of the promises of prosperity that Deng Xiaoping had made with the policies of 'reform and opening'.[3]

Basic Measurements gives us a glimpse into what became an intense concern with questions of standardisation (*biaozhunhua*) as part of the industrialisation of furniture production in China before and after the economic reforms of 1978. For this chapter, it is the starting point to think about how crucial everyday goods were thought of and produced during the early Mao era (1949–76). The concern with standardisation in post-war China was anything but new. Different practices of standardisation in architecture, goods manufacturing, and other areas, as Lothar Ledderose writes, can be found throughout imperial China.[4] After 1949, standardisation then also became a cornerstone of the Chinese Communist Party (CCP)'s project to industrialise the country. At the general level, the case for standardisation seemed clear: standards facilitated mass production, could unify technical components and measurements, and render them comparable. They could help systematise the amount and kind of raw materials used to produce goods and economise materials, energy, and time used in production. Standardised designs might also reduce time spent on planning and production while enabling higher outputs. Finally, technical standards could make quality control more manageable, thereby promising to improve the products people used in their daily lives. To party planners, standardisation was thus a central element in the developmental project of building socialism, and there were numerous examples of standardisation in early-PRC (People's Republic of China) industrial production; from electronics, sewing machines, bicycles, and other consumer goods to agricultural tools, factory machinery, railways, cars, building components, and much more.[5]

[3] Xiaobing Tang, 'Decorating Culture: Notes on Interior Design, Interiority, and Interiorization', *Public Culture*, Vol. 10, Issue 3, 1998, p. 530–48; Jie Li, *Shanghai Homes: Palimpsests of Private Life*. New York: Columbia University Press, 2014; Jennifer Altehenger, 'Modelling Modular Living: *Furniture and Life* Magazine and 1980s Interior Design in China', *Journal of Design History*, Vol. 35, Issue 2, June 2022, 151–67.

[4] Lothar Ledderose, *Ten-Thousand Things: Module and Mass Production in Chinese Art*. Princeton: Princeton University Press, 2001.

[5] On the history of different light industry goods and their development see the series edited by Yun Shen, *Gongye sheji Zhongguo zhi Lu*, Dailian: Dalian ligong daxue chubanshe, 2017. Karl Gerth discussed their circulation as desired objects in *Unending Capitalism: How Consumerism Negated China's Communist Revolution*, Cambridge: Cambridge University Press, 2020. Literature on the history of

In contrast to several of these examples, the story of Mao-era furniture standards is far less clear cut. To be sure, woodworkers had long considered the question of standardisation of measurements, techniques, tools, materials, and workflows. Into the twentieth century, the standard reference work for carpentry was the fifteenth-century manual *Lu Ban Jing*, which provided measurements and instructions on material choice for dozens of furniture items.[6] After 1949, however, it was unclear whether and how furniture production would be industrialised, whether standards were necessary, and how they should work. Even though furniture manufacturing had a long and distinguished history in China,[7] and many central party leaders enjoyed good furniture in their daily lives, this did not mean that they were focused on furniture mass production as an elementary part of national policymaking. Furniture occupied an odd position within the larger universe of goods at first – neither elevated to prestige socialist consumer object like wristwatches or bicycles nor afforded much of a media presence.[8] Research institutes and factories created furniture prototypes, discussed standards, and compiled manufacturing guidelines in an attempt to improve supplies, yet their resources were inadequate, and their impact often limited to the local or regional. Although the government repeatedly affirmed its intent to mechanise furniture production, and made some investments in technological improvements, most manufacture remained manual or semi-mechanised until the 1980s. China's furniture manufacturing sector thus never produced enough to meet demand.

Those involved in furniture manufacturing were uncertain how their work fit into the party's vision for the creation of a socialist society, giving rise to many exchanges about whether, why, and how to standardise furniture production, at which level (local, provincial, regional, or national), and who should determine which standards, according to which criteria. State planners, light industry officials, local government officials, architects, designers, and woodworkers knew that the furniture people owned, or were able to obtain, would be integral to their experience of the 'transition to socialism', influencing their perception of themselves and

standards beyond China is vast, but some indicative works include: Nils Brunsson and Bengt Jacobsson, *A World of Standards*. Oxford: Oxford University Press, 2002; Stefan J. Link, *Forging Global Fordism: Nazi Germany, Soviet Russia, and the Contest Over the Industrial Order*. Princeton: Princeton University Press, 2020; Haruhito Shiomi and Kazuo Wada (eds), *Fordism Transformed: The Development of Production Methods in the Automobile Industry*. Oxford: Oxford University Press, 1995; JoAnne Yates and Craig N. Murphy, *Engineering Rules: Global Standard Setting Since 1880*. Baltimore: Johns Hopkins University Press, 2019;

[6] Klaas Ruitenbeek, *Carpentry and Building in Late Imperial China: A Study of the Fifteenth-Century's Carpentry Manual* Lu Ban jing. Leiden: Brill, 1996.

[7] Christine Ho discusses debates surrounding the question of craft and light industry in 'Design and Handicraft', in Jennifer Altehenger and Denise Y. Ho (eds.), *Material Contradictions in Mao's China*. Seattle: University of Washington Press, 2022, 63–83.

[8] Chen (ed.), *Zhongguo jiaju shi*. Beijing: Zhongguo qinggongye chubanshe, 2009. Furniture is also occasionally covered in the rich account of Yun Shen, *Zhongguo xiandai sheji guannian shi*. Shanghai: Shanghai renmin meishu chubanshe, 2017.

others, and of the political system that governed their lives.[9] Concerned about more than aesthetics, they connected standardisation to questions of the management of daily life, resource allocation, and the organisation of labour and space. This chapter traces two examples of these discussions. The first focuses on the city of Tianjin, a former treaty port some 80 miles southeast of Beijing. During the 1950s, a group of municipal officials responsible for the management of industry and commerce worked to standardise furniture manufacture as part of their reorganisation of the city's carpentry businesses. For them, standardisation was a means to regulate labour, wages, and key resources, and break away from capitalist profit-oriented furniture production. The second example moves to the national magazine *Architectural Journal* (Jianzhu xuebao) where, between the late 1950s and early 1960s, designers and architects debated how furniture standardisation could help meet people's residential needs.

Both examples show that the material environments of the Mao era, at the level of policy and manufacture, resulted from diverse ideas about what a socialist society and its material culture might look like. China's central authorities may not have turned furniture standardisation and mass production into a symbol of 1950s post-war recovery in the same manner as governments in other countries did.[10] Nevertheless, people across the country linked furniture standards – in discourse and practice – to the party's promise of giving, or at least striving to give, 'the masses' equal access to similar living standards.

Furniture for a 'new society'

Propaganda posters indicated which items artists considered part of the ideal interior of an urban home for officials and workers in the early 1950s.[11] In the image 'Did you figure it out?' (Figure 8.1) a worker sits at the home table trying to work out a calculation. He and his wife live in one room. They do not have a lot of furniture, but they have a wooden bed, stool, side table, and a table probably used for

[9] Jie Li, *Shanghai Homes.*

[10] This included the Utility Furniture Program in the UK, discussed in Harriet Dover, *Home Front Furniture: British Utility Design, 1941–1951.* London: Scholar Press, 1991, design in East and West Germany, discussed in Katrin Schreiter, *Designing One Nation.* Oxford: Oxford University Press, 2021, and the increased production of light industry goods especially after 1956 in the Soviet Union, see Susan Reid, 'Communist Comfort', *Gender and History*, Vol. 21, no. 3, 2009, pp. 465–98; Christina Varga-Harris, *Stories of House and Home: Soviet Apartment Life during the Khrushchev Years.* Ithaca: Cornell University Press, 2015; Natalya Chernyshova, *Soviet Consumer Culture in the Brezhnev Era.* London: Routledge, 2013;.

[11] The CCP's classification system labelled most who lived in the countryside as 'peasants' although a majority of rural inhabitants did not engage in agricultural labour. On this process among papermakers in rural Sichuan see Jacob Eyferth, *Eating Rice from Bamboo Roots: The Social History of a Community of Papermakers in Rural Sichuan, 1920–2000.* Cambridge, MA: Harvard University Asia Centre, 2009, 123–4.

Figure 8.1 'Did you figure it out?', by Lin Shuzhong and Li Mubai, 1954, Landsberger Collection PC-1954–00

eating, as a desk, and as a workplace to sew, and a source of heating and warm water. Given that they have a sewing machine, theirs is a comfortable life.[12] The image 'Moving into a new house' (Figure 8.2), meanwhile, shows a young worker family readying their newly built apartment. The family is moving into one of the prestigious 'new villages' (*xincun*) built during the 1950s, with several pieces of new wooden furniture and quite a bit of spare floor space. Next to a radio and cupboard with two drawers and two doors, they too have a wooden stool used both for sitting and to climb onto the cupboard. These items – the bed, table, cupboards, chairs, and stools – appear in different combinations throughout posters of the decade and after. None were entirely new designs, but many became closely associated with life in the PRC.

The furniture trade and industry that manufactured such furniture was diverse and colourful. In the initial years after the PRC was founded, furniture was made much as it had been in decades before: in small workshops and furniture factories, many of which were family run. There were also larger urban mills founded during the Qing dynasty, in the early Republic, and during Japanese occupation. In northern regions, everyday furniture was often made using solid hardwood felled in local forests. Meanwhile in southern regions, carpenters also made furniture out of bamboo and rattan, which were local and affordable materials well-suited to the humid climate. Many carpenters had their own techniques, tools, and designs, appropriate for the kinds of timber they worked with.

People looking to acquire furniture had several options. Before 1949, department stores in larger cities sold a mixture of foreign-made furniture, 'Chinese-style' furniture, and 'Western-style' modern furniture produced in China.[13] There were also big furniture companies that had fashionable showrooms where customers could browse sample items,[14] though much of this was reserved for the more affluent citizens. Otherwise, people bought their furniture from local shops or carpentry workshops, from travelling salespeople who came to their neighbourhoods, or at nearby markets. Sometimes stores were affiliated to larger factories, some of which had machinery and did not produce all their furniture by hand. The majority of furniture though was still made manually and in smaller operations.

Most stores, workshops, and local factories continued their work during the early 1950s and many customers were satisfied with the furniture available. As postwar inflation stabilised, more people could afford to buy. Not everyone saw this diverse landscape of furniture production as something positive, however; municipal officials thought it was inefficient, uneconomical, difficult to manage, and a relic of China's past that should be overcome. They wanted to integrate workshops,

[12] Antonia Finnane, *How to Make a Mao Suit: Clothing the People of Communist China*. Cambridge: Cambridge University Press, 2023. See also, Andrew Gordon, *Fabricating Consumers: The Sewing Machine in Modern Japan*. Berkeley: University of California Press, 2012.

[13] Elizabeth Lacouture, *Dwelling in the World: Family, House, and Home in Tianjin, China, 1860–1960*. New York: Columbia University Press, 2021.

[14] For this kind of furniture store in Tianjin see ibid., 172–4.

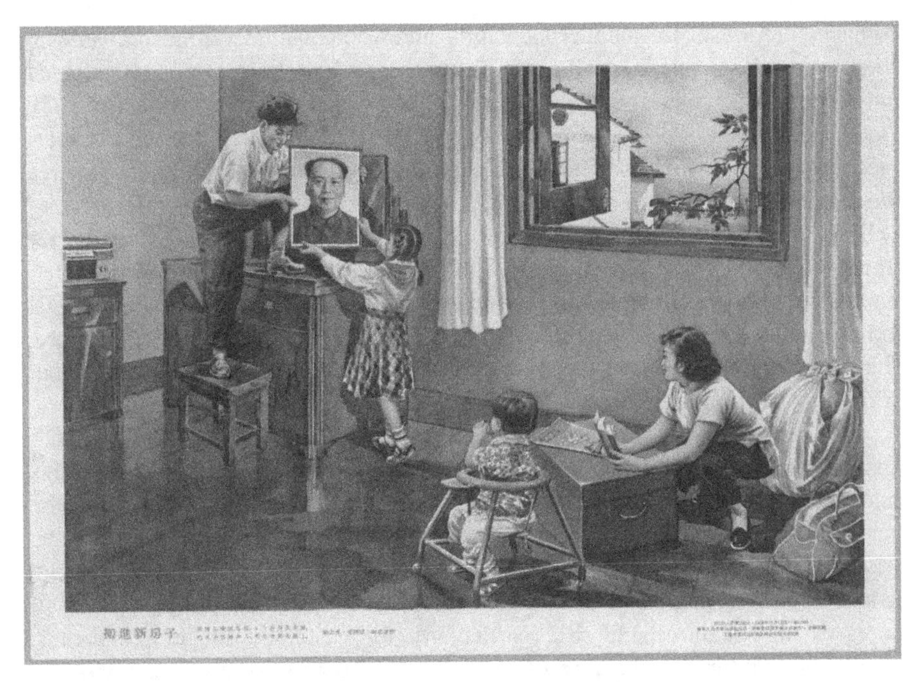

Figure 8.2 'Moving into a new house', by Xie Zhiguang, Shao Jingyun, and Xie Mulian, 1953, IISG BG D29/682, International Institute of Social History (Amsterdam)

factories, carpenters, peddlers, and store owners into a centralised planning process. Standardising products and their makers through a transformation (*gaizao*) of industry was supposed to 'serve the people', making more and more affordable furniture available to everyone. It was a process all trade sectors underwent during the 1950s, though the timeline depended on how crucial to socialist construction party-state authorities thought they were.[15] In the case of the furniture industry, government officials aimed to merge furniture workshops into bigger cooperatives and small factories into large joint state-private managed enterprises or state-owned factories with the resources to develop into fully mechanised units. These could produce furniture in large batches and at the speed required in an industrialising country.

The transformation of industries produced a few well-known factories in China's major cities. In Beijing, this included the Beijing Timber Mill and Beijing Northern Suburbs Mill, which were both established in 1952 and 1956 respectively

[15] On the history of 'transformation' see the articles in the special issue 'The Great Spoliation: The Socialist Transformation of Industry in 1950s China', *European Journal of East Asian Studies*, Vol. 13, No. 2, 2014; and also Xiaocai Feng, 'Rushing Toward Socialism: The Transformation and Death of Private Business Enterprises in Shanghai, 1949–1956', in *The People's Republic of China at 60: An International Assessment*, ed. William C. Kirby. Cambridge, MA: Harvard University Asia Centre, 2011.

from a merger of several smaller local mills and wood manufacturing companies.[16] There were also a select few flagship factories specially built in the 1950s, such as Beijing's Guanghua Timber Mill.[17] Guanghua was the flagship project of CCP Vice-Chairman and People's Liberation Army (PLA) chief, Zhu De who chaperoned the mill's construction. Guanghua extended over some 710,000 square metres and was equipped with an own product design room, machinery design rooms, as well as laboratories for physics, chemistry, and material sciences.[18] Called integrated wood processing factories, these mills processed large quantities of timber, manufactured boards of all kinds, researched and developed materials and designs, and produced as well as distributed a variety of timber goods, including furniture. They showcased the young government's accomplishments, curating occasional exhibitions of furniture prototypes to highlight their work and products.

An equally important element of transformation – and far more widespread – was the reorganisation of the many small private furniture workshops and stores that had long furnished people's homes. Transformation in this case included the standardisation of wages, working hours, prices, and designs. Local governments presented this as advantage and safeguard for ordinary customers. From early on, central directives instructed local governments to consider the ubiquity of small stores as an uncontrollable risk, particularly for customers of limited means. Regarding furniture, as the case of Tianjin will illustrate, local officials worried that a competitive market environment would allow savvy, dishonest carpenters to make unrealistic promises to their customers and then rob them of their hard-earned wages by not delivering goods of the anticipated quality.[19] These were exactly the kind of practices state socialism promised to eradicate, especially from 1953 when the government carried out the Five-Anti Campaign to clamp down on bribery, tax evasion, theft of state property, and cheating on government contracts.

Standardising furniture to manage labour: The case of Tianjin

Tianjin was a significant urban centre with a sizeable furniture industry. Almost four years after the PRC was founded, the Tianjin municipal government estimated that there were some 846 small wood manufacturing units, mostly privately-run,

[16] Beijing gongyezhi jiancaizhi bianweihui (ed.), *Beijing gongyezhi – jiancai zhi*. Beijing: Zhongguo kexue jishu chubanshe, 1999, 249–51.

[17] 'Mucai jiagong ji jiaju gongye' in *Jianguo yilai de Beijing chengshi jianshe ziliao*, Vol. 7, 'Jianzhu cailiao', 1994, 64.

[18] Guanghua Mill was placed under the leadership of Wang Kai (1917–2006) who had obtained an MA in Forestry from Michigan University and returned to China in 1947. Wang Shen, 'Huainian fuqin – 'Yang mujiang' Wang Kai', *Zhongguo linye chansheng*, No. 6, 2011, 44–5, 44. On the layout of the mill see Peng Liang, *Zhong-Xi fang sheji jiaoyu moshi de bijiao yanjiu*, 159.

[19] Tianjin Municipal Archives (TMA) X0077-C-001893-010, Tianjin shi renmin zhengfu gongshangju, 'Muqi jiaju shengchan gaijin gongzuo baogao', December 1953.

employing nearly 3,500 people.[20] Together with the employees of bigger factories, Tianjin therefore likely had well over 5,000 people working in wood manufacturing, of which many would have been involved in furniture production. Within only a few years after 1949, this municipal furniture industry quickly turned into a major target for socialist transformation and standardisation.

When the city's officials first began their surveys of local furniture businesses, their reports identified several practices they considered in tension with socialist theory and practice. Private furniture businesses were accused of underreporting on the amount of furniture they produced and overreporting on the number of workers they hired. Both were strategies to manipulate account books in companies' favour. Reports also accused carpentry businesses of 'theft of labour and raw materials' (*tougong jianliao*), saying they had economised on materials and sold customers pieces that only seemed of a high quality but really were not.[21] A third practice government officials were suspicious of was one businesses commonly used to remain profitable: restricting employment, using temporary labour, and determining wages case by case.

Many of these practices could indicate the kind of dishonest business operations officials were so worried about, but there were alternative explanations for these practices as well. In the case of so-called 'theft of labour and raw materials', for instance, the problem was often that many small enterprises did not acquire and store large amounts of raw material at any time. They bought materials when needed at local markets. Material quality could therefore fluctuate, depending on what was available.[22] Some carpenters also used materials that state-owned mills sold off as waste. This material was not always poor, but, when it was so, Tianjin's bureaucrats saw the practice as dishonest towards customers.[23] Similar explanations existed in the case of temporary labour. Many businesses had few permanent employees, were family run, and therefore did not operate on a transparent wage system for family members who worked for the business. They often also used temporary labour to help with larger commissions, but officials decried this reliance on temporary labour as excessive. In one report, the smaller Xinhua workshop served as a case in point. The workshop had six apprentices plus about two dozen regular employees. In early 1953, it had accepted a commission for 2,000 sets of school desks and chairs, to be delivered within two months. To cope, the workshop managers hired 200 temporary workers. City officials maintained, however, that temporary workers presented a risk to product quality. The unknowing customer paid the normal price, assuming a skilled worker made their furniture.[24] This charge was not without its ironies. During the Mao era, many state-owned factories relied

[20] Ibid.

[21] On the accusation of 'theft of labour and raw materials' see Eyferth, *Eating Rice from Bamboo Roots*, 121.

[22] TMA X0283-C-001058-005, Tianjin shi di san shangyeju, 'Muqi jiaju hangye xiaozhuan', September 1961.

[23] Lin Gang, 'Zai jiaju zhizao chang li', *Renmin ribao*, 17 January 1957, 2.

[24] TMA X0077-C-001893-010, 'Muqi jiaju shengchan gaijin gongzuo baogao'.

on temporary labour often for the same reasons that private businesses had: it kept costs low.[25] In this particular context, however, officials equated temporary labour in a private business with inadequate skills and interpreted Xinhua's decision to take a commission they could not manage in-house as an example of capitalist greed and labour exploitation.

Municipal officials were not wrong to worry that private businesses' flexible approach to working hours, prices, and designs gave them some advantages over state-owned factories. For example, many small shops vied to offer the best possible service. As one report dryly remarked, carpenters could be 'nimble' in their service attitude, going the extra length for customers, which was something state-owned factories could or would not do. Some gave six-month warranties, or they were more willing to visit customers' homes to make repairs when needed, even long after the sale.[26] Shops were usually open twelve hours a day and did not have firm working hours, so when orders were plenty, they would work longer and if needed they would call on family members to help. They were also accommodating about designs: many worked to order, often offering unique designs specific to individual workshops and their craftsmen. By contrast, state-owned factories were not meant to offer such services – or at least not to the same extent. Their product portfolios were more restricted, often already limited to producing larger quantities of selected designs, and they offered less variety and fewer adjustments to suit customer needs.[27]

Questions of wages, labour time, material use, and product specifications were thus closely connected, in businesses' day-to-day operations and in the eyes of city officials. Wages across employees in different workshops doing similar jobs were often uneven and sometimes a source of discontent. Businesses used wages to attract talented carpenters, but from the government's perspective everyone making similar chairs should earn the same wage. This wage should be determined by whether their labour was unskilled, semi-skilled, or skilled. That was easy for government officials to demand, but difficult for businesses to implement. Similarly looking chairs were not necessarily similar in make and could require varying levels of expertise and technique, a factor official reports seldom accounted for. The next problem was that too many design options and too much furniture variety on offer with different producers made it impossible for state officials to standardise prices in any simple way. If one chair was unlike the next, how would their respective prices be aligned? This had become clear, for example, when a district public affairs office ordered chairs for its rooms. The public affairs officials thought the price they agreed on was for chairs that had backrests measuring 2.7 *chi* (one *chi* equalling roughly a third of a metre). The company instead manufactured backrests measuring 2.6 *chi*; a difference of circa 3cm. The officials felt they had been

[25] On temporary labour in Tianjin see also Jeremy Brown, *City versus Countryside in Mao's China: Negotiating the Divide*. Cambridge: Cambridge University Press, 2012.
[26] TMA X0283-C-001058-005, 'Muqi jiaju hangye xiaozhuan'.
[27] TMA X0077-C-001893-010, 'Muqi jiaju shengchan gaijin gongzuo baogao'.

cheated and filed a complaint, but the company argued that this was how they always made their chairs. Presumably, nobody had thought to ask what the exact measurements would be, assuming that the category of 'chair' meant the same thing to everyone. In the atmosphere of the time, city officials took this as an example of cheating and profiteering, which was something they argued better quality control could have averted. After all, the manufacturer had saved a substantial amount of material making shorter chairs.[28]

These problems were not specific to Tianjin; officials in other cities encountered similar concerns. In Tianjin, they tried to deal with them systematically, hoping to simultaneously address issues of 'quality control' and 'efficiency' (as they understood it). In late 1953, the Municipal Office of Industry and Commerce produced a city-wide policy that would integrate the standardisation of furniture designs and processes of production. There were multiple steps to this process of policymaking. First, officials from the Office of Industry and Commerce organised a discussion meeting with selected members of labour and management from the city's furniture industry. In this, they followed the mass line: new policies should originate in consultations with representatives of those groups of the 'masses' whom the policy concerned. During the meeting, participants established how different workshops operated, the prices they charged, how much workers earned, and what kind of materials they used. Next, they organised a larger conference on wood furniture production and economisation. At the end of the conference, participants agreed on an 'improvement plan' to be overseen by a production improvement committee in charge of supervision and investigation of all furniture producers. Finally, they assembled a team of seven people from different factories, all of whom were versed in technical drawings. This team produced a set of 64 furniture items, including tables, desks, and shelving for offices and schools, as well as beds, wardrobes, tables, and other items for private spaces such as homes and dormitories. Compiled over a month, drawings were as precise as possible, including exact exterior, interior, and surface measurements, construction details, and specifications of which and how much material to use.[29]

With this material in hand, they then set out on the next phase to determine the relationship between these objects and the labour needed to make them. Officials consulted with selected municipal districts that had in turn collected data about work patterns, personnel, and wages from their districts' furniture manufacturers. They determined the amount of 'standard work hours' each of these furniture pieces would require and how much each piece should cost. The calculations presumed an eight-hour workday for a permanent employee as one standardised *gong* ('one workday').[30] How much an actual hour of work on each furniture item was worth

[28] Ibid.

[29] Ibid.

[30] The unit of 'one workday' stems from the imperial period and, as Lothar Ledderose explains, it measured a day lasting from sunset to sundown in order to measure 'human labour in the same way that material was measured'. Ledderose, *Ten-Thousand Things*, 137.

in this metric depended on whether it was made by a skilled worker, apprentice, or temporary worker, and whether the product was made manually or using machines. Eight hours of skilled manual work equated with one 'workday'. An apprentice who did eight hours of manual work had worked 0.75 'workdays'. Eight hours of skilled work involving machinery came to 1.13 'workdays', and for an apprentice it was 0.85 'workdays'. The amount of 'workdays' needed for each item of furniture was then also specified: a chair would need one 'workday', a double bed 2.5 'workdays'. The price would be calculated accordingly, with allowances for criteria such as wood quality.[31] Linking furniture measurements and material specifications to labour hours, level of skill and type of labour, the plan combined payment by piece with an eight-hour workday.

In these plans, Tianjin officials approached the transformation of local industry as a modular process of standardisation that connected labour, material, and design. It is unclear how far this plan was implemented and to what effect. In many aspects, the furniture items listed aligned with later records of technical standards in other localities: one-drawer and two-drawer desks, square dining table, square stool and rectangular bench, large wardrobe with two doors or with two doors and drawers, single and double beds, all kept simple with little embellishment. When it came to aspects such as standardised wages, however, the plan was soon outdated. Structures of production and wages shifted across the 1950s and 1960s, in the furniture world as across all trades. In summer 1961, Tianjin's Third Municipal Bureau of Commerce reported that the payment by piece, or 'piecework wage', embedded in an eight-hour workday had resulted in higher output in many cooperatives in 1956. Workers earned a decent average salary of about 80 yuan per month and could increase this if they manufactured more pieces. Officials thought that this system had mostly worked well.[32] During the Great Leap Forward (GLF), Mao then heavily criticised the payment by piece system, and it was eliminated. Everyone now received a standard wage; on average 65 yuan per month, depending on skill levels.[33] This was lower than earlier wages, reflecting the fact that the government considered furniture, as part of light industry, less significant than heavy industry. Productivity was no longer relevant to income, and reports noted that this resulted in an increase in the number of 'unhappy workers', leading to an official estimate that overall work output had halved compared to the early 1950s.[34]

Reports suggest that, over the course of the 1950s, different attempts to standardise products and production led to adverse results. For one, service quality decreased. Officials reasoned that this was because people had more incentive to produce, but less incentive to sell and to make furniture available to customers. To

[31] TMA X0077-C-001893-010, 'Muqi jiaju shengchan gaijin gongzuo baogao'.
[32] TMA X0283-C-001058-005, 'Muqi jiaju hangye xiaozhuan'.
[33] Ibid.
[34] Ibid.

improve official oversight, the unity of production and retail sales, characteristic of small commercial workshops, though also found in larger factories, had been severed. Those who produced furniture now had less contact with customers. Those who sold it had little connection to the products they sold. Officials also worried that workers paid no attention to how much material they used; a fact they blamed on workers no longer caring about the net costs and overall profitability of items. They had calculated that the ratio of wood waste in timber processing was up by a third. This was a poor result especially after party leadership and local governments had spent much time advocating better usage of wood waste. It also meant that fewer items of furniture were produced from the available material.[35]

The city's residents, meanwhile, often complained about a lack of variety and quality of whatever was on offer. Many had to wait a long time to receive new furniture as part of the work-unit distribution system, and many carpenters either had less incentive to meet customer demand or found it difficult to do so within the given restrictions. Fewer products and less variety also met with more demand. In summer 1960, officials at the Tianjin Municipal Department Store warned that furniture shortages were being exacerbated by the fact that there was a sizeable group of people who had benefitted from the reorganisation of the work force during the 1950s and now had more 'purchasing power' that they wanted to invest in new furniture.[36]

Municipal officials' search for ways to address these production shortages exposed some of the paradoxes of their earlier standardisation attempts. One obvious solution to the problem of shortage was to encourage people to consult repair services.[37] This saved labour and it was cheaper for customers, but the structure to enable or incentivise this kind of repair work was no longer in place. The fact that people found it difficult to have furniture repaired was a result of the transformation of carpentry workshops. Repairs had been a traditional and established service provided by carpenters into the early 1950s, but they declined during the first five years after 1949. Where previously Tianjin had had some 424 repair businesses with 766 employees who could service homes and fix furniture, by 1961 there were only 19 repair services with 175 employees. Desperate to fix furniture shortages, in early 1961, the Tianjin Third Municipal Bureau of Commerce decided to allow people who had once run repair shops not just to return to doing so, but to receive increased individual payment for their services once more.[38] In the case of Tianjin, then, a by-product of attempts to plan and standardise at different levels was a temporary return to some of the structures in place already during the early 1950s.

[35] Ibid.
[36] TMA X0279-C-000414-018, Tianjin shi baihuo gongsi, 'Guanyu muqi jiaju shishang gongying ji huoyuan fenpei banfa de qingshi', June 1960.
[37] Joshua Goldstein, *Remains of the Everyday: A Century of Recycling in Beijing*. Oakland: University of California Press, 2020.
[38] TMA X0283-C-001058-005, 'Muqi jiaju hangye xiaozhuan'.

Standardising furniture to make space: The debate in the *Architectural Journal*

In their plan for the transformation of Tianjin's furniture industry, officials did not mention how the 64 furniture pieces they had designed could be arranged in different spaces, nor did they consider the link between design, healthy living, and productivity. Their collective method nonetheless resembled the approach to standardisation via design that people in urban planning, architecture, and design institutes took when they considered the role of furniture in one of the most pressing projects of the Mao era: the construction of residential housing.

In the second half of the 1950s, more architects and designers began to think about furniture design as a constitutive element of life in socialism. During the GLF, furniture came under the wider remit of plans to technologise and modernise everyday life. That architects and furniture designers should work together, with many architects also designing furniture, was nothing unusual. It had been common practice in Republican China and globally. After 1949, however, architecture and furniture were separated administratively: architecture came under the auspices of the Ministry of Construction, furniture became a concern of the Ministry of Light Industry and the Handicraft Administration, and the timber industry was overseen by the Ministry of Forestry. The late 1950s therefore saw attempts to reconnect people working in different areas as part of the state's policy of 'integrated use' (*zonghe liyong*) of the country's resources. The policy, as Sigrid Schmalzer has highlighted in her analysis of Mao-era systems thinking, applied to all aspects of industry, handicraft, and life more generally, as it called on institutions and individuals to strive to maximise usage of all resources – objects, people, spaces, materials, and manufacturing capacity.[39]

Integrated use found specific application in conversations about furniture and housing. The editors of *Architectural Journal*, which launched publication in 1955, classified furniture as essential to the integrated use of residential spaces. For them, this meant working out how to design furniture that was economical to produce and how to incorporate it into urban 'small-space residential housing design' (*xiao mianji zhuzhai sheji*). Contributors to *Architectural Journal* who put forward proposals were often based in major universities or research centres. Shanghai's Tongji University and Beijing's Tsinghua University had departments for design research. The Central Academy of Crafts and Design, established 1956 in Beijing, opened an interior design research unit shortly after. City and regional governments, meanwhile, set up design institutes throughout the 1950s.[40]

[39] Sigrid Schmalzer translates *zonghe liyong* as 'integrated use' and explains the concept in her talk 'Connecting the Dots: A History of Systems Thinking in Chinese Agricultural Science and Politics' (www.hkihss.hku.hk/en/events/mmea-lecture-by-sigrid-schmalzer-20210303/). Joshua Goldstein translates it as 'comprehensive use' in the context of Mao-era recycling, where the emphasis was on fully using and reusing every bit of material possible. Joshua Goldstein, *Remains of the Everyday*.

[40] The infrastructure of design institutes is discussed in Charlie Q.L. Xue and Guanghui Ding, *A History of Design Institutes in China: From Mao to Market*. London: Routledge, 2018. Esp. chapter 2, 'The emergence of state-owned design institutes'.

A growing number of articles in *Architectural Journal* called readers' attention to the relevance of furniture and its standardisation in architectural practice during the late 1950s. Authors highlighted what they considered an increasingly pressing question: how to reduce the minimum amount of residential space per capita by maximising how space was used. Doing so had become a necessity in major urban centres because there was not nearly enough housing for the many people who needed it. With housing in excess of demand, architects and planners spent the early 1950s trying to figure out how much residential space people needed. One question was what constituted reasonable need. The other how to use this assumption of reasonable need to determine who should receive how much space while construction of new housing was ongoing. As two authors, Ye Zugui and Ye Zhoudu, wrote in February 1958, 'small-space' living 'saves resources, maximises investment, decreases costs, and economises land use'.[41] *Architectural Journal* was thus full of articles about how to divide the interior of new residential buildings to fit in a maximum of units and accommodate as many residents as possible.

This was not the first time architects tried to work out standardised ratios for space per capita. Prior to the GLF, a consensus had emerged among many contributors to the journal that 4m^2 per capita should be considered the ideal in any housing unit. Compared to similar space per capita calculations in other countries, 4m^2 per person was extremely small. In the UK in 1961, for example, the Parker Morris Committee's report on housing needs recommended a minimum of 17.5m^2 per person in a four-person flat, noting that the 'present average', considered too low, was 15.5m^2.[42]

Yet even this 4m^2 average was more aspiration than reality for most Chinese at the time, and architects knew this. By the GLF, however, with pressure on space higher than ever and many people waiting for housing, some looked for ways to work around the 4m^2 average. For some, furniture was the key to more planning flexibility. Ye Zugui and Ye Zhoudu, for instance, argued in early 1958 that the 4m^2 ideal should not be seen as written in stone. It all depended on how space was used and what kind of furniture was necessary. Eight people in 15m^2 or five in 8m^2 might not be much space per person, they wrote, but it was possible with good furniture placement, provided one used bunk beds, slim furniture designs, and worked with an appropriate rectangular floor layout. Their point of departure was per capita furniture need, not per capita space. It was a radical proposal for integrated use, not least because they also maintained that forcing people into extremely tight spaces would save on heating in winter (and one might simply use fans in the summer heat). Still, their contribution highlighted questions central to this chapter: what role would furniture play in housing planning? How might one determine per capita

[41] Zugui Ye and Zhoudu Ye, 'Guanyu xiaomianji zhuzhai sheji de jinyibu tantao', *Jianzhu xuebao* (JZXB), 1958, Issue 2, 30–6, 30.

[42] Ministry of Housing and Local Government, 'Homes for Today and Tomorrow'. London: Her Majesty's Stationary Office, 1961, 33–5.

furniture needs? What would this mean for how and what kind of furniture was manufactured?

For architects, the initial obstacle to standardisation was old furniture. Many new-built apartments could not accommodate a full set of furniture, including a bed, chair, table, cabinet, and wardrobe.[43] Some residents of coveted apartments in Shanghai's newly built Caoyang Village were unhappy because the space was too small for the things they already owned.[44] In 1959, interior architect Zeng Jian (1925–2011), then based at Shanghai's East China Industrial Architecture Design College, criticised that many architects thought about space per capita, building material, and issues of construction, few about what sort of furniture people might put into the rooms they built.[45] In the first years of *Architectural Journal* after 1955, some architects had included furniture arrangements in their drawings of individual units, to show how to maximise by fitting a bed, table, chairs, and perhaps a chest of drawers, wardrobe, or bookcases into one room. Yet they seldom specified furniture dimensions, making the inclusion of furniture more symbolic than practicable. Most did not add furniture at all – it was left up to residents to figure out what of the furniture they already had they could fit in.

Such inattention to the details of measurement and placement was not just a problem for those who already had furniture. In 1959, the Beijing Municipal Mill criticised that because there was no standardisation of batch production and little coordination between architects and furniture producers, even the new furniture they manufactured would often not fit.[46] They argued that failure to coordinate contravened the GLF goal of making residential housing 'practical, economical and where possible attractive'. It was also wasteful, running counter to the ethos of integrated use.

The project of aligning furniture with housing raised the same question that Tianjin's officials had had to contend with earlier: what should the starting point for standardisation be? In the case of the journal, it was anthropometry and the sizes and proportions of human bodies. To establish average measurements for common furniture items, one needed to agree on average measurements of potential users. In 1959, the magazine published excerpts from the Ministry of Building and Construction's internal reference material, which outlined the average height of an adult male Chinese citizen. Drawing on collective research conducted by the Beijing Medical College, Shanghai No. 1 Medical College, the Shandong Province Medical College, Fudan University's Anthropology Teaching and Research Group, they had established that the average adult man in socialist China measured 169cm in height (210cm with arms raised above the head) and 42cm

[43] For critical reader letters on this topic see, for example, Zhang Qingming, 'Shengchan huodong jiaju', *Renmin ribao*, 7 September 1956, p. 2.

[44] Jian Zeng, 'Duogongneng jiaju sheji', *JZXB*, 1959, Issue 6, pp. 32–3, 32.

[45] Ibid.

[46] Beijing mucai chang shejike, 'Jiaju sheji', *JXZB*, 1959, Issue 3, 19.

from shoulder to shoulder.[47] Having consulted furniture factories in cities including Beijing, Shanghai, and Wuhan, they recommended that the seating area of a chair should be 40cm from the floor, and backrest about 90cm (80cm for a foldable chair). A standard table should be 78cm high and about 90–100cm wide. The average size for a double bed would be 140x195cm and for a single 80x195cm.[48]

Such an approach had weaknesses. For one, these measurements were on the small side, and many Chinese people were taller. Anthropometrical research continued and a few years later, for example, the Ministry published elaborate calculations of average height and circumferences as part of their attempts to create architectural standards, now including women at an average height of between 153 and 158cm.[49] These revised compilations also tried to standardise according to assumptions about regional variations, or what they called 'local standard body height'. It was a challenging project, built on Republican-era legacies of anthropometry designed to strengthen Chinese nation and race.[50] Researchers in 1962 presumed that people's average height might be determined by the province in which they lived.[51] At that moment, determining regional variations might have seemed a promising way of centrally planning just how much material would need to be allocated to different provinces and localities for furniture production. These calculations, however, had difficulties accounting for historical patterns of migrations and for contemporary domestic migrations taking place across different regions in China as part of the CCP's campaigns and industrialisation projects.

Other architects agreed that users should be the point of departure for standardisation. In April 1959, in one of the early calls for a user-focused approach, Zhou Ming criticised that furniture designs at times thought more about individual items, and how to optimise their production, material usage, and sizing, than about how people lived, how furniture might be placed, and how measurements might correspond to common preferences for placement of different items. In other words, the size of a table might not just be determined by how the table might be used, but by where in the room it might be placed.[52] For Zhou, where to place the bed was the first question, from which all other design choices would then follow.

Zhou's argument echoed across many other contributions where the bed was the anchor for interior layouts, at the heart of the family's one-room home. Beds were indispensable even in designs that presumed one might do without most

[47] Jianzhu gongcheng bu jianzhu kexue yanjiuyuan gongye yu minyong jianzhu yanjiushi, 'Renti gebu chicun bili', *JZXB*, Issue 8. 1959, 21–4. The 169cm average for an adult man was lower than figures proposed during the Republican period.

[48] Ibid., 'Zhuzhai zhong changyong de jiaju chidu', *JZXB*, 1959, Issues 9 and 10. 1959, 76–9.

[49] Jian Qin and Yiqing Hua, 'Renti chidu de yanjiu', *JZXB*, 1962, Issue 10, 21–3, 21.

[50] Jia-chen Fu, 'Measuring Up: Anthropometrics and the Chinese Body in Republican Period China', *Bulletin of the History of Medicine*, Vol. 90, No. 4, 2016, 643–71, and David Luesink, 'Anatomy and the Reconfiguration of Life and Death in Republican China', *Journal of Asian Studies*, Vol. 76, No. 4, November 2017, 1009–34. I am grateful to David Luesink for his helpful suggestions.

[51] Qin and Hua, 'Renti chidu de yanjiu', 21.

[52] Ming Zhou, 'Zhuzhai sheji Zhong de jige shiyong wenti', *JZXB*, 1959, Issue 4, 37.

furniture items, such as Ye Zugui and Ye Zhoudu's 1958 layout. In March 1961, the architect Dai Nianci (1920–91) of the Beijing Industrial Architecture Design Institute recommended that architects organise rooms around two pivots: the bed and prospective residents' profession. Beds might then be placed with headboard to the wall extending into the room, or with head and one side of the bed pushed into the corner against two of the unit's walls.[53] The bed was also of concern to Zeng Jian who by 1961 had joined the interior design team at the Beijing Industrial Architecture Design Institute, and his colleague Yang Yun. They argued that the configuration that allowed 'entrance on both sides' of the bed might be 'convenient' but that placing the bed in a corner was better. This was a question not merely of how individual items of furniture were used, or their size, but also of 'traffic' and 'movement' across the room – a bed in the corner opened up more floor space. It also, Zeng noted, accorded with Chinese customs: 'old-style' beds were usually only open on one side, and a Northern *kang* was only accessible from one side.[54]

These contributions did what Zhou had called for: they based standardisation on investigations into people's homes. In 1960, their teams visited people's homes in Beijing to study furniture use. Furniture was now seen in connection with how people used spaces and with the kind of spaces people had. Zeng and Yang, for example, wrote that simply thinking about floor space was insufficient, one would have to think about room layout as well as placement and number of doors and windows.[55] They argued, as Zeng had done before in an article on modular furniture, that planners should move away from a focus on individual furniture items to be placed in the room, and towards integrating built-in furniture such as wardrobes or overhead cupboards (esp. over the bed), fold-out tables, sofa beds, and other multifunctional designs.[56] Such built-in furniture could be part of the original layout of the apartment, and not something residents would have to construct or acquire, thus making it an element to be factored into building budgets rather than a concern for residents later on.

Meanwhile, building on local investigations in Shanghai, Zhu Yaxin of Shanghai Tongji University's College of Design, proposed a quantitative approach to thinking about furniture as a component of standardisation in 1962. Zhu and a team of researchers started from the problem of how to translate residents' experiences living with furniture, their self-expressed needs, and their habits of using furniture in everyday life into quantifiable standardised unit layouts. Having gathered data on what furniture people had and how they used it in Shanghai residential homes, the team determined a set of 'most commonly used' furniture styles they had encountered during their visits. They took the measurements of these furniture items and

[53] Nianci Dai, 'Guanyu zhuzhai biaozhun he sheji zhong jige wenti de taolun', *JZXB*, 1961, Issue 3, 2–5, 5.

[54] Jian Zeng and Yun Yang, 'Guanyu zhuzhai neibu de sheji wenti', *JZXB*, Issue 10, 1961, pp. 12–14, 12.

[55] Ibid.

[56] Zeng, 'Duogongneng jiaju sheji'.

the average amounts of such furniture found in homes as a basis for their calculations. This, Zhu wrote, made up the average 'furniture surface area' in Shanghai homes. Then they calculated what they called a 'furniture coefficient' (*jiaju xishu*); the percentage furniture occupied in room footage overall, relating it also to the number of residents that occupied a unit. If three residents shared one double bed – for instance parents and children – that changed the calculation. The equation thus allowed readers to calculate the ratio of residents to apartment and furniture surface area. They determined that if furniture occupied more than 60 per cent of the surface area, the average resident would feel crowded (though it helped if they had two rooms to organise room 'traffic'). If furniture occupied less than 40 per cent, and people had all the furniture they needed, they had more space than necessary.[57] This approach was premised on the measurements of Shanghai furniture, but it could be replicated across the country if other teams used the same methods to calculate a furniture coefficient for their locality. For academics such as Zhu and his colleagues in other institutes, the case for standardisation was thus clear. The question remained, however, which parts were best standardised at the local level and which at the national level; which was a quandary not resolved for several years.

Conclusion

Debates of this kind continued throughout the Mao period, eventually leading into projects such as the Ministry of Light Industry's *Basic Measurements* and 1980s housing reforms.[58] The two case studies highlight the early PRC as a decades-long experiment in trying to build socialism, mostly under conditions of austerity and material shortages. In this context, the absence of comprehensive national standards for furniture production should not be seen as an example of failed state planning or inadequate modernisation, but rather as the result of a discussion concerning the purpose, desirability, and extent of standards, and the connection between national management and local needs. Debates about and shifts in techniques and strategies to standardise furniture – from components to materials and design – were projects to determine how best to turn political visions of socialism into material practice.

Although aesthetics were important, standardisation through design was not primarily focused on new socialist aesthetics. The concern was much more about how to try and ensure that goods were equitably produced, paid for, that they used as few material resources as cleverly as possible, and that they enabled similar living conditions for people across the country. For some, the solution was to have standards specific to different localities, and to retain maximum flexibility in resource use. Others thought it was better to have standards at the national level so as to ensure that resources were used the same everywhere.

[57] Yaxin Zhu, 'Zhuzhai jianzhu biaozhun he xiaominji zhuzhai sheji', *JZXB*, 1962, Issue 2, 25–9.

[58] Jennifer Altehenger, 'Modelling Modular Living'.

At the ideological level, Maoism therefore did not equate to a formalised or total vision for material life. A comprehensive vision that had no space for experimentation would have precluded the possibility of discovering better ways to standardise, and it would have contravened the project of moving between theory and practice in an ongoing quest to materialise socialism. In practice, the absence of comprehensive standards contributed to the challenge of manufacturing at scale in all but a limited number of large factories until the 1980s.

9

Mass Muralism and Mass Creativity in the Early PRC

CHRISTINE I. HO

When wu zuoren's *Peasant Painter* was exhibited as part of the Chinese works of art at the prestigious 1958 *Art of Socialist Countries* exhibition in Moscow, the painting conveyed two concepts of socialist creativity that were arguably divergent in symbolism and ideology (Figure 9.1).[1] On the one hand, *Peasant Painter* is a highly polished work of high socialism; finished in an art studio by a Beaux-Arts-trained painter who had studied in Paris and Brussels between 1930 and 1935, the oil painting is confident in its command of spatial recession, finely modelled portraiture, and naturalistic disposition of bodies and objects within space. Its skill represented socialist China's accommodation of Soviet socialist realism through institutional structures as well as style, exemplifying socialist realism's claims to an international lexicon for visualising united revolutionary struggle and history. On the other hand, *Peasant Painter* might be understood to be conveying the tenets of Maoist artistic practice. Departing from the international vision for socialist realism, *Peasant Painter*'s subject is depicted working outside the elite, academic, and studio-oriented practice that produced narratives of revolutionary history. Instead, socialist creativity is figured as rural and innocent, playful and enthusiastic: the peasant painter examines with pleasure a wall painting, his contribution the latest addition to a series of other murals that appear, partially obscured by the irregular staggering of the walls, but also partially visible, revealing images of towering stalks of corn, stands of sunflowers, and overgrown cabbages.

In part documentary in intent, the mural depicted within the painting was widely extolled and published as one of the achievements by a peasant named

[1] *Shehui zhuyi guojia zaoxing yishu zhanlanhui: Zhonghua renmin gongheguo zhanpin mulu* (np: np, c. 1958): cat. no. 77. On Wu Zuoren, see *Xueyuan yu yishu: Wu Zuoren bainian danchen jinian zhan*. Beijing: Zhongguo meishuguan, 2008.

Proceedings of the British Academy, **267**, 167–188, © The British Academy 2025.

Figure 9.1 Wu Zuoren, *Peasant Painter*, 1958. Oil painting, 114 x 146cm. From *Zhonghua renmin gongheguo zhanlanpin mulu* (n.p., c. 1958).

Wang Fuwen during the Pi County mural campaign during the 1958 Great Leap Forward (GLF) (Figure 9.2).[2] The mural depicts three furnaces, fanned by the mythical figure Sun Wukong, the Monkey King's stature exaggerating the expanse of the conflagration; below is a poem written by a Ding Shan, which reads, 'The coal-burning place can attack two million *jin* at once, like Sun Wukong overtaking the volcanic mountains!' In the version reproduced in the *Art of Socialist Countries* catalogue, Wu Zuoren's painting is accompanied by a brief discussion of the mural painting campaign, and its inclusion in the exhibition was a moveable proxy for a site-specific work and a stand-in for mass creativity of Maoist China writ large, requiring the artist to observe some responsibility to fidelity while replicating the peasant's mural.

Nonetheless, Wu Zuoren has taken liberties with the mural's slogan and figure, which is lightly revised around the face to allow for immediate iconographic recognition. In correcting the peasant painting, the painting acknowledges the conflicting modes of artistic technique and visuality that are presented as an undifferentiated

[2] 'Sun Wukong danzuo huoxianshan', *People's Daily* (August 30, 1958): 5th edition; in color in *Wenyibao* no. 18 (1958), 3, and *Meishu* no. 10 (1958): unpaginated.

Figure 9.2 Wang Fuwen, *Flame Mountain*, 1958. Mural, dimensions unknown. From *Wenyibao* no. 18 (1958), 3.

whole: the painting stages a meeting between academic and amateur technique, oil realism making possible the idealised peasant portrait, even while seamlessly imitating the mural's crude disjunctions of scale and irregular, uncertain, and almost primitive strokes. The mural is daubed, as the painting so explicitly demonstrates, with a paint brush dipped in tinted whitewash slopped around in rice bowls with water poured from a teapot, yet we are continually aware that nothing of the sort has been used in the painting's careful stroking forth of the amateur painter's brawny, ruddily radiant good health. One might be hard-pressed to locate the precedence of creativity in either mode, for where the easel painting succeeds through technical skill and academic refinement, the mural operates within a richly metaphorical register regardless of material paucity, employing legendary folk heroes to animate production quotas of the GLF with playful yet martial attributes.

Peasant Painter's depiction of the mural campaign becomes ever more ambivalent when one considers the version currently in the National Art Museum of China. This version has been repainted, removing the more famous mural in favour of a more conventional composition that stereotypes peasant painting, sandwiching a tractor driver between rows of wheat stalks and sunflower beds, and imitating the effect of peasant painting with a childishly scrawled sun. Lost in the repainted version is the insertion of playful imagery as figments of the imagination conjured forth from the peasant's repertory of folk tales, substituting a more workmanlike mural that has a ploddingly illustrative effect. In the version that represents Wu Zuoren's corpus, one might think of the reworked painting on the wall as subtly reinstating a return to a preference for the trained over unschooled hand, the precedence of the professional artist over the anonymous one.

Peasant Painter embodies the tensions that have surrounded contemporary debates over the legacies of Maoism in the global history of post-war art. Revisionist histories of post-war and contemporary art have underscored the disparate and often contradictory legacies of Maoist visual culture, legacies that are far from foreclosed and definitive, despite presumptions of the period's unceasing monotony. Where exhibitions of post-war art have sought to expand definitions of modernism by incorporating Mao-era painting and sculpture in order to emphasise the diversity of socialist realism beyond the Soviet centre, narratives of revolutionary visual culture have prioritised the visual ephemera of Maoism from China to the global South, from peasant painting, Red Guard prints, to big character posters.[3] Maoist art and visual culture has tended to be siloed within two distinct realms, divided between fine arts and mass arts, each entailing distinct consequences for interpreting the significance of Maoism within its global embrace. To understand Maoist art constituted within the domain of the fine arts is to see socialist painting as a cosmopolitan and modernist project, embedded within debates over the status of realism and its aesthetic effects in forging revolutionary consciousness; to understand Maoist artistic practices as part of global revolutionary culture is to see mass art as part of, and antecedent to, grassroots campaigns and guerrilla movements of post-colonial solidarity.

Reconsiderations of Cold War media histories have also shifted the balance from the elite production of official state culture in favour of mass art practices. The historiography of Soviet studies has long foregrounded the production of revolutionary consciousness through experiments in amateurism and the culture of

[3] Okwui Enwezor, Katy Siegal, and Ulrich Wilmes (eds.), *Postwar: Art between the Pacific and the Atlantic, 1945–1965*. New York: Prestel Pub, 2016; David Joselit, *Heritage and Debt: Art in Globalization*. Cambridge: The MIT Press, 2020, pp. 56–69; Flavia Frigeri and Kristian Handberg (eds.), *New Histories of Art in the Global Postwar Era: Multiple Modernisms*. New York: Routledge, 2021; Jacopo Galimberti, Noemi de Haro Garcia, and Victoria H. F Scott (eds.), *Art, Global Maoism and the Chinese Cultural Revolution*. Manchester: Manchester University Press, 2020.

agitation.[4] A developed body of cultural theory has illuminated the entwinement of avant-garde and mass agitational practices through, for example, the theorisations of media form and collective transformation in proletarian photography, while the propaganda posters have been examined for their creation of 'medium intimacy'.[5] Meanwhile, scholarship in China studies recently has seen a new, careful consideration of mass art practices that were enacted through media experienced collectively, from theatre, film, photography, song, radio, to museums.[6]

Current instrumentalisation of socialist visual culture compels a reconsideration about the historical significance of Maoist art. Until recently, scholarship on mass art (*qunzhong yishu*; also referred to as 'amateur art' (*yeyu yishu*), 'peasant painting' (*nongmin hua*), 'worker-peasant-soldier art' (*gongnongbing meishu*), and very rarely as 'proletarian art' (*puluo yishu*)) tended to sympathise with the plight of the professional artists who were pressed into service as uncredited technicians rectifying works made by the unskilled hands of workers, peasants, and soldiers.[7] Yet, at the same time, a double disappearance has been effected: as histories of post-war modernism and contemporary art have increasingly incorporated global Maoism, these anonymous artists have rarely been acknowledged in histories of Maoism, despite avowals to reframe histories from the grassroots. This chapter takes up this problem by examining the subject of *Peasant Painter*, the national mass mural campaign of 1958. Perhaps the most ambitious and radical mass art campaign of the Seventeen Years, the mass mural 'fever' (*bihua re*) sought to extend Maoism's radical promises, not only making non-artists the subject of art, but to transform them into agents of artistic creativity. This attempt to create artistic subjectivity, however, destabilised and underscored the hierarchies and values of the artworld: the mural campaign saw widespread dissension over the parameters of creativity and how Maoism would be made by artists. To make Maoist mass art was not merely to establish the category of propaganda, but to recognise the very social power held by art – not as an index of representational truth but of its wielding of taste as an instrument of cultural capital – and so to question the very values of the artworld itself.

[4] Margarita Tupitsyn, *The Soviet Photograph, 1924–1937*. New Haven: Yale University Press, 1996; Lynn Mally, *Revolutionary Acts: Amateur Theater and the Soviet State, 1917–1938*. Ithaca: Cornell University Press, 2016.

[5] Maria Gough, 'Radical Tourism: Sergei Tret'iakov at the Communist Lighthouse', *October* 118 (2006), 159–78; Robert Bird *et al.* (eds.), *Vision and Communism: Viktor Koretsky and Dissident Public Visual Culture*. New York: The New Press, 2011.

[6] Denise Y. Ho, *Curating Revolution: Politics on Display in Mao's China*. Cambridge: Cambridge University Press, 2017; Ying Qian, 'When Taylorism Met Revolutionary Romanticism: Documentary Cinema in China's Great Leap Forward', *Critical Inquiry* 46, no. 3 (2020), 578–604; Jie Li, *Utopian Ruins: A Memorial Museum of the Mao Era*. Durham: Duke University Press Books, 2020; Laurence Coderre, *Newborn Socialist Things: Materiality in Maoist China*. Durham: Duke University Press, 2021.

[7] Jonathan Spence and Ann-ping Chin, *The Chinese Century*. New York: Random House, 1996, p. 177.

Mass art in the early People's Republic (PRC)

Studies of mass art in China have largely extrapolated from the Hu County, Hebei Province peasant painting movement during the late Cultural Revolution. Ellen Johnston Laing's touchstone study of Hu County peasant painting exposed interactions between amateur and professional artists in what had been loudly trumpeted, and received, as the fullest potential of Maoism: peasants who had so benefitted from socialist redistribution of cultural capital and knowledge that they had become artists themselves.[8] Building upon Laing, Ralph Croizier demonstrated peasant painting's evolution into export art, tracking its afterlives under Reform-era capitalism, while more recent studies have emphasised the contexts of international solidarity, underscoring the receptiveness of international leftists to Maoist soft power.[9] Central to these studies is the unabashed hypocrisy of state rhetoric, the distance between theoretical celebration of the reversal of cultural hierarchies proclaimed in Mao Zedong's seminal 1942 discussion of art and literature at Yan'an, and the working reality of on-the-ground practices, as well as a clear-cut division between state instrumentality and constraints upon artistic freedom – both on the part of the professional artist who must subordinate themself to a peasant voice, and the peasant painter who must sell their work to promote a state vision.

Hu County painting was simultaneously a representative and extraordinary development within the longer history of worker-peasant-soldier art as the defining feature of Maoist cultural practice. Enthusiastic reception of peasant painting and subsequent scholarly critique homogenised other attempts in the prior three decades to create the radically distinct forms of expression, institutions of production and exhibition, and systems of appraisal that underlaid the category of mass art. In its broadest conceptualisation, mass art was creative activity, encompassing the humanistic spheres of literary, musical, theatrical, and visual production, conducted by workers, peasants, and soldiers who lacked the cultural capital necessary for full participation in what Arthur Danto has called the artworld. As this provisional definition suggests, the very constitution of a mass art required a wholesale revision of the values of art-making itself; an art made by and within the values of the 'masses' required a cognitive conversion into Danto's famous formulation: 'To see something as art requires something the eye cannot decry—an atmosphere of artistic theory, a knowledge of the history of art: an artworld'.[10]

[8] Ellen Johnston Laing, 'Chinese Peasant Painting, 1958–1976: Amateur and Professional', *Art International* 17, no. 1 (1984), 2–12, 40, 64; Laing, *The Winking Owl: Art in the People's Republic of China*. Berkeley: University of California Press, 1988, pp. 31–2.

[9] Ralph Croizier, 'Hu Xian peasant painting: From revolutionary icon to market commodity', in *Art in Turmoil*, ed. Richard King. Vancouver: University of British Columbia Press, pp. 136–65; Emily Williams, 'Exporting the Communist Image: The 1976 Chinese Peasant Painting Exhibition in Britain', *New Global Studies* 8, no. 3 (2014), 279–305.

[10] Arthur Danto, 'The Artworld', *The Journal of Philosophy* 61, no. 19 (1964), 571–84.

When the first Literary and Art Workers' Conference was held in July 1949, its leftist attendees arrived intending to perpetuate mass art campaigns begun in Yan'an.[11] In both Yan'an and treaty-port metropolises in Shanghai, artists and art critics had first- and second-hand knowledge of Soviet proletarian leisure clubs, the spaces that provided factory workers with resources to perform, paint, and poeticise.[12] By the early 1940s, art workers had carried out short-term experiments in founding amateur art groups in northeastern China, such as one amateur art group headed by the cartoonist Hua Junwu.[13] Yan'an cultural workers described a larger plan to create handbooks that would assist rural residents in producing visual propaganda in the absence of professionally-trained artists. Among these activities was compiling 80 circulars that demonstrated the principles of human anatomical drawing, which they cited as one of their major achievements in 1940.

Soviet amateur art handbooks provided the basis of mass art pedagogy, which Chinese cultural workers both built upon and questioned for their relevance to rural values. In January 1950, the recently reconstituted Lu Xun Academy of Fine Arts published a teaching manual titled *Short Term Art Training Class Handbook*, created from field experience during a six-month art course for factory workers in northeast China.[14] Divided into two sections, on fundamental artistic techniques (copying, sketching, basic human anatomy, basic perspective, composition, colour, illustration, narrative cartoons, and posters), and the tasks of functional art (blackboard design, art typography (*meishuzi*), enlargement of leaders' portraits, to the design of meeting rooms), the Handbook was dogged with a pervasive desire to create a mass art that both observed Soviet precedent, while creating a new convention for national mass art. In a section on copying, for example, the authors suggest basing works on Soviet examples, such as transforming *Soviet Red Army Defeats German Enemies* in order to create *Dongbei Air Union Defeats Japanese Enemies* (Figure 9.3). On the other hand, the authors criticised using Vera Muhkina's monumental 1937 steel sculpture *Worker and Kolkhoz Woman* as the basis for *Painting the Chinese People Constructing New China*, suggesting that the figures' physiognomy were too specific to the Soviet revolution, not suitable for the Chinese masses (Figure 9.4).

An early example of similar handbooks that would follow in subsequent decades, the Lu Xun Academy's *Short Term Art Training Handbook* teaches amateur art as a set of fundamental skills that would furnish the constant thrum of revolutionary education throughout the Maoist visual and material everyday surround. Correspondingly, the goal of the art lessons in the *Handbook* serve three primary

[11] Zhao Shuli, 'Tan qunzhong chuangzuo', *Wenyibao* 1, no. 10 (1949), 6.
[12] On Soviet workers' art programmes, see Evgeny Dobrenko, *The Making of the State Writer: Social and Aesthetic Origins of Soviet Literary Culture*. Stanford: Stanford University Press, 2001.
[13] Wang Peiyuan, *Yan'an Luyi Fengyunlu*. Guilin: Guangxi shifan daxue chubanshe, 2004.
[14] Luyi meishu bu (ed.), *Jianyi meishu gongzuo shouce*. Shenyang: Xinhua shudian, 1950.

Figure 9.3 'Dongbei Air Union Defeats Japanese Enemies', from Luyi meishu bu, ed., *Jianyi meishu gongzuo shouce* (Shenyang: Xinhua shudian, 1950).

Figure 9.4 'Painting the Chinese People Constructing New China', from Luyi meishu bu, ed., *Jianyi meishu gongzuo shouce* (Shenyang: Xinhua shudian, 1950).

channels of socialist education: updating the village or factory's public blackboard (*heiban bao*), painting political leaders' portraits, and designing meeting halls. The creation of these ephemeral environments, now lost, cannot be overstated: a presumption of low literacy rates meant that the village needed to receive news and information by pictorial means, created by someone who had some experience with drawing, composition, and colour. Alongside the expansion of radio and cinematic programmes in the countryside, the expanse of the blackboard wall, divided into several sub-units, decorated with auspicious imagery and small stylised portraits, and written in art typography (*meishuzi*) became the interface for communication, where knowledge of daily slogans, awards, and campaigns were disseminated and collectively examined. A representative and underrecognised architectural feature of Maoist China's designed environments, the public blackboard recognised that the architecture of the village was a communicatory agent, in which designated public walls could be used to transmit messages.

Illustrations created for exercises throughout the *Handbook* underscore the paradoxes of transplanting urban art academy knowledge into workers' lives. Its lessons employed everyday objects plucked from mundane environments and prosaic circumstances, such as workers working on a lathe and picking through scrap steel, to analyse scale, proportion, and perspectival angles. Beyond hand-eye coordination as a problem of somatic discipline, the *Handbook* also lays out lessons on portraying revolutionary iconography and, in so doing, reveals the challenges of teaching the implied cognitive processes of art. The difficulty of concrete example and metaphorical expression is confronted in a lesson on posters, which explains that posters are temporally-fixed signs that employ symbolic form in appealing colours to explain an immediate political or social problem. The authors propose an example of creating a thematic poster: how should a poster depict Sino-Soviet friendship? As it guides the amateur art worker through steps of conceptualising a poster, the *Handbook* acknowledges the difficulties of revolutionary knowledge. First, the worker must identify symbols of China (Chinese flag, Qianmen, Chinese people); next, they should specify symbols of the Soviet Union (Soviet flag, Soviet people, Kremlin); third, they must pinpoint symbols of friendship (shaking hands, walking forward in unison, two overlapping heads facing the same direction) (Figure 9.5). Once combined, the synthesis of acceptable symbols would convey a satisfactory vision of 'Sino-Soviet friendship'.

At the same time, not all combinations were equally successful: the *Handbook* warns of a political mishap in a poster of a Chinese and Soviet girl performing folk dances, who appear to be inadvertently trampling flags under prancing feet. Posters, in other words, required pre-existing knowledge of the conventions of pictorial significance beyond the worker's ken. While the Lu Xun Academy cultural workers attempted to standardise the creation of popular art, or self-generated propaganda, the fuzziness of the term that they insisted upon, 'art', and its attendant implications of cognitive transformation and symbolic code could not be wholly subsumed into the schematic processes and forms provided by the *Handbook* and other similar manuals.[15]

[15] Lu Xun meishu xueyuan, *Qunzhong meishu shouce*. Shenyang: Liaoning huabaoshe, 1958; Meixie Guizhou fenhui chouweiyuan, Guizhong sheng qunzhong yishuguan (eds.), *Duanqi meishu xunlianban jiaocai*. Guiyang: Guizhou renmin chubanshe, 1959.

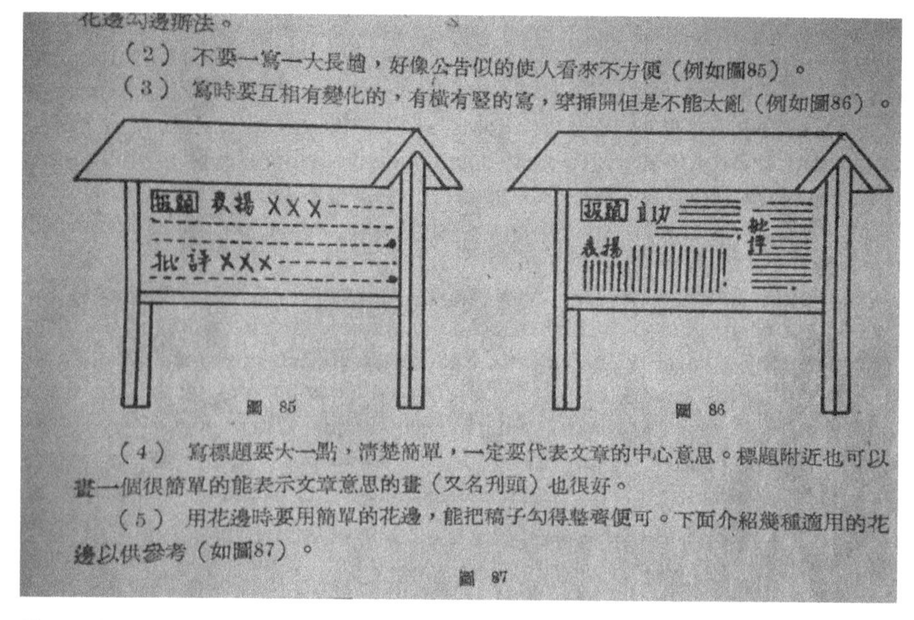

花邊勾邊辦法。

（2） 不要一寫一大長趟，好像公告似的使人看來不方便（例如圖85）。

（3） 寫時要互相有變化的，有橫有豎的寫，穿插開但是不能太亂（例如圖86）。

圖 85　　圖 86

（4） 寫標題要大一點，清楚簡單，一定要代表文章的中心意思。標題附近也可以畫一個很簡單的能表示文章意思的畫（又名刊頭）也很好。

（5） 用花邊時要用簡單的花邊，能把稿子勾得整齊便可。下面介紹幾種適用的花邊以供參考（如圖87）。

圖 87

Figure 9.5 'Blackboard design', from Luyi meishu bu, ed., *Jianyi meishu gongzuo shouce* (Shenyang: Xinhua shudian, 1950).

Theorising amateurism and amateur creativity

Appraising the creative act as a function of making propaganda, despite attempts to regulate its performance, was already embedded in the first exhibitions of amateur art in the PRC. Among the many exhibitions celebrating Chinese Communist victory in the late months of 1949 was a worker's art exhibition organised in Beijing. Ten art workers organised art groups in factories and industrial plants, numbering approximately 20 members, to make wall newspapers (*qiangbao*), posters (*zhaotiehua*), illustrations (*chatu*), cartoons (*lianhuan hua*), art typography (*tu'an zi* or *meishu zi*), sketches (*xiesheng*), woodblock prints (*muke*), and to design leisure clubs. In December, these art groups exhibited their works in one of the earliest amateur art exhibitions, organised by the Beijing municipal art worker's association and the municipal Propaganda and Education Association, showing works with titles such as *Production is A Tool to Exterminate Robber Chiang (Kai-shek)* for one to two days in participating factories, from the Beijing quilt and clothing factory, Renli carpet factory, Qishibing factory, streetcar factory, People's publishing plant, telegraph station, Housheng match factory, Xinhua printing plant, Qinghe clay factory, Shijingshan steel factory, and Shijingshan electrical plant.[16]

[16] Hu Man, 'Beijing gongren yeyu huazhan ji', *Wenyibao* 1, no. 7 (1949), 25–6.

In an exhibition review, the critic Hu Man reiterates the central propositions of worker's art as acts of emancipation. As Hu explains, creating posters on vaccination such as *Everyone Prevents an Epidemic*, or on economising electricity, such as *Save, Then Save More: Change Large Bulbs for Smaller Bulbs*, achieved the goals of amateur art by engaging workers in self-education, while carving woodcuts from photographs of Marx, Lenin, Stalin, Mao Zedong, and Zhu De allowed workers to recognise their political leaders. Worker's art was much more than the mere creation of propaganda; amateur art addressed the problem of visual art's inherent passivity, presuming that visual art was an inert object received by an apathetic viewer who required interpretation. As such, worker's art turned art from object into making, into an active process of participation and engagement.

Two central issues figured prominently in Hu's review: first, citing visitor feedback, he identified works that lacked ideological content and technical facility. Responding to these criticisms, Hu countered that the standards of professional artists should not be applied to amateur artists. Even if they required more guidance on artistic technique or night school to expand their political and cultural resources, amateur artists had used 'their own hands and heads to reflect their own production and lives'. As in Land Reform theatre, the aestheticisation of propaganda production would entail sensuous contact with political knowledge.[17] Consequently, the work of the cultural cadre was not to apply standards of a fetishistic art regime to amateur art. Amateur art advanced two radical features that contravened earlier tenets of artistic judgement: it removed technical facility as a marker of excellence, and it reconfigured standards for appraising experimentation in art-making. Hu, among other Marxist critics, sought to justify these as the characteristics of amateurism, the fullest expression of a specific form of creativity ascribed to class identity and social origin. In other words, labouring classes were endowed with a purity that extended to a noble, genuine, and indomitable source of creativity.

Perhaps one of the key historical events that sanctioned ideas of amateur creativity was the promotion of Taiping Heavenly Kingdom murals.[18] First 'discovered' as art in 1952 100 years after their creation, the murals commissioned by Taiping rulers during their occupation of southeastern China were described as visually exemplifying the proto-revolutionary movement.[19] These murals, of fragmentary landscape paintings and auspicious animals such as cranes, deer, peacocks, and mandarin ducks, were painted without a sense of consistent perspective in a style that was characterised as naive, awkward, and childlike.[20] The Taiping murals of

[17] Brian DeMare, *Mao's Cultural Army: Drama Troupes in China's Rural Revolution*. Cambridge: Cambridge University Press, 2015.

[18] Audrey Spiro, 'Paintings of the Heavenly Kingdom: Rebellion and Its Conservative Art', unpublished paper, College Art Association (1992); Jonathan D. Spence, *God's Chinese Son: The Taiping Heavenly Kingdom of Hong Xiuquan*. New York: W.W. Norton & Company, 1996, p. 188.

[19] Luo Ergang, *Taiping tianguo yishu*. Nanjing: Jiangsu renmin chubanshe, 1959.

[20] Wan Qingli, *The Century was Not Declining in Art: A History of Nineteenth-Century Chinese painting*. Taipei: Xiongshi tushu gufen youxian gongsi, 2005, pp. 174–7.

Nanjing and elsewhere were not only taken as evidence that political rebels could be keen patrons of art, in spite of decimating temples and other objects of cultural heritage in their wake, but that the Taipings recognised the artistic worth of the unnamed 'painting artisans' (*huajiang*) who had been disparaged as mere labourers in the hierarchy of the fine arts. The Taiping murals set historical and conceptual precedent for amateur artists, positing an innate urge for self-expression and collective decoration, even as their paintings were technically unskilled and underappreciated by the official artworld.

Between 1949 and 1952, virtually nothing that could be understood as approaching the category of 'fine art' was produced in the early PRC. Instead, art academies were reconstituted in military formation that were indebted to Yan'an-era mobilisation campaigns, with teams of artists forming amateur art clubs and study groups with farmers and factory workers. Under the leadership of the Ministry of Propaganda, each province subsequently opened a Mass Art or Culture Gallery (*qunzhong yishuguan*), which managed the relationship with the county-level cultural stations (*wenhua zhan*), frequently by converting ancestral temples or other former religious spaces into secular cultural spaces. The earliest was the conversion of the Imperial Ancestral Temple in Beijing into the Worker's Culture Palace, completed by International Worker's Day in 1950.[21] By 1953, according to the statistics provided in the first Five-Year Plan, the number of cultural galleries nationwide had reached 2,455, and there were 4,296 Cultural Stations. Following the launch of the Second Five-Year Plan in January 1958, most provinces had established Mass Art Galleries in time for the Spring Festival holiday in disused spaces such as temples, where exhibitions, as one typical report proclaimed, 'did not show the works of famous artists … but are full of the simple, awkward works of peasants'.[22] Alongside performances of dance, theatre, music, and political meetings, cadres also organised annual exhibitions of amateur-produced art, assisted by internal circulars on orchestrating mass art campaigns.[23]

The launch of the first Five-Year Plan in 1953 saw a dramatic reversal of the balance between amateur and fine art. Mass art, as process and object, enacted the politics of a mobilisational revolutionary state, but the technocratic goals of the first Five-Year Plan required a different cultural programme, one that could articulate the national identity of a bureaucratic socialist state through professional accomplishment, rather than amateurism.[24] Establishing a fine-art programme with Soviet expertise entailed not only the institutionalisation of a Soviet-style pedagogical system throughout the art academies, but a renewed emphasis on the technical ability to produce material-intensive easel paintings and familiarity with the socialist realist

[21] Chang-tai Hung, 'A Political Park: The Working People's Cultural Palace in Beijing', *Journal of Contemporary History* 48, no. 3 (2013), 556–77.

[22] 'Xiangcunli de meishu zhanlanguan', *Meishu* no. 6 (1958), 9.

[23] Zhejiang qunzhong yishuguan, *Qunzhong meishu gongzuo tongxun*, begun October 1956.

[24] Julia Andrews, *Painters and Politics in the People's Republic of China, 1949–1979*. Berkeley: University of California Press, 1994.

canon, knowledge that returned cultural power and authority to trained artists. Where academy artists had suspended prior artistic projects in favour of producing New Year's prints, for which they effected a deliberately naive style associated with rural visual traditions, they now returned from the field to studios to create an official state art, which was one that would be judged by international standards.[25] Mass-oriented activities continued in other forms, as professional artists continued to regularly go down to the countryside as part of their official duties and incorporated 'going to the countryside' (*xiaxiang*), or immersion in rural life, into their creative process, but amateurism was largely sidelined, as the state-building work of the fine arts took precedent over amateur creativity.

Mass muralism and the Pi County campaign, 1958

During the GLF, amateur art's marginalisation within the art world was confronted in radical form and practice, presaging mass art's ascendancy during the Cultural Revolution. Beginning in early 1958, art academy artists and village artists alike participated in a nationwide mural campaign, one that saw every public wall as potential for creating large-scale paintings, often spanning between two to three metres in length. The height of the GLF's mural campaign was concentrated between March and December 1958 as cadres competed with each other to create 'mural counties' (*bihua xian*).[26] As certain centres, such as Pi and Shulu counties in Jiangsu Province, acquired a reputation as a 'mural county', the murals constituted a key part of their local identity and would be promoted well into the early 1980s.[27] Stemming from the blackboard drawings that featured in village centres, the mural campaign sought to create an immersive visual environment for the continual agitation of the GLF's mobilisational politics.[28] At the same time, the mural campaign, and its promotion on a national level, also saw itself as the convulsive overthrow of what was seen as a retrenchment of conventional art world hierarchies. It was at both levels that Wu Zuoren's *Peasant Painter* acknowledged, even as it covertly denied, the dynamics of fine and mass art.

A key event was the artist Gu Yuan's *xiafang* visit in January 1958 to Dazhai village, Zunhua County, approximately 160 kilometres away from Beijing. Without

[25] Yan Geng, 'A New Earthly Paradise: Appropriation and Politics in Li Keran's Representation of Beihai Park', *Archives of Asian Art* 64, no. 1 (2014), 75–92.

[26] Andrews, *Painters and politics in the People's Republic of China,* pp. 225–7. Studies include Fu Xiaoyan, 'Shehui shijian zhong de Pixian "nongminhua huodong" yu "nongminhua"', (MA thesis: Central Academy of Fine Arts, 2007); Li Weiming, 'Dayuejin yundong zhong de nongmin bihua—Yi Jiangsu Pixian nongmin bihua wei zhongxiu', *Meishu xuebao* no. 4 (2009), 23–8.

[27] Feng Xianling, Ding Shangqian, 'Nongjia xihua huanletu—fang nongmin huaxiang Jiangsu Pixian', *Renmin ribao* (October 3, 1982), 2nd edition.

[28] Krista Van Fleit Hang, 'More, Better, Faster—the Ming Tombs Reservoir and a different path for Maoist culture', in *Literature the People Love: Reading Chinese Texts from the Early Maoist Period (1949–1966)*. New York: Palgrave Macmillan, 2013.

prior academic art education, Gu began his studies of printmaking in 1939 in the ad hoc circumstances of the Yan'an Lu Xun Art Academy, and Gu consolidated his capital as part of the Yan'an-trained faction of the socialist artworld through appointments at the People's Art Publishing House and at the Central Academy of Fine Arts.[29] In Dazhai, Gu elaborated upon a local custom of painting upon walls for local celebrations, including the Spring Festival, and painted a series of murals, daubed in red earth and coal dust with a house painting brush, which affected a consciously naive, crude, yet charmingly stylised quality attributed to folk art. Instead of apotropaic symbols, Gu applied the village mural style to GLF slogans. In one example, three peasants cross the Yangtze River seated on incongruous vessels of oversized sweet potato, corn, and sorghum; the masts are comprised of a hoe, shovel, and rake; and underneath, a slogan reading, 'A *mu* of land produces a thousand *jin* of the 'three freshnesses', potatoes, corn, and sorghum; with an abundant harvest of the three, the year's harvest is so good that it crosses the Yellow River to the Yangtze' (Figure 9.6). Gu's 32 murals were compiled in a modelbook, in which the artist Fang Ji described Gu Yuan as following in the footsteps of anonymous artisans of temple murals and Sino-Japanese War propaganda muralists.[30]

Hebei Province then became one of the foremost centres of the mass mural movement. To advance GLF cultural programmes, the Ministry of Culture convened the National Village Mass Culture and Art Work Conference in Beijing from 20 to 30 April 1958.[31] At the conference, cultural cadres reported on their successes. The head of the worker's club in Houqianzhuang, Gao Xueqian, reported on one of the first model cases that were carried out by cultural cadres in Changli County, Hebei Province, when 164 murals had been completed within three days; Huang Deyu, the chairman of the Changli County propaganda bureau, eclipsed this report by describing his county's mural campaign between 8 to 20 March, when 65,000 murals were produced over 12 days. Another major centre for the mural campaign was Shulu County (present-day Xinji), Hebei province, which boasted 95,000 murals, produced by 150,000 people.[32]

It was, however, Pi County, Jiangsu Province, that became the most promoted, and acclaimed, centre of mass muralism, the premier 'mural county' or a 'satellite' of the art world. Before the GLF, Pi County had already been recognised as a successful model for cultivating mass cultural activities, especially two cultural galleries in Xuzhou and Huaiyang counties, where teachers and students from nearby Nanjing Art Academy had provided art training courses. Like Gu Yuan, Nanjing Art Academy students had observed that the Pi County peasants would paint ploughs,

[29] *Gu Yuan yishu de neizai jingshen* (Changsha: Hunan meishu chubanshe, 2016).

[30] Gu Yuan, *Nongcun bihua fanben*. Tianjin: Tianjin meishu chubanshe, 1958.

[31] 'Shengchan dayuejin, wenhua yishu jinjin gen—ji quanguo nongchun qunzhong wenhua yishu gongzuo huiyi', *Meishu* no. 5 (1958), 20–1.

[32] Hebei sheng wenhuaju, *Hebei bihua xuan*. Baoan: Hebei renmin chubanshe, 1959; Guo Jun, 'Zouchu huashi, kaizhan bihua gongzuo', *Meishu* no. 6 (1958), 8; 'Renren jidong shou, cuncun shi hualang: Shulu Pixian liangxian bihuahua', *Renmin ribao* (August 18, 1958), 6th edition.

Figure 9.6 Gu Yuan, 'Rely on the Three Freshnesses to produce 1,000 *jin* from a single *mu* of land', 1958. From *Nongcun bihua fanben* (Tianjin: Tianjin meishu chubanshe, 1958).

harrows, and cattle around entryways to symbolise abundant harvests.[33] The genesis of the mural campaign in Pi County was Guanhu Township, where the local Party secretary Liu Zhaoqi and the cultural gallery director Han Shaoxian had organised a conference on peasant painting.[34] The Party Secretary Liu later recounted his doubts about the mural:

> As for mural painting, at the beginning I didn't really pay attention to it. Station chief Han said to me we should paint murals, and I asked, 'Why?' 'To express the Great Leap Forward.' I thought that it wasn't important. Afterward he asked a few more times, and I still thought, does painting count as work? … I heard county Party Secretary Zheng discuss mural painting, describing murals as a propaganda tool that should be used to the fullest, and it was then I truly opened my mind. Now I fully understand that painting is a kind of thought work, that wherever murals are plentiful, propaganda efforts have been vibrant, all work is vibrant; wherever mural paintings are few, then propaganda efforts have been poor, and altogether, the work even poorer.

Despite the stated doubts above, such was the cultural cadre's success in Guanhu that its population of 2,000 achieved 2,500 murals, reportedly completing over 500 posters in one night. Their campaign expanded outward, eventually totalling 23,000

[33] Nanjing yizhuan meishuxi lilun zu jiti baogao, 'Pixian nongminhua yanjiu', *Yishuxue yanjiu* no. 11 (2012), 464–96.
[34] 'Gongchanzhuyi yishu de mengya', *Meishu* no. 9 (1958), 3.

murals and 15,000 posters in a count taken at the end of July 1958, with participating artists numbering 108,000 people, of which 6,000 were peasants.[35] By the first two weeks of August, Pi County's production numbers had quadrupled and tripled, reaching 100,000 murals and 45,000 posters; the numbers continued to grow by the time an exhibition of the mural painting movement was organised in Beijing's Beihai Park and the headquarters of the Chinese Artists' Association at the end of August, with photographs standing in for 60 murals and nearly 200 posters.[36]

The quantification of production numbers was one of the foremost considerations for cadres when they reported to the central Ministries of Culture and Propaganda. We can compare the numbers that individual fine-art artists pledged in 1958 to achieve with the numbers reported by the mural campaigns: in Beijing, which concentrated a greater proportion of the nation's artists, cartoonists committed to 58,000 cartoons, oil painting groups promised 6,000 paintings, sculptors guaranteed 1,507 works and 120,000 words worth of theoretical texts, printmakers 2,112 prints, and ink painters presented a package of 5,812 paintings, 8 books, and 160,000 words worth of historical writing.[37] All of the above groups reported a membership smaller than 80 artists. If untrained artists could be encouraged to participate in a local art campaign, cultural cadres would easily exceed the promised production quota of professional artists.

Attempting to reach the target ratio of one mural to every five or six households was a feat of mobilisation. It required that the county-level cultural galleries organise their cadres to visit each commune, attempting to locate someone who knew how to make images or art, which was, according to accounts, at most one or two people per commune. If there was no suitable volunteer in the commune, then the cadres had to find a person who was literate. After a viable candidate had been located, they participated, sometimes unwillingly, in a month-long art course of the kind described in the previous section, learning to conceptualise imagistic content of an acceptable type, supplied with *lianhuanhua* and pictorials to copy, and taught basic techniques to realise their inventions through drawing, colour, and composition. They then returned to their communes to begin producing murals and posters, as well as teaching others, but, as numerous cadres noted, these artists who had undergone a crash course were frequently reluctant to expose themselves to judgement comparing themselves to professional artists.

Both Hebei and Jiangsu provinces benefitted from their proximity to major academies, the Central Academy of Fine Arts and the Nanjing Art Academy (now

[35] These numbers vary dramatically depending upon the source; I have given the most conservative numbers here. The catalogue edited by Lu Shaofei gives the totals for mid-August as 34 participating villages, 414 communes with 1,800 art groups, 15,000 amateur artists, 105,000 mural paintings, and 78,000 posters.

[36] *Gongnongbing huaji* (Beijing: Renmin meishu chubanshe, 1958). See also the description of the official delegation's visit, in Luo Shuzi, 'Canguan Pixian Chefu xiang bihua—meishu lilun jiaoyanzu tongxun', *Yishuxue yanjiu* no. 11 (2012), 450–1.

[37] 'Meishu jie Dayuejin', *Meishu* no. 3 (1958), 4.

the Nanjing University of the Arts).[38] As students were dispatched to the countryside, they were given the briefest preparation, handed a packet of mimeographed readings.[39] In Beijing, groups of students travelled under the slogan, 'Turn the capital's outskirts into socialism's new art gallery', during which time they described their struggles with enlarging drafts, selecting pigments for outdoor use, and preparing suitable walls.[40]

Despite the role of art academies, the murals themselves, due to their histories of folk production and origins in village customs and traditions, sought to channel an agrarian purity, thereby becoming the medium of the revolutionary countryside. A cursory look at the remaining reproductions of murals suggests an undifferentiated and uninteresting morass of political slogans about rural industrialisation, communal living benefits, and abundant harvests. In comparison with earlier forms of amateur art, however, one of the striking features of the mass mural movement was a new emphasis on creating a distinctively agitational peasant form of expression, which was one that diverged from the folk workshop traditions that characterised New Year's prints.[41] The peasant mural, created through village campaigns, would return to the city and ruralise the urban, and prevail as a feature of agrarian expression during later peasant painting campaigns. When students and their professors travelled to villages to produce murals, they brought with them modelbooks and examples of Soviet caricature. Nevertheless, their collected images of political leaders stabbing their enemies with spears, rifles, and swords, provided few precedents for the images of folk industry that became associated with the mass mural.[42] Rather, it was not in Soviet or academic models but the outlandish exuberance and quirky combinations of Pi County murals, and the work of cultural cadres to identify and celebrate them as expressions of amateur creativity, that created the visual identity of a distinctively Maoist form.

Debating aesthetic subjectivity

To refer to the peasant murals as the blandly conventional 'sprouts of Communist art', as many period commentators parroted, is to miss the particular strangeness of their idiosyncratic presence on the national art stage. Unlike earlier mass art

[38] *Meishu ganbu xiachang xiaxiang shenghuo suxie* (Beijing: Renmin meishu chubanshe, 1958).

[39] Nanjing yizhuan meishuxi huihua zhuanye xuesheng jitichuangzuo, 'Pixian sanshi ri chugao (jiexuan)', *Nanjing yishu xueyuan xuebao* no. 12 (2012), 450–62. See also Mo Ji, 'Yiduan lishi, yige shenhua—du *Pixian sanshiri* you gan', *Nanjing yishu xueyuan xuebao* no. 2 (2005), 241–9.

[40] 'Jiang Zhongxing, Fan Zeng, 'Women zenyang zai kan danxiang hua bihua', in *Xuanchuanhua yu bihua*. Beijing: Renmin meishu chubanshe, 1958.

[41] James A. Flath, *The Cult of Happiness: Nianhua, Art, and History in Rural North China*. Vancouver: University of British Columbia Press, 2004, pp. 134–49.

[42] Lu Shaofei, *Sulian zhengzhi fengci xuanchuan hua*. Beijing: Renmin meishu chubanshe, 1958, part of the Qunzhong meishu huaku.

campaigns that were largely viewed as propaganda, the mass murals were instead featured, and promoted, as fine art. Murals from Jiangsu and Hebei provinces were photographed and sent to Beijing in a special exhibition of rural art campaigns, accompanied by a catalogue titled *Worker-Peasant-Soldier Art*; recorded in short-lived specialised mass art publications, including Shanghai's *East Wind* and the Central Academy of Fine Art's *Mass Art* journal; and rhapsodised over in special issues of the national publications *Fine Arts*, *Literary Gazette*, to *Cartoons*.[43]

Challenging conventional definitions of fine art, the peasant murals singled out for national promotion were untutored and disjunctive, crudely drawn in quirky proportions, and often conveyed their messages by means of compositions organised around repetitive elements that avoided any requirements of perspectival consistency.[44] Unlike Hu County peasant painting, the impoverishment of materials also featured heavily in the impression of the resourcefulness of peasant painting. With a limited supply of art materials, as well as the nationwide constriction of commodities in the countryside, peasants made use of basic materials in their environment, such as red silt, red sand from coal pits, chimney soot, white lime, and crushed leaves to make green pigments, and pig bristles as brushes.[45]

Unlike the crisp clarity of Gu Yuan's modelbook, the peasant murals were outlandishly unorthodox and wildly scrappy. The featured image of the mass mural campaign that began this chapter, Wang Fuwen's mural *Flame Mountain*, disregards conventional ideas about rational spatial relations in favour of graphic simplicity that dramatises through repetition and iteration (Figure 9.2). So, too, does another mural by a peasant named Zou Wuquan from Xiao County, Anhui Province, also monumentalise its subject, the harvest of grapes, through limited means, alternating between the organic dynamism of trellised lines and clusters of splodged dots (Figure 9.7). Although other amateur artists created more naturalistic or technically skilled murals, it was these – a soldier steering an oversized soybean pod with strings attached to its leaves, or an overgrown, anthropomorphised carrot and taro with linked arms, scurrying into a storehouse – that were celebrated for their inexpert rawness, in which imagination triumphed over competence to animate a peasant's interior subjectivity.

Few writings on socialist art are as strange as the appreciative essays that Communist art critics produced about the murals as aesthetic objects.[46] Most prolific among these responses was the critic Wang Zhaowen, who wrote five essays (*suibi*) to promote the phenomenon, published across the cultural nexus of *Literary Gazette*, *People's Daily*, and *Fine Arts*, albeit in different voices appropriate to

[43] *Dongfeng* was only published for two years between 1958 and 1959, and *Qunzhong meishu* published by the Tianjin Art Publishing House for six issues in 1959.

[44] Lu Shaofei, *Jiangsu Pixian nongmin bihua xuanji*. Beijing: Renmin meishu chubanshe, 1958.

[45] 'Jiangsu Pixian nongmin huazhan mingri kaimu', *Renmin ribao* (30 August 1958), 7th edition.

[46] Hua Junwu, 'Bai qunzhong weishi', *Renmin ribao* (30 August 1958), 8th edition; Zuo Hai, 'Lun Zhongguo dazhonghua', *Renmin ribao* (30 August 1958), 8th edition; Li Qun, 'Nongmin zoushang le huatan', *Renmin ribao* (30 August 1958), 8th edition.

Figure 9.7 Zou Wuquan, 'Produce 15,000 this year, guarantee 30,000 next year', 1958. Mural, dimensions unknown. From *Gongnongbing huaji* (Beijing: Renmin meishu chubanshe, 1958).

the respective audiences of the publications.[47] One of the most important Marxist critics of the early PRC, Wang Zhaowen left his native Sichuan to study sculpture at the National Hangzhou Art Academy in 1932, then arrived in Yan'an in 1940, where he became known for carving a monumental Mao Zedong relief sculpture; in the early PRC, Wang edited *People's Art*, which later became *Fine Arts*, the mouthpiece of the official art world.[48] With his training in Western-style art academies and as a leading figure of ideological correctness, Wang could not failed to have understood the difficulty of integrating the peasant mural into the artworld.

It fell to Wang Zhaowen to establish a new criteria for appreciating amateur art. That peasant murals lacked a pre-existing canon and accrued history of aesthetic terminology to evaluate the success of the movement was a stumbling block for many artists and administrators.[49] As Wang writes, art cadres 'cannot see the good

[47] Wang Zhaowen, 'Waihang neihang', *Wenyibao* no. 18 (1958), 23–6; Wang Zhaowen, 'Jigao yeda de qunzhong meishu', *Renmin ribao* (30 August 1958), 7th edition.
[48] Ping Jian, *Wang Zhaowen zhuan*. Yinchuan: Ningxia chubanshe, 2009.
[49] Cai Pu, 'You nianhua tan dazhong de xinshangfa', *Renmin meishu* no. 2 (1950), 43–5; Yang Keyang, 'Kan bihua you gan', *Dongfeng* (1959), 30, Gao Ziliang, 'Guanyu nongmin hua de tigao wenti', *Dongfeng* (1959), 32–3, Zheng Yinnong, 'Quan dangzhai dongshou, shihua man qiantou', *Dongfeng* (1959), 34.

in peasant painting, negate the good in peasant painting … Why can they not see the most basic good of peasant painting? … Because the characteristics of the art cannot rely upon a naturalistic eye to meet it'.[50] In order to write the mass mural into fine art, Wang does not default to the presumption that class status determines its excellence; this, he knows, is an indefensible response that cannot impart to mass art a clear sense of value beyond identity. In response to those who think that amateur works are 'perfect, that the technique is even better than professionals, everything is excellent', he writes, 'Actually, several works show that the hands of the masses do not answer to their wishes, that what they conjure up they cannot paint according to their wishes. So this kind of praise is also fallacious'.[51]

Instead, Wang Zhaowen's criticism enacts an eye that is untouched by the values of the existing artworld, one that performs aesthetic judgement uncorrupted by academic training. Wang's Communist eye could appreciate the peasant paintings for their successes and failures, even as he sought to couch the amateur mural into a familiar language of art history. Writing on Wang Fuwen's *Flame Mountain*, Wang Zhaowen compares the peasant's mural favourably by denigrating a well-known ink painter, writing, 'This type of composition is much bolder by comparison with the landscape painter Huang Binhong's empty compositions … Who knows what artist or what school the peasant artist followed for the composition'.[52] In disparaging Huang, Wang comments upon *Flame Mountain*'s originality and purity free of exterior influence or historical consciousness. To incorporate *Flame Mountain* within the realm of established canons thus inspires Wang to draw two analogies: the first, by means of a Chinese painting theorem that refers to the use of negative space, or 'The more a scene is hidden, the loftier the heights that it attains; the more a scene is exposed, the more trivial its achievements'. In the second, Wang compares the mural with the primitive, unfinished qualities of Han dynasty paintings that lacked anatomical knowledge.[53]

Awkwardness, in Wang Zhaowen's essays, becomes a virtue, a feature associated with amateur creativity that differentiates peasant painting from professional production, marking its purity. Wang frequently characterises the murals as childishly awkward (*zhizhuo*), which bespeaks the naivete and wholesomeness of peasant painters. 'Why', he writes, 'is childish clumsiness also beautiful? I think that because they are serious about their creation, that they are earnest, they break conventions, they don't have established rules'.[54] To demonstrate this, he turns to the meandering strokes of Zhou Wuquan's *Grape Harvest* mural, writing that it triumphs over other works because of its artlessness and innocence: 'In Peking opera, it's similar to how the *dan* must have sloping shoulders, his costume must use long belts with swooping

[50] Wang Zhaowen, 'Gongnongbing meishu, hao!' *Meishu* no. 12 (1958), 12.
[51] Wang, 'Gongnongbing', 11.
[52] Wang Zhaowen, 'Wanzheng bu wanzheng', *Renmin ribao* (4 November 1958), 7th edition.
[53] Wang, 'Wanzheng bu wanzheng'.
[54] Wang, 'Gongnongbing', 14.

lines, the pattern of his embroidered robe must emphasise the relationships between emptiness and fullness … .'[55] These qualities of untrammelled spontaneity, precocious earnestness, and unconventional sources of imagery underscored the foreign obsessions, technical mania, and narrow-mindedness of fine art practitioners.

Mass creativity, throughout Wang Zhaowen's several essays on peasant murals, signifies the ascendancy of revolutionary romanticism over revolutionary realism because of the very ungainliness, imprecision, and clumsiness of mass muralism's forms.[56] Peasant murals achieved revolutionary romanticism because the untrained hand allowed for unfettered human spirit to flow freely over the imposition of esoteric knowledge, dogmatically implemented. In arguing for the aesthetic subjectivity enacted by peasant murals, Wang took a position that was opposed to other artists who also participated in amateur art campaigns, but saw their role as paternalistic and dogmatic, enlightening the peasants by introducing them to a wider span of cultural references, often foreign in origin, and concepts of beauty, often requiring skilful expertise.[57] Rather than merely transcribing the exterior appearance of things, as professional artists did, Wang and his Yan'an colleagues argued that mass murals channelled future-oriented utopian thought, leaping past dogmatism and urban elitism. Mass muralism, then, became the apotheosis of cultural Maoism in its most radical form, its childlike spontaneity and rural primitivity standing in for a renewed strain of radical tradition that could topple the bureaucratic artworld, both its institutions and reactionary class consciousness.

Maoism and Cold War creativity

Wang Zhaowen's promotion of peasant creativity and mass muralism was mounted at a time when critics and artists engaged in overt, and covert, debates over the role of creativity in making Maoist culture. Maoism has mostly been thought of in terms of its *lack* of creativity – characterised by rigid ideological parameters, lockstep conformity, and limitations on creative experimentation – that manifested the authoritarianism of the early PRC.[58] Debates over mass creativity were not limited to disputes over creating a new artistic culture for the early PRC, however, and may arguably be more broadly understood as part of anxieties over the promotion of creative activity as a key arena of contestation in Cold War culture. Within the contexts of Cold War media histories, the making of mass art has stood in contradistinction with the polemics of democracy that drove parallel attempts to

[55] Wang, 'Gongnongbing', 16.

[56] Wang, 'Waihang yu neihang', pp. 23–6.

[57] Vivian Y. Li, '"How-To" Make Art in Communist China: Professionalizing Amateur Artists', in *The Oxford Handbook of Communist Visual Cultures*, eds. Aga Skrodzka, Xiaoning Lu, and Katarzyna Marciniak. Oxford University Press, 2020, pp. 223–47.

[58] Laikwan Pang, *The Art of Cloning: Creative Production during China's Cultural Revolution*. London: Verso, 2017.

propagandise the exercise of freedom associated with individual creativity in arenas as diverse as Abstract Expressionism, children's art, design, and photography.[59] Yet both democratic and Communist countries wielded the language of creativity to describe the successes of their political project, and these discourses of creativity were occasionally consonant in overlooked junctures. Chief among these junctures was the description of childlike creativity as a marker of egalitarianism, a discourse that met surprisingly in a 1942 exhibition of wartime drawings by Chinese children, exhibited at the Museum of Modern Art in New York.[60] In this way, mass creativity's role in defining Maoism's cultural resources might be understood as a forerunner to, rather than the antithesis of, creativity as a hallmark of post-socialist China's knowledge economy.[61]

The mural campaign temporarily upended the hierarchies of the art world. Originating in poor counties and villages that sought to exceed the material resources, knowledge, and connections concentrated in cities, the rural underpinnings of the mural campaign undermined the centrality of urban intellectuals to shaping national cultural programmes. From their very assumptions to the work that had gone into laying out a stylistic identity of socialist China during the first Five-Year Plan, artists were sidelined by the mural campaign and its primary agents, ambitious Propaganda Department cadres. Instead, the effect, or goal, of the mural campaign was to solidify worker-peasant-soldier art as a genre unto itself, one that could not be directly challenged because of the classism that was attached to its definition. One of the results of the mural campaign was that its aesthetic effects and rural practices would in turn move back into the city. The practices and reception of the mural campaign presented a full-scale theorisation of mass art in toto, instrumentalising the infrastructure of amateur art training courses and exhibition spaces that had been established in the previous years, but more critically, began to propose the practice of amateurism as possessing a distinct aesthetic subjectivity, one that would become continually rearticulated, revaluated, and rethought within Cultural Revolution visual culture.[62]

[59] Jamie Cohen-Cole, *The Open Mind: Cold War Politics and the Sciences of Human Nature*. Chicago: University of Chicago Press, 2016, 35–62; Serge Guilbaut, *How New York Stole the Idea of Modern Art: Abstract Expressionism, Freedom, and the Cold War*. Chicago: University of Chicago Press, 1983; Greg Barnhisel, *Cold War Modernists: Art, Literature, and American Cultural Diplomacy*. New York: Columbia University Press, 2015; Amy Ogata, *Designing the Creative Child: Playthings and Places in Midcentury America*. Minneapolis: University of Minnesota Press, 2013.

[60] Fred Turner, *The Democratic Surround: Multimedia and American Liberalism from World War II to the Psychedelic Sixties*. Chicago: University of Chicago Press, 2015, 184–90. The exhibition is cited in John Blakinger, 'MoMA's Child Artists: The Politics of Creating Creative Children', in, *Modern in the Making: MoMA and the Modern Experiment, 1929–1949*, ed. Austin Porter and Sandra Zalman. Bloomsbury Publishing, 2020.

[61] Lily Chumley, *Creativity Class: Art School and Culture Work in Postsocialist China*. Princeton: Princeton University Press, 2016.

[62] See, for example, *Gongnongbing huabao*, published for 244 issues in Zhejiang between 1967 and 1975.

10

An Unlikely Moment of Revival? The Return of Gods in Early 1960s Zhejiang

XIAOXUAN WANG

THE RELIGIOUS QUESTION has been one of the major issues in nation-building (*jianguo*) facing modern Chinese regimes, including the Communist Party. Since its early years, two aspects of religion have always concerned Chinese Communist Party (CCP) leaders: religion as cultural phenomenon and as social institution. On the one hand, as a cultural phenomenon, religion was primarily considered as a matter of beliefs or a thought issue (*sixiang wenti*). Mao himself believed that widespread 'superstition' among peasants could only be solved by persuasion and education.[1] On the other hand, the overthrow of social institutions underpinning religious beliefs was a much more urgent task. Divine authority (*shenquan*) and lineage authority (*zuquan*), in the eyes of Mao, were among four authorities (or 'four ropes') that obstructed the emancipation of peasants.[2]

By the early 1960s, after a string of sociopolitical campaigns to transform the 'old society' of pre-1949 China, those of the landlord class – traditional leaders and major patrons of local religion – had been entirely wiped out from local society, along with their political and economic powers. The vast majority of temples and churches that had survived the Land Reform Campaign (1946–53) were either closed, severely damaged, or converted in the frenzied campaigns of the Great Leap Forward (GLF, 1958–62). A full-fledged revival of religion seemed improbable. But the Chinese authorities were concerned, if not surprised, to find that religion was making a comeback throughout the country. Many reports in *Neibu Cankao* (Internal Reference), an internal publication by the state-owned Xinhua Agency, in the early 1960s painted a picture of re-invigoration of various religious activities, from reconstruction of temples and lineage halls, ritual activities, fortune-telling,

[1] Mao Zedong, *Mao Zedong Xuanji, di yi juan* (Selected works of Mao Zedong, volume one). Beijing: Renmin chubanshe, 1991, 33.

[2] Mao, *Mao Zedong Xuanji*, 32.

Proceedings of the British Academy, **267**, 189–208, © The British Academy 2025.

to reorganisation of 'reactionary teachings and sects' (*fandong huidaomen*) and Christian house meetings.[3] These activities spread widely across different regions of the country, from Liaoning in the northeast to Guangdong in the south and Yunnan in the southwest. According to the statistics of 22 counties in Liaoning alone, 4,182 'feudal superstition professionals' (*fengjian mixin zhiye fenzi*) remained active in 1963 – most of them spirit mediums.[4] In big cities like Tianjin, the number of religious practitioners who congregated in still-open religious sites in 1963 doubled that of 1959, and the investigators noticed the same trend across each major religious tradition.[5]

What does the short surge of spiritual activities in the aftermath of the GLF and the Great Famine (1959–61) reveal about the conditions of religion at the time? What then, in retrospect, can the early 1960s religious resurgence tell us about the seemingly surprising return of religion after Mao? Until the phenomenal religious revival of the early 1980s, the country seemed to be only 'a short leap from' secularisation.[6] One premise of the chapter is that Maoism was not just a one-way process through which the state extended its bureaucracy and power into society. What grew organically together with the penetration of the Maoist state was also the making of its subject: the governed learned to live with – not just resist – the shifting political environments as they self-adjusted, regrouped, and found niche in which to survive. In this vein, the experience of religious communities like the story told here should also be part of how Maoism was made.

This chapter showcases the experience of religious communities in Wenling County, Zhejiang Province of the southeast coast, through the religious revival in Mount Fang in the aftermath of the GLF. Drawing mainly on local state archives in Wenling and surrounding counties, this chapter traces the interactions and networks of a variety of groups, from ordinary religious practitioners to ritual specialists of many different traditions as well as grassroots cadres, to illustrate the making and demise of the Mount Fang revival. As I show in this chapter, the efforts to perpetuate their religious traditions led to the (sometimes forced) collaboration of various groups, especially between Buddhist/Daoist monks and spiritual mediums. This gave rise to a highly unusual ritual economy that saw the flourishing of temple and ritual activities in the rather remote Mount Fang area. This ritual economy was short-lived. Yet its emergence during the Mao era and amidst the deep crises of local religious life suggests the potential for revival and indeed the limits of the power of the Maoist state. Equally, if not more importantly, this short-lived ritual

[3] *Neibu Cankao* (Internal Reference), 24 January 1962; 29 June 1962; 25 December 1962; 8 February 1963; 18 June 1963; 25 June 1963; 26 July 1963; 8 October 1963. See also Adam Chau, 'Popular Religion in Shannbei, North-Central China', *Journal of Chinese Religions* 31(2003): 41–2; Stephen Jones, *Plucking the Winds: Lives of Village Musicians in Old and New China*. Leiden: CHIME Foundation, 2004, 135–8; Henrietta Harrison, *The Missionary's Curse and Other Tales from a Chinese Catholic Village*. Berkeley: University of California Press, 2013, chapter 6.

[4] *Neibu Cankao* (Internal Reference), 15 August 1963.

[5] *Neibu Cankao* (Internal Reference), 30 July 1963.

[6] Rebecca Nedostup, *Superstitious Regimes: Religion and the Politics of Chinese Modernity*. Cambridge, MA: Harvard University Asia Center, 2009, 279.

economy points to the flaws and the ironies of the revolution as the Maoist rule often, albert inadvertently, revitalised traditional idioms while profoundly weakening the economic base and political structure of traditional rural societies.

Surviving a decade of turmoil following 'liberation'

A coastal county in Taizhou Prefecture, eastern Zhejiang, Wenling borders Yueqing County of Wenzhou Prefecture to the south. Although fishing was (and still is) the mainstay of the local economy, Wenling, like most places in China at the time, went through Land Reform and collectivisation after 1949. Political atmosphere in the county was tense in the 1950s. After the Communist guerrilla forces took over Wenling, local Kuomintang (KMT) armies retreated to nearby Jigu Island, where they continued to skirmish with the Communist forces until the mid-1950s. Afterwards, Wenling, especially some of its islands, remained at the forefront of the military standoff across the Taiwan Strait between the People's Republic of China (PRC) and the Republic of China (ROC) .[7]

In spite of the tense political atmosphere, religious practices were not absent from public view in Wenling in the early 1950s. The 1955 official register recorded 304 Buddhist monks, 505 Buddhist nuns, and 323 Daoist priests, many of whom were still providing services, such as doing the rite of feeding 'flaming mouths' (*fang yankou*),[8] performing Daoist rites (*zuo daochang*), or fortune-telling (*kanxiang*). Some Protestant communities also continued to meet publicly and develop new converts, ignoring warnings from officials. As a county government report pointed out, religious activities significantly increased with the rise of infectious diseases in 1955. That year, epidemics, especially measles, were spreading across the county. At least 300 young children reportedly died of measles. Many people sought help from efficacious women (*linggu*) and flower(/tree)-reading women (*kanhuanu*), two major types of female spirit mediums in the region. In Muyu Township, 20–30 spirit mediums were estimated to provide services on a daily basis.[9]

The GLF beginning in 1958 threw religious activities into a moment of complete disruption.[10] To religious communities in Wenling, the Great Leap dealt a huge

[7] For a history of the Taiwan Strait standoff between the PRC and the ROC, see Michael Szonyi, *Cold War Island: Quemoy on the Front Line*. Cambridge: Cambridge University Press, 2008.

[8] The ritual of feeding flaming mouths is a Buddhist ceremony most commonly performed to save the dead from hell.

[9] 'Wenling Xian dangqian zongjiao huodong qingkuang baogao' (Report on current religious activities in Wenling County), 16 April 1955, Wenling City Archives, 7-1-8: 54–6.

[10] For recent scholarship on the Great Leap Forward and the Great Famine, see Frank Dikötter, *Mao's Great Famine, The History of China's Most Devastating Catastrophe, 1958–1962*. New York: Bloomsbury, 2010; Kimberley Ens Manning and Felix Wemheuer (eds.), *Eating Bitterness: New Perspectives on China's Great Leap Forward and Famine*. Vancouver: University of British Columbia Press, 2011; Felix Wemheuer, *Famine Politics in Maoist China and the Soviet Union*. New Haven: Yale University Press, 2014; see also Felix Wemheuer and Wenyu Jing's chapter in this volume, 'Seeing Like the Maoist State: Perceptions of Peasant Resistance in Official Documents'.

blow, in terms of both organisation and material foundations. Numerous religious properties were seized as raw materials for industrial developments such as building backyard furnaces. Local cadres in Wenling also took the occasion of the Great Leap to mobilise the masses to debate the existence of gods and encourage activists to denounce religious faiths in order to suppress religious communities. Consequently, churches and temples were forced to close and their properties were 'donated', as a county official report noted, to other institutions such as local brigades or communes. Clergies were forced to leave religious orders and return to secular life. Before the Great Leap, 32 churches were still active. By April 1959, however, only eight churches remained open with just around 400 people regularly attending meetings. The Wenling County government planned to retain five Christian churches and one Buddhist temple while confiscating all other temples and churches.[11]

In addition to suppression of religious activities, political campaigns after 1949 also made the bond between religion and community more fragile than ever. As political campaigns worsened existing feuds and grievances against religious institutions and communities, they ultimately provided excuses and opportunities for attacking religious sites. During and after the Land Reform Campaign, some villagers took furniture and other facilities from village temples at will, as they considered temples as landlords and therefore targets of the Land Reform. People sometimes simply moved into temples and churches or turned them into meeting venues of the village government.[12]

Crucially, however, as the following story of Mount Fang demonstrates, the bond between religion and community was much more tenacious than it might appear to the Communist government. The campaigns of the GLF had rather complicated and indeed contradictory impacts on religious life. Instead of just alienating local communities from their religious institutions, for many religious leaders and practitioners in local communities, the tense political atmosphere of the Great Leap and the sense of uncertainty in fact ended up working to reinforce such a bond. In other words, the socialist system actually 'made' a religious community through its GLF policies.

The formation of a religious revival in Mount Fang

The revival of religious activities in Mount Fang appears to have begun quietly towards the end of the Great Famine. It was only during the Socialist Education Movement of the mid-1960s, however, when local authorities in Wenling began to identify and investigate the revival in Mount Fang as an incident of so-called

[11] 'Xianwei tongzhan bu guanyu 1958 nian zongjiao gongzuo zongjie ji jinhou gongzuo yijian de baogao' (County department of united front's report on religious work in 1958 and plans for the future), 22 April 1958, Wenling City Archives, 1-1-446: 22–7.

[12] 'Guanyu wo xian zhixing zongjiao zhengce de jiancha baogao' (Examination report on the implementation of religious policies in our county), 8 December 1956, Wenling City Archives, 17-1-11: 49–53.

'restoration of feudal superstition'. Mao believed that there appeared to be acute social issues in rural China in the few years after the GLF. A nationwide Socialist Education Movement was staged to re-indoctrinate socialist thoughts and rectify revisionist elements in politics, economy, ideology, and grassroots organisations. In the practices of socialist education, resurfacing religious activities such as the revival in Mount Fang were often targets of investigation and attack.

Mount Fang (or *Fangshan*, also known as Mount Fangyan or simply *Fangyan*) is located in the borders of Wenling, Taizhou, and Yueqing County of Wenzhou. Though on the fringes of two prefectures, Mount Fang was not inaccessible. In the 1960s, provincial roads connecting Taizhou and Wenzhou ran through the hill-sides of Mount Fang, from which travellers could also, via inland waterways, reach Songmen Port (Wenling) and Haimen Port – the main port of Taizhou Bay.

The mid-1960s religious revival in Mount Fang was centred in two sites: Cloud Heaven Temple (*Yunxiao si*) and Jade Toad Palace (*Yuchan gong*).[13] Cloud Heaven Temple, a Buddhist temple located in Shangyang Brigade, Xinjian Commune, Daxi District, was allegedly first built in 1820. Like most Buddhist monasteries, the main building was the Great Hall of Powerful Treasure (*Daxiong Baodian*) with the Hall of Great Compassion (*Dabeige*) in the rear row. Most curiously, however, the front hall of the temple was dedicated to a popular deity Great Lord Yang (*Yangfu Dashen*), one of the most influential local deities in southern and eastern Zhejiang. One would also find a place to worship Lord Yang in Jade Toad Palace, a Daoist temple less than a mile from Cloud Heaven Temple. Located in a small village called Lingtou Brigade (Huwu Commune of Yueqing County, which had been part of Wenling until 1951), Jade Toad Palace traces its origin back to Priest Chen Tiyang who first arrived at Mount Fang in 1851. Jade Toad Palace still had 41 rooms and 2 caves after the Great Leap (see Figure 10.1 for today's Jade Toad Palace).

In the late 1950s and early 1960s, Abbot Kefang and Abbess Liu Peihua were respectively in charge of Cloud Heaven Temple and Jade Toad Palace.[14] Kefang was from Baiyantan, a small village just a few miles from Mount Fang. He had been the abbot of the temple since 1944 but left in 1953 (perhaps out of fear for the political atmosphere). Before returning in 1958 at the request of Commissar Xie Xianye of Shangyang Brigade (according to Kefang himself), he moved around several Buddhist temples in Ningbo Prefecture, north of Taizhou. In 1961, Kefang became the abbot again after the passing of a critically ill Kezhong, the abbot of Cloud Heaven Temple at that time.[15] Abbess Liu Peihua was also a native

[13] Mount Fang in this chapter refers to both the mountains and the religious communities consisted of Cloud Heaven Temple and Jade Toad Palace. Cloud Heaven Temple is known among the locals as 'Fangshan Temple' (*Fangshan si*). Similarly, Jade Toad Palace is more commonly referred to by the locals as 'Goat Horn Caves' (*Yangjiao dong*).

[14] Kefang is a Dharma name. We do not know what Kefang's birth name is.

[15] 'Guanyu Fangshan, Yangjiaodong zongjiao mixin huodong qingkuang de diaocha baogao' (Investigation report on the situation of religious superstition activities in Mount Fang and Goat Horn Caves), 20 June 1964, Wenling City Archives, 17-1-80: 242; 'Seng Kefang jiaodai cailiao' (Monk Kefang's confession materials), 7 March 1965, Wenling City Archives, 17-1-118: n/a.

Figure 10.1 The main cave of Jade Toad Palace today (Courtesy of Junliang Pan)

but from Xiaoqiu Brigade, Huwu. Her experience before 1949 was rather murky. Some information in government documents hints that she was perhaps an adherent of a certain redemptive society.[16] She used to reside in Rising Prosperity Temple (*Fuxing gong*), Lingtou, but transferred to Jade Toad Palace in 1951 when Land Reform was taking place in Wenling.[17]

Before 1960, both Cloud Heaven Temple and Jade Toad Palace were desolate. During and after the Land Reform, many statues of divinities had been demolished even though they were supposed to be protected as temples of officially sanctioned Buddhism and Daoism. Villagers from nearby Shangyang Brigade kept taking furniture and other facilities from Cloud Heaven Temple. Aside from Kefang there were just one or two monks living in the temple that still had 31 rooms. Nevertheless, Kefang and Peihua all seemed determined to carry on the operations of their temples. Kefang recruited three monks between 1962 and 1964, including two middle school graduates. Liu and five male priest colleagues collectively recruited five

[16] 'Diaocha jianju cailiao' (Investigation and confession materials), January 1965, Wenling City Archives, 17-1-118: n/a.

[17] 'Guanyu Fangshan, Yangjiaodong zongjiao mixin huodong qingkuang de diaocha baogao' (Investigation report on the situation of religious superstition activities in Mount Fang and Goat Horn Caves), 20 June 1964, Wenling City Archives, 17-1-80: 242.

to six young, mainly female, Daoists after 1960. Although Cloud Heaven Temple and Jade Toad Palace were once forced to close down, they were officially given permission to reopen by Zhejiang Province Bureau for Religious Affairs after the Great Leap. Thanks to both the legal status and remote location of the two temples, they soon became a place where outside religious communities, especially followers of local deities, sought refuge. In addition to Liu Peihua herself, several colleagues of Liu had previously served in other temples across Wenling.

The most conspicuous of Mount Fang's status as a sanctuary of religious activities was the arrival of many spirit tablets (*shenwei*, literally divine seat), often sent by spirit mediums – a curious fact that occurred after the repressive religious policies of the Great Leap. Spirit tablets are tablets inscribed with names of deities – an object extremely important in Chinese territorial cult because it symbolises the residence of deities.[18] Buddhist and Daoist temples in China have the tradition of hosting local deities since the former is also part of local communities. In other words, Buddhist/Daoist temples are also village temples as they were often founded by and served the same group of patron families. Therefore, those from Xinjian Commune where Cloud Heaven Temple is located sent spirit tablets of their deities to the temple and the latter not surprisingly allowed the spirit tablet to stay.

It is unusual, however, that deities sought refuge in areas distant from their home villages/neighbourhood. In the case of Marshal Fang whose spiritual tablet was sent to Cloud Heaven Temple, for instance, the deity is from Chengnan (County seat south) District, far from Mount Fang. One of the most important dimensions of traditional Chinese organisations is their territoriality, so the deities they worshipped and the spiritual mediums who served the deities were also territorial, meaning that they mainly served population in a certain area. The displacement of spirit mediums and/or their deities from their home villages/neighbourhood can only suggest that the local religious ecology was greatly disturbed – something that one should not be surprised to see in the aftermath of the Great Leap. Before Kefang's return in 1958, the community in a nearby village sent to Cloud Heaven Temple a spirit tablet of their main deity: the Emperor of Jiang San (*Jiangsan Dadi*), a popular deity in some areas of Wenling. Since Kefang came back, he personally received at least three spirit tablets. Spirit medium Lin Ying of Gu'ao Brigade, Hengshan Commune, Chengnan District, sent Marshal Fang (*Fang Xianggong*). Someone from Xinjian Commune sent Great King of Chen Jiu (*Chen Jiu Dawang*). In late 1961, Xiaomeiyu of Kantou Brigade, Xinjian Commune, sent The Goddess of Dragon Mother.[19] Other than these three, though, we do not know how many more spirit tablets that Kefang had personally received.

[18] We do not have the details about how spirit tablets were sent: the trip to Mount Fang, the reception ceremony, or other details. However, given the size of them – usually a piece of wood and a seat, it should not be difficult to carry them without being noticed.

[19] 'Seng Kefang jiaodai cailiao' (Monk Kefang's confession materials), 7 March 1965, Wenling City Archives, 17-1-118: n/a.

Although the official investigations only mentioned a few deities whose spirit tablets were sent to Mount Fang, it is possible to identify through the investigations names and basic backgrounds of a total of 27 spirit mediums who were active in the Mount Fang revival. Most mediums were in fact not from Daxi, the district where Mount Fang is. The presence and activities of spirit mediums from across the county in Mount Fang suggest that the number of spirit tablets sent to Mount Fang was likely way more than what official documents revealed. Those who sent spirit tablets initially might only hope to find a quiet provisional home for their deities. However, the arrival of many spirit tablets soon stirred up a chain of changes in Mount Fang's religious ecology when both Kefang and Peihua sought to revitalise their temples.

In Mount Fang and elsewhere in Zhejiang, many communities attempted to restore or rebuild temples and ancestral halls after the end of the Great Famine,[20] despite the fact that the restoration plans were often short of money due to the Land Reform and collectivisation. In particular, the Land Reform removed major traditional patrons of communal religion and took away almost all land possessions of religious institutions. Temple reconstruction initiatives sometimes were able to use the money of village collectives with the permission of brigade or commune cadres.[21] People would also go door to door collecting donations. Ritual activities (and local opera performances) were another critical, if not the most important, channel to collect funds for temple restoration. Buddhists and Daoists in Mount Fang were especially motivated in this regard because of the need to repair their large religious complexes.

How do you establish a steady supply of patrons to use your ritual services? It did not take long for Kefang and Peihua to realise that they could take advantage of the worship of Great Lord Yang in their temples and spirit mediums coming to Mount Fang. This proved to be a sensible move. Although spirit mediums, Buddhists, and Daoists were often competitors, collaborations between spirit mediums and Daoist priests or between spirit mediums and Buddhist monks/nuns have also been a longstanding tradition since the late imperial era. In the lower Yangtze or Jiangnan region in the Ming and Qing dynasties, for instance, most of those *xiangtou* (literally, 'incense head') who organised ritual services and pilgrimage to sacred Daoist or Buddhist sites were often spirit mediums themselves.[22] Mediums,

[20] 'Tangxia Qu dangqian de jiejidouzheng' (Current situation of class struggle in Tangxia District), 15 December 1962, Ruian City Archives, 49-14-16: 64; 'Pizhuan Zhengfa dangzu guanyu zhizhi fengjian canyu de gezhong fubi huodong de yijian' (Approving and circulating Party committee of [county] political and legal office's notification on stopping restoration activities of various residual feudal forces), 5 May 1963, Rui'an City Archives, 1-15-69: 18.

[21] 'Guanyu Donglian Gongshe fengjian mixin huodong qingkuang de diaocha baogao' (Investigation report on the situation of feudal superstition activities in Donglian Commune), 6 May 1963 Yueqing City Archives: 1-15-27: 42.

[22] Vincent Goossaert, 'Daoism and Local Cults in Modern Suzhou: A Case Study of Qionglongshan', in *Chinese and European Perspectives on the Study of Chinese Popular Religions*, ed. Philip Clart. Taipei: Boyang, 2012, 199–228; Wang Zhenzhong, 'Huayun jinxiang: minjian xinyang, chaoshan xisu

it was also observed, asked clients to seek the help of Daoist priests 'exorcising the baleful stars' (*rangxing*) or Buddhist monks 'appeasing the soul of the dead' (*chaodu*). They would later 'get a share of money from Daoist priests and Buddhist monks'.[23]

To spirit mediums and pilgrims, Cloud Heaven Temple and Jade Toad Palace did not seem to be particularly influential and attractive as Buddhist/Daoist sacred sites. The two temples were relatively young. The main cave (*Yangjiao yi dong* or Goat Horn Cave No. 1) of Jade Toad Palace was built in 1915, less than 50 years old by the 1960s. The 1896 county gazetteer did not mention these two temples at all.[24] What really motivated pilgrims instead was the worship of Lord Yang. Its presence in both Cloud Heaven Temple and Jade Toad Palace suggests the rising influence of the deity in the area since the nineteenth century. It is hard to know the history and scale of pilgrimage to Mount Fang for the worship of Lord Yang before 1949. Official investigations reveal, however, that during the Mount Fang 'superstitious activities', the number of pilgrims to these two temples around the birth date of Lord Yang was way more than other dates. Many pilgrims carried fish and smoked meats instead of vegetarian food, according to official investigations and confession letters of spirit mediums and pilgrims.[25] This is yet another bit of evidence that people were more likely going there to visit Lord Yang rather than solely to burn incense at a Buddhist or Daoist temple. To put it differently, these Buddhist or Daoist ritual service providers far more likely benefitted from the worship of Lord Yang than vice versa.

The following stories come from a confession letter of a medium called Dong Xiaohua,[26] written on 17 January 1965, at the request of local officials who were sent to investigate the Mount Fang 'superstition activities'. It shows how Buddhists/Daoists and spirit mediums began to work together to bring pilgrims to Mount Fang as well as the role of Lord Yang. Dong of Shangdong'ao, Daxi Commune, was a spirit medium of a deity named Liu Liu. Her explanations for the troubles of clients, according to herself, were usually possession by demons. For demons to depart

yu Ming Qing yilai Huizhou de richang shenghuo' (Pilgrimages to Mount Jiuhua and Mount Qiyun: Folk Beliefs, Pilgrimage Custom and Daily Life in Huizhou since the Ming and Qing Dynasties), *Difang wenhua yanjiu* 2.2(2013): 38–60.

[23] Gu Chuanjin (ed.), *Puxi xiaozhi* (A concise history of Puxi). Shanghai: Shanghai guji chubanshe, 2008, 11–12.

[24] The year 1896 is a year during the Guangxu reign of the late Qing dynasty. The next Wenling county gazetteer was compiled in 1992 in which Cloud Heaven Temple appears as the youngest one of five major Buddhist monasteries in the county.

[25] 'Diaocha jianju cailiao' (Investigation and confession materials), January 1965, Wenling City Archives, 17-1-118: n/a. Many brigades, official reports accused, simply went to Mount Fang to have banquets, which, it was revealed, involved alcohol and meat including pork, fish, and goose meat. See 'Dui Fangshan, Yangjiaodong mixin huodong qingkuang de diaocha he jinhou yijian de baogao' (Report on investigation of superstition activities in Mount Fang and Goat Horn Caves and plans for the future), 6 June 1964, Wenling City Archives, 79-1-562: 101.

[26] 'Diaocha jianju cailiao' (Materials from investigation and informants), 17 January 1965, Wenling City Archives, 17-1-118: n/a.

the bodies of the affected, she would advise to offer sacrifices. When she met Liu Peihua during the birthday celebration of Great Lord Yang in Jade Toad Palace in May 1961, Liu tried all means and persuaded her to tell clients to employ the ritual services of Jade Toad Palace to solve their problems. Per Peihua's persuasion, in the next few years, Xiaohua often told clients that their illness or troubles could only be solved by the Great Lord (Yang) of Goat Horn Caves (*Yangjiao Dong Dashen*). She would then take them to Daoist priests of Jade Toad Palace to perform ritual healing. Some clients of spirit mediums were also told that they had to send 'donation for [temple] reconstruction' (*suyin*, literally construction silver) to Mount Fang before their illness was cured or miseries disappeared.[27]

With the assistance of figures like Dong Xiaohua, in the few years since 1960, Cloud Heaven Temple and Jade Toad Palace received tens of thousands of pilgrims each year. The 18th of the 5th lunar month is the birthday of Great Lord Yang, the most important date in Mount Fang. On that day (24 June) of 1964, at least 4,000 people went to celebrate in Mount Fang, according to official estimates. Between 1 January and 10 July 1964, 13,159 people stayed overnight in the lodges on top of Mount Fang,[28] the lodge of Cloud Heaven Temple's register shows. Some pilgrims had to stay overnight either because they would not be able to return home the same day or because of ritual requirements. If we take into consideration those who did not stay overnight, the total number of visitors to Mount Fang in the period 1960–4 could be considerably larger.[29]

Spirit mediums who were investigated all denied receiving payment from the temples of Mount Fang for their roles. It might be possible that some of these spirit mediums indeed did not receive money from the temples of Mount Fang. However, they must have gotten something in return, otherwise the collaboration between spirit mediums and Buddhists or Daoists would not have been able to run smoothly and bring so many visitors to Mount Fang. We may speculate that, since some spirit mediums placed spirit tablets of their deities in Mount Fang, they would have to bring clients there to perform ritual services. In return, they asked clients to seek help of Buddhist and Daoist ritual specialists in Mount Fang or use other services in Cloud Heaven Temple and Jade Toad Palace.

Looking beyond economic calculation, however, we can clearly see how the hostile political reality had forced spirit mediums, Buddhist monks, and Daoist

[27] 'Diaocha jianju cailiao'.

[28] 'Guanyu Fangshan, Yangjiaodong zongjiao mixin huodong qingkuang de diaocha baogao' (Investigation report on the situation of religious superstition activities in Mount Fang and Goat Horn Caves), 20 June 1964, Wenling City Archives, 17-1-80: 242–3.

[29] As so many visitors poured into Mount Fang, they even stimulated other economic and social activities – side products of the Mount Fang revival that also caught the attention of local authorities. Almost every day more than 30 people were begging in the nearby area of Cloud Heaven Temple and Jade Toad Palace, an official report noticed. A female beggar from Daxi commune said that she could earn as much as three to four yuan in a good day. Theft and gambling also reportedly increased. 'Guanyu Fangshan, Yangjiaodong zongjiao mixin huodong qingkuang de diaocha baogao', 245.

priests to collaborate in order to weather the difficult days under Mao. Spirit mediums sending spirit tablets of deities to Mount Fang was apparently a preventative act to evade potential political storms. The financial needs of Cloud Heaven Temple and Jade Toad Palace similarly stemmed from the damage that political campaigns did to religious sites. In this vein, the creation of pilgrimage networks in Mount Fang was indeed extraordinary. The newly created pilgrimage networks, together with the arrival of many spirit tablets, suggest that the Mount Fang religious revival was by no means the resumption of old traditions. Though the enduring influence it brought to the local religious landscape was still not identifiable at the time, such an influence would come to light in the post-Mao era, to which I will return at the end of the chapter.

The emergence of a ritual economy

What was also extraordinary about the Mount Fang religious revival is that the symbiosis between spirit mediums and these two temples in the end did much more than helping local deities survive political storms of the day. It formed the cornerstone of the brief religious revitalisation in Mount Fang in the few years after the Great Leap. As a result of responses to harsh political environments, a small-scale ritual economy emerged out of increased interactions and communications between religious leaders and specialists in Mount Fang and outside communities, especially those spirit mediums who sought to carry on business.

In the chain of Mount Fang ritual economy, one finds three groups of people: *xiangke* (literally, 'incense guest') or pilgrims, ritual brokers, and Buddhist and Daoist ritual specialists. Pilgrims, according to official investigations, are rural 'basic masses' (*jiben qunzhong*) from Wenling, neighbouring Huangyan, Yuhuan, Linhai, and Yueqing (Wenzhou), but there were also a considerable number of grassroots cadres and factory workers among the visitors.[30] Ritual brokers were mainly those spirit mediums but also included contacts of Cloud Heaven Temple and Jade Toad Palace across the county. Ritual brokers were in charge of mobilising and organising pilgrimage to Mount Fang. Viewing this from the division of labour perspective, it was mainly spirit mediums who were ritual brokers, but they were also ritual service providers. This then raises an important feature of the practices at Mount Fang: it was a ritual economy containing two types: those of spirit mediums and the salvation rituals of Buddhists/Daoists.

Most of the Buddhists and Daoists who performed rituals in Mount Fang were not residential Buddhist monks or Daoist priests. In the case of Cloud Heaven Temple, most monks or household monks (*yeheshang*) were hired from various nearby town[ship]s, who earned commission from performing rituals for Cloud

[30] 'Dui Fangshan, Yangjiaodong mixin huodong qingkuang de diaocha he jinhou yijian de baogao', 100–1.

Heaven Temple. Sometimes monks from the outside simply paid Kefang in order to use Cloud Heaven Temple for their own ritual services. Such was the case of those from Jinjila Brigade of Muyu Town[ship] who recited scriptures in Cloud Heaven Temple and gave Kefang 20 per cent of the total income.[31] In any case, providing ritual services in Mount Fang at that time must have been financially viable if not lucrative because some monks were even willing to pay their brigade cadres in order to serve at Cloud Heaven Temple.[32]

Another way to look into this ritual economy at Mount Fang is the volume of money being circulated. First of all, the income from lodges was considerable: between the first and 26th of the 5th lunar month in 1964, the lodge of Cloud Heaven Temple earned about 1,200 yuan. The archival documents contain no information regarding the income of Jade Toad Palace, but Kefang told investigators that Jade Toad Palace's income was at least three to four times theirs. This was perhaps not entirely baseless, as the scale of Jade Toad Palace was bigger than Cloud Heaven Temple. Jade Toad Palace also had more clergies and spirit mediums collaborating with them, which helped to generate more traffic. In addition to the lodges, the two temples also received income from selling items such as oracle slips (*qianshi*), incense, candles. Cloud Heaven Temple also sold many 'flaming mouths' certificates (*yankou die*) for 2–3 yuan, likely a ritual document that people burnt at home for protection or blessing.

The most important source of income for Cloud Heaven Temple and Jade Toad Palace was none other than the ritual services, which curiously flourished in the aftermath of the heavy human loss of the Great Famine. Among the ritual services performed, two types of rites stood out: *yankou* and *daochang*.[33] Cloud Heaven Temple performed the rite of feeding flaming mouths about 130 times in 1963, and 120 times in 1964, according to Kefang's estimate. An incomplete income account shows that Cloud Heaven Temple gained 1227.8 yuan from performing the ritual of feeding flaming mouths between the first and 26th of the 5th lunar month in 1964. Cloud Heaven Temple also offered *bao'anshou yankou* (the rite of feeding flaming mouths for blessing of peace and longevity) for the well-being and prosperity of families. The ritual of feeding flaming mouths is a quite expensive service, given the average income at that time. The average individual annual income in Wenling in 1962 was 99 yuan,[34] but two versions of the rite of feeding flaming mouths in

[31] 'Fangshan, Yangjiaodong wupo shenhan deng goujie qingkuang' (The situation of collusion among witches, spirit mediums, and others in Mount Fang and Goat Horn Caves), 6 June 1964, Wenling City Archives, 17-1-118: n/a.

[32] Household monk Xie Maojing of Kantou Brigade for instance was not allowed to do service in his village. He ended up paying Lin, the head of brigade security, in order to leave for Mount Fang. See 'Fangshan, Yangjiaodong wupo shenhan deng goujie qingkuang'.

[33] *Daochang* literally means Daoist rites. We do not know exactly what kind of Daoist rituals were offered in Jade Toad Palace. The investigators did not seem to know the nature of *yankou* and *daochang* and their basic differences. Sometimes they used *yankou* to refer to ritual services of Jade Toad Palace.

[34] Wenling xianzhi bianzuan weiyuanhui, *Wenling Xianzhi* (Wenling County Gazetteer). Hangzhou: Zhejiang renmin chubanshe 1992, 110.

1963, for instance, cost respectively 12 and 16 yuan,[35] which was about the income of one and a half to two months for one person in the county. So it was very common for people who were eager to use the service of the ritual of feeding flaming mouths but short of money to pool together, sometimes involving dozens of individuals, to jointly buy one performance of the ritual of feeding flaming mouths.

What can explain the large demand and proliferation of the ritual of feeding flaming mouths? Nothing in official investigations or confession letters directly or explicitly tied this to the human loss of the Great Famine. Yet one cannot stop connecting the proliferation of the ritual of feeding flaming mouths, at least in part, to the collective mentality of local communities in the aftermath of the GLF. Premature death in the entire Taizhou Prefecture during the Great Famine (1959–60) was estimated to be 10,000, which was not a huge number compared to other regions of the country.[36] Nevertheless, in a small region such a loss of population could have driven up anxieties and fears among local residents about the bad things that the ghosts of those who died prematurely might visit upon living people.[37] In this vein, the surge of the ritual of feeding flaming mouths can be viewed as a collective response to appease the souls of the dead and prevent them from harming the living.[38]

Ritual services were not the only motor in the ritual economy of Mount Fang revival. The temples spent the money from ritual and other services on repairing and building temples, roads, and other facilities, which then brought more pilgrims to Mount Fang. In such a way, ritual economy and temple restoration/expansion reinforced each other to sustain the religious revival between 1960 and 1965. Many pilgrims in fact were attracted to Mount Fang by the cause of helping to repair temples and statues of divinities. Since 1960, both temples underwent reconstruction, and Jade Toad Palace even added two caves. With the collaboration of nearby brigades, the two temples built two roads (from Jade Toad Palace to nearby places) and expanded the road connecting Cloud Heaven Temple and Jade Toad Palace, which in total cost about 7,400 yuan.[39] In Little Dipper Cave, less than a mile from Jade Toad Palace, Abbess Liu opened a new branch of Jade Toad Palace in 1964 with the permission of local brigade and commune cadres.

[35] Two versions of the rite of feeding flaming mouths in 1963 costed respectively 12 and 16 yuan. In 1964, Cloud Heaven Temple offered four versions of rites of feeding flaming mouths. Each cost respectively 8, 10, 12, and 14 yuan.

[36] Cao Shuji, *Dajihuan—1959 nian—1961 nian de Zhongguo renkou* (Chinese Population between 1959 and 1961). Hong Kong: Shidai guoji chuban gongsi, 2005.

[37] Ralph A. Thaxton, *Catastrophe and Contention in Rural China: Mao's Great Leap Forward Famine and the Origins of Righteous Resistance in Da Fo Village*. New York: Cambridge University Press, 2008, 305–6.

[38] Similar concerns of wild ghosts after the Great Famine were observed in Yi community in Yunnan. See Erik Mueggler, 'Spectral Chains: Remembering the Great Leap Famine in a Yi Community', in *Re-envisioning the Chinese Revolution: The Politics and Poetics of Collective Memories in Reform China*, eds. Ching-Kwan Lee and Guobin Yang. Stanford: Stanford University Press, 2006, 64–5.

[39] 'Guanyu Fangshan, Yangjiaodong zongjiao mixin huodong qingkuang de diaocha baogao', 244.

The economic dimension of temple activities in Mount Fang is especially striking in the early 1960s when frantic industrialisation and collectivisation of the Great Leap had left most villagers on the verge of starvation. Hill Gates argues that Chinese folk culture is a product and embodiment of a 'petty-capitalist mode of production', because of the extraordinary economic dimension of traditional Chinese religion.[40] Similarly Adam Chau uses 'religious enterprise' to stress the service-provision aspect of a temple.[41] The appearance of a 'religious enterprise' like Cloud Heaven Temple and Jade Toad Palace in the early 1960s suggests the capacity of what we might call an 'idiom of petty capitalism', which was never fully destroyed even when traditional economic structures had been largely destroyed.

In the Mount Fang case, the readiness of people to pay money for often very expensive ritual and other related services is especially telling, considering how poor people were back in the 1960s. What is equally telling is the importance of material interest in smoothing temples' relations with local communities and mobilising cadres and villagers as well as ritual specialists. Continuing to trace the social foundation underpinning the ritual economy in Mount Fang reveals an extensive *guanxi* network in nearby brigades, communes, and town[ships] that took Cloud Heaven Temple and Jade Toad Palace many years to form. In the building of *guanxi* networks, exchange of material interest appeared to be quite important if not the most important, as official reports concluded, serving as a social lubricant. For a long while, Cloud Heaven Temple did not get along with cadres and members of nearby Lantianhu Brigade of Xinjian Commune. The brigade not only tore down two houses of the temple during the period of Land Reform but also took many furniture from the temple. Though Kefang claimed that he was invited to return by a cadre of Lantianhu, he held strong grievances against Gu Xiaofu, the Party secretary of Lantianhu, especially regarding the ownership of some abandoned wood in a small temple near the Little Dipper Cave. When Cloud Heaven Temple took that wood, Gu went to ask Kefang to pay for it. The rapprochement between the temple and the brigade only came after the temple's activities started picking up. Cloud Heaven Temple lent money to some members and cadres of Xinjian Commune and the temple sent 100 yuan to Lantianhu Brigade as grain ration so that the temple could be exempted from participating in production. The brigade also considered that ritual activities were conducive to its members since they could earn extra income by helping the temple build houses and roads.[42] In fact, it was essential to maintain good relations with the local community and especially with the cadres, the revival of temple activities was almost impossible without their support, both administratively and financially.

[40] Hill Gates, 'Money for the Gods', *Modern China*, 13.3(1987): 259–77; *China's Motor: A Thousand Years of Petty Capitalism*. Ithaca: Cornell University Press, 1996.

[41] Adam Chau, *Miraculous Response: Doing Popular Religion in Contemporary China*. Stanford: Stanford University Press, 2008.

[42] 'Guanyu Yunxiaosi Yuchangong de fengjian mixin huodong he fushi ganbu qingkuang de diaoca baogao' (Investigation report on Cloud Heaven Temple and Jade Toad Palace's superstition activities and their corrupting of cadres) 23 January 1965, Wenling City Archives, 17-1-118: 4.

Compared to Kefang, Liu Peihua of Jade Toad Palace was much more socially skilful and successful. Her circle of acquaintances included the Party secretary of Lingtou Brigade (Huwu Commune), Director of Huwu Commune, chiefs of police and tax offices in Dajing Town, and Director of Daxi Commune. Yet her social circle was not just limited to cadres in nearby places. She had warmly hosted Huang Yitao, the Vice County Head of Yueqing, and cadres of Wenling County UFWD and Civil Affairs Offices. Her acquaintances even included a Vice County Head of Huangyan, a county to the north of Wenling. When the investigators went to ask about the conditions of Jade Toad Palace, Liu Fabing, Party secretary of Lingtou Brigade, not only immediately jumped to the defence of the Daoist temple but even informed Liu Peihua of the investigation on the same day.[43] As part of Peihua's plan to rebuild Little Dipper Cave as a branch of Jade Toad Palace, she received written consent from the cadres in Lantianhu Brigade and Lingtou Brigade, including Gu Xiaofu, the Party secretary of Lantianhu. They even sent one *mu* and more of housing land and five *mu* and more of forest land to the Little Dipper Cave. Perhaps as *quid pro quo*, when Jade Toad Palace rebuilt the Little Dipper Cave in March 1963, Lantianhu Brigade undertook the bulk of the construction and Gu supervised the construction himself. In February 1964, Gu agreed to transfer the household register of Daoist Ruan Meifu from Huangyan County to his brigade. Ruan was thereby able to settle in and manage the affairs of Little Dipper Cave. The investigators believed that the existence of cadres who patronised and protected Jade Toad Palace lent Liu Peihua courage to openly resist the investigation, which explains why the information about Jade Toad Palace in official reports is far less extensive than that of Cloud Heaven Temple.

Grassroots cadres providing protection and support for the temple revival in Mount Fang might have had economic incentives, but they were also motivated by their faiths. As mentioned above, cadres were also consumers of ritual services, directly or indirectly – in other words, they were also part of the pilgrim group. If we take this into consideration, Mount Fang's ritual economy in fact had two economic chains: money not only flowed from pilgrims to the spirit mediums and temples, but it also flowed from the latter to the former via cadres providing protection for religious activities in Mount Fang.

A sudden end and the ensuing story

During the second half of 1963 to early 1966, the Chinese government promoted socialist education in rural areas and some cities. In order to implement central policies on the Socialist Education Movement, Wenling County government convened a study group of religious leaders in June 1964 and in the same month,

[43] 'Guanyu Yunxiaosi Yuchangong de fengjian mixin huodong he fushi ganbu qingkuang de diaoca baogao', 4–6.

the county government launched a formal investigation into religious activities in Mount Fang. Though the local authorities were very likely aware of the existence of pilgrimage to Mount Fang given its scale, no sign indicates that any major investigation or action had been done prior to the June 1964 investigation.

In early March 1965, after three investigations conducted between June 1964 and January 1965, Wang Zhilian, an officer from Wenling County Party UFWD and several other officials from Daxi District and Xinjian Commune went to visit Cloud Heaven Temple. According to Kefang's petition letters, Wang told them that they were restoring 'feudal superstition'. The government would by no means allow them to stay in the temple anymore. The government planned to eventually eliminate religion but in Wenling they first wanted to make an example of the Cloud Heaven Temple. Kefang said that he was really frightened and did not know what to do. When he was told to submit an application (to leave the temple), the officials were not fully satisfied with what he wrote. In the end he had to record each sentence that Wang dictated to him.[44]

Official investigations provided no information on whether or how the many cadres – protectors and collaborators in Mount Fang revival – were disciplined, but the punishments for the spirit mediums, Buddhists, and Daoists were fairly light considering the harsh political environment of the day: none got criminal penalties. Some spirit mediums, Buddhists, and Daoists were seemingly only asked to write self-criticisms. Most assets of Cloud Heaven Temple and Jade Toad Palace were confiscated, and Liu Peihua and other senior Daoists were sent to the temples of Mt Yandong of Yueqing or temples where they originally worked. The young Daoists were sent to factories or their places of origin. All Buddhists were sent back to their home villages.

The officials might think that the punishments for those directing the Mount Fang revival were already very light – after all no one was sent to prison – but the fallout from their handling of the Mount Fang incident lasted until the eve of the Cultural Revolution, suggesting a delicate situation in the local politics of religion at the time. Kefang refused to leave from the beginning. In April 1965, shortly after returning to his village of birth, Kefang began to send petition letters to governments of various levels, from Wenling County People's Committee to Zhejiang Provincial Office of 'Letters and Visits' and the State Council. Kefang complained that Wang Shulian, the county official who told him to leave the temple, did not handle the Mount Fang case appropriately, in accordance with state policies and rules. He claimed that Wang only intended to coerce Buddhists into returning to non-cleric life and disregarded religious freedom. In the meantime, Kefang also filed a suit in Wenling Court against Gu Xiaofu, Party secretary of Lantianhu Brigade, with whom he had long feuds.

In January 1966, Kefang went to Hangzhou in person to deliver a letter to Zhejiang Provincial Commission of Religious Affairs. In March of the same year,

[44] 'Seng Kefang shensushu' (Seng Kefang's petition letters), 12 April 1965, Wenling City Archives, 17-1-118: n/a.

Kefang even returned to Cloud Heaven Temple and stayed until 18 July when he was taken by members of Xinjian Commune to parade through the streets. In 2 March 1967, at the height of the Cultural Revolution, Kefang once again returned to Cloud Heaven Temple. It is not clear how long he stayed in the temple this time, but, during the Cultural Revolution, most buildings were destroyed, and the remaining structures were used as facilities for a ranch. Kefang returned in 1983 when he began his second endeavour to rebuild Cloud Heaven Temple and he passed away there in 1998. Thanks to Kefang and others' efforts, Cloud Heaven Temple was fully rebuilt and even expanded, covering more than 3,000 square metres with more than 50 rooms.[45] Kefang's life after the brief Fangshan religious revival demonstrates, once again, the tenacity and adaptability of religious communities in finding space to cohabitate with Maoism and Communist rule in general.

Conclusion: The ironies of revolution

A brief but unusual revitalisation of religion appeared in Mount Fang, Wenling, at a seemingly unlikely moment in the aftermath of the Great Famine, thanks to a string of actors and factors. The ambition of Abbot Kefang and Abbess Liu Peihua to restore 'incense fire' or prosperity to their temples was one cause. Yet the invigoration of ritual activities would not have happened so quickly and successfully without the arrival of many spirit tablets brought by local communities and spirit mediums who sought refuge for their deities in Mount Fang. The accommodation of deities from outside communities in the temples of Mount Fang provided an opportunity that soon connected ritual services of spirit mediums and Buddhists/ Daoists, creating a hybrid ritual economy that hummed along for a few years before the forced closure of the temples in 1965. The Mount Fang ritual economy is characterised by the presence of many grassroots cadres. Without their administrative and financial support, a ritual economy of this scale would not have been possible and the deep involvement of cadres in the ritual economy demonstrates the limits of the Maoist state's penetration. On the one hand, the state could not even fully transform the beliefs of its principal local agents;[46] many of Mount Fang's visitors and donors were cadres, and on the other, the state had to rely on those cadres, who often sought to conspire with other local people to gain economic or other benefits – for themselves or for their communities as they learned to govern and be governed. The Mount Fang religious revival shows that the systematic destruction of the sociopolitical and economic structure of the 'old society' by the Maoist revolution did not mean that traditional social organisations and networks disappeared entirely. Their

[45] The life of Abbess Liu Peihua after Jade Toad Palace was closed is unclear. Yet the reconstruction and expansion of the palace began in 1985.

[46] Steve A. Smith, 'Local Cadres Confront the Supernatural: The Politics of Holy Water (*shenshui*) in the PRC, 1949–1966', *The China Quarterly* 188(2006): 999–1022.

visible structure was largely gone but sometimes a less visible part remained.[47] The knowledge of traditional social forms and other cultural 'tool-kits'[48] such as beliefs, rituals, and idioms of petty capitalism were deeply entrenched and diffused in everyday practices. They could be evoked at times of need, implying that religious revival was always possible, even during the Mao years. From this perspective, we should be less surprised by the flourishing of religious life immediately after the Cultural Revolution.

The irony is that Maoist rule often created the conditions for traditional idiom to be evoked even though it had fundamentally devastated the economic foundation and established leadership of rural organisations. It has been observed elsewhere in China that 'the irrationality and experienced immorality of certain state actions led villagers to turn over even more to traditional norms and forms'.[49] The making of the religious revival in Mount Fang reveals the various ways that religious practitioners reacted to and coped with extraordinary social-political and economic conditions created by the GLF and its aftermath. For instance, political mobilisation and actions to close down and demolish religious sites during the Great Leap forced followers of local deities to resort to the temples of Mount Fang. The large demand for and proliferation of the ritual of feeding flaming mouths to save souls of the dead in the Mount Fang revival could also be read as a collective response to the trauma of the Great Famine directly stemming from the Great Leap.

Consequently, the early 1960s religious revival in Mount Fang did not last long, but it is certainly not something that left no tracks in its path. If we trace the history of Jade Toad Palace, for instance, it only occupied one cave at the start in the late nineteenth century. By the time Liu Peihua was in charge in the 1960s, the temple, including the main cave, expanded to four caves where the family of Lord Yang and other local deities were worshipped. Since the reconstruction of Jade Toad Palace in the 1980s, the worship of the Yang family in Jade Toad Palace has expanded into seven caves.

The most striking scene that visitors today will discover is in the second floor of the palace's main cave (*Yangjiao yi dong*) (see Figure 10.2), where more than 40 statues of local deities with known and unknown origins, including Lad Liu, Master Wu, General Zheng, General Hu, and General Huang, make the place look almost like a pantheon.[50] Similar scenes can be seen in other caves. It should not be hard to

[47] Michael Szonyi, 'Lineages and the Making of Contemporary China', in *Modern Chinese Religion II: 1850–2015*, eds. Jan Kiely, Vincent Goossaert and John Lagerwey. Brill, 2015, 433–87.

[48] Ann Swidler, 'Culture in Action: Symbols and Strategies', *American Sociological Review* 51.2(1986): 273–86.

[49] Edward Friedman, Paul G. Pickowicz, Mark Selden, and Kay Ann Johnson, *Chinese Village, Socialist State*. New Haven: Yale University Press, 1991, 269.

[50] Pan Junliang, 'Yandangshan he daojiao' (Yandang Mountain and Daoism), Conference Proceeding: 'The First Japan-France Chinese Religious Researchers Conference: Holy Land in Chinese Religion: Cosmology, Geography, Body Theory' (Dai 1-kai Nihon Furansu Chūgoku shūkyō kenkyūsha kaigi: Chūgoku shūkyō ni okeru seichi — uchū-ron-chi rigaku karada-ron), 2014.

Figure 10.2 A corner of Goat Horn Cave No. 2 (Courtesy of Junliang Pan)

speculate that some of the deities whose spirit tablets were sent to Jade Toad Palace during the Mao years were permanently settled there even after political circumstances have subsequently changed.

Looking back on Jade Toad Palace's history, the 1960s revival, one can argue, is in fact a critical period of reinvention and development: the Daoist temple vastly expanded its influence by accepting numerous outside deities while essentially capitalising on the popularity of the worship of Lord Yang. The 1960s revival for Jade Toad Palace was certainly not a period of disruption that the Maoist era is often associated with in terms of religious life. In this vein, the Mount Fang case invites us to rethink what the Mao years meant for religion. It should be safe to say that, in certain areas like Mount Fang, the engagement with Maoism did engender something new and innovative. The collaboration of spirit mediums with Buddhist/Daoist clergies may be a tradition, but the particular kind of religious networks and landscape created in the early 1960s is new and has a significant influence on the present. If we turn to other religious traditions, house churches are perhaps an even better example. House churches as we know today are largely a result from the efforts by Chinese Christians to adapt church life to Maoist rule. In Wenzhou, Zhejiang, the political storms of the Great Leap forced Christians to completely turn to house meetings, which then dramatically brought about a series of far-reaching organisational transformations.[51] As exemplified by the temples of Mount Fang and the house churches of Wenzhou, the socialist system of the Mao era paradoxically catalysed the (re)making of religious communities, fostering religious innovations and transformations that continue to shape China's religious landscape today.

[51] Xiaoxuan Wang, *Maoism and Grassroots Religion: The Communist Revolution and the Reinvention of Religious Life in China*. New York: Oxford University Press, 2020, chapter 4–6.

Part IV

Expertise and Revolutionary Epistemology

11

How Geography Won the Battle and Geographers Lost the War

SHELLEN X. WU

> The giant tsunami of the people's revolution has overturned Old China and liberated geography. Inevitably, following the coming wave of economic construction we must look to create a high tide of cultural construction. Geography has a wide-open future. Geographers too must make progress. They have begun to dismantle the old thinking of individualism and dogmatism. They must earnestly and industriously work for the people and build a people's geography.[1]

THE 1950 INAUGURAL issue of the journal *Geographical Knowledge* (*Dili Zhishi*) set down the ideological foundations for the field of geography in a newly established Communist China. Geography, the preface to the issue intoned, is a sociological as well as scientific discipline. Not just an academic field, geography impacted the daily life of average citizens. In the pre-revolutionary Old China, the editors argued, geography was held captive by the ruling classes and kept separated from the masses. The people's revolution, however, has washed away Old China and, in the process, liberated geography. Going forward, therefore, geographical research would rely on the masses to reconceptualise the discipline. The strident tone of the inaugural editorial creates the illusion of a radical break with the past, giving the impression that henceforth a new geography would develop as befitting a New China. In the subsequent decade the pages of *Geographical Knowledge* showcased examples of progress from around the country in the building of a New China, from dam construction to terraced agricultural efforts. This depiction of dramatic reconceptualisation of the field of geography, however, intentionally concealed significant continuities in the spatial and territorial imagination in twentieth-century China across political divides.

[1] *Dili Zhishi,* Vol. 1, Issue 1 (1950), p. 1.

Proceedings of the British Academy, **267**, 211–231, © The British Academy 2025.

In recent years Sinologists have pointed to continuities in various aspects of state and society between the Republican era and post-1949 'New China'. A modern conceptualisation of Chinese territoriality follows this narrative – arising during the late Qing, taking shape during the Republican period, and finally becoming fixed during the early years of the People's Republic (PRC). Both the Kuomintang (KMT) regime in the Republican period and their communist successors after 1949 subscribed to a territorial vision of China that incorporated the imperial expanse in the national geo-body, which Thongchai Winichakul defined as 'a man-made territorial definition which creates effects – by classifying, communicating, and enforcement – on people, things, and relationships'.[2] The concept of the geo-body has deeply influenced modern China historians, who have connected the concept to the rise of Chinese nationalism.[3] Elsewhere, I have argued that territorial conceptions in modern China were concomitant with its entry into geo-modernity, as defined by clearly demarcated borders and the use of the latest science and technology to develop borderlands through agriculture and the intensive exploitation of natural resources.[4] The emergence of the Chinese geo-body and its embrace of geo-modernity took place concomitantly during a turbulent period of transition between empire and the formation of the nation-state.

From this perspective, the PRC is a success story – a unique case of a modern nation that has retained the territorial extent of the last empire. The individuals who helped to create this notion of an inviolable territorial body of modern China, however, fared far less well. This chapter examines these two opposing trajectories – of the discipline of geography, which became essential to defining the Chinese geo-body, and of the geographers, who found themselves imminently dispensable to the nation – by moving both forwards and backwards in time from this starting point in 1950. The case study of geography further complicates our understanding of the 1950s and both the continuities and ruptures that occurred during the founding decade of the new regime. The divergent fates of geography and geographers showcase the contingent and seemingly arbitrary nature of how a discipline and its practitioners became targets of political attacks. It is also a story

[2] Thongchai Winichakul, *Siam Mapped: A History of the Geo-Body of a Nation*. Honolulu: University of Hawaii Press, 1994, p. 17.

[3] See Robert Culp, *Articulating Citizenship: Civic Education and Student Politics in Southeastern China, 1912–1940*. Cambridge, MA: Harvard University Asia Center, 2007, p. 74; Peter Zarrow, 'The Importance of Space', in *Educating China: Knowledge, Society, and Textbooks in a Modernizing World, 1902–1937*. Cambridge: Cambridge University Press, 2015, chapter 7, pp. 214–45; William A. Callahan, 'The Cartography of National Humiliation and the Emergence of China's Geobody', *Public Culture*, Vol. 21, Issue 1 (2009), pp. 141–73; Bill Hayton, 'The Modern Origins of China's South China Sea Claims: Maps, Misunderstandings, and the Maritime Geobody', *Modern China*, Vol. 45, no.2 (2019), pp. 127–70.

[4] Shellen Wu, *Birth of the Geopolitical Age: Global Frontiers and the Making of Modern China*. Stanford: Stanford University Press, 2023, p. 2.

of how the categorisation of knowledge mattered greatly – in many cases an innocuous decision later meant the difference between life and death – for individuals navigating the new regime.[5]

The many geographies of China

At the time of the inaugural issue of *Geographical Knowledge*, the journal was published by the Geosciences Society (*dixue hui*), with the editorial committee based in the geography department at Nanjing University. The term *dixue* served as an umbrella term that covered the full range of subjects in the Earth sciences from the late Qing, incorporating traditional epistemologies that included geography (*dili*) and newly introduced sciences from the West like geology (*dizhi*).[6] Within *dixue*, attention particularly focused on control over the country's natural resources. By the end of the Qing dynasty, the initial official and merchant interest in developing Chinese mines to fuel nascent industries had turned into outright alarm against what was seen as an insatiable foreign encroachment on Chinese territorial sovereignty in a competition over natural resources, particularly coal and precious metals.[7] In 1909, a group of like-minded intellectuals founded the China Geoscience Society (*Zhongguo dixue hui*) in Tianjin. In 1910, one of the founding members of the society, Zhang Xiangwen (1867–1933), and other members published the first Chinese periodical devoted to geography and geology, *Geoscience Magazine* (*Dixue zazhi*).[8]

In addition to founding the *Geoscience Magazine,* Zhang served as the President of the China Geographical Association. In his later years, he wrote *Collected Works on Earth Sciences* (*Dixue congshu*) and other textbooks on human geography. Zhang represented a generation of scholars who had grown up inculcated in the values of the classics and advancement in the civil examinations. That world collapsed precipitously in the first decade of the twentieth century. In Zhang and others like him, including Bai Meichu (1876–1940), who served as the editor-in-chief of the *Geoscience Magazine* for 25 years during its early decades of publication, the intellectual values of the late Qing continued to inform the study of geography in

[5] This was true not only for geographers but also statisticians. In the 1950s, pointed attacks on social statistics were particularly targeted at the state statistics bureaucracy. The statistician Xu Baolu (1910–70), who pioneered probability study and mathematical statistics in China, on the other hand suffered no repercussions because he was based in the Department of Mathematics at Peking University rather than employed by the state statistical apparatus. Arunabh Ghosh, *Making it Count: Statistics and Statecraft in the Early People's Republic of China*. Princeton: Princeton University Press, 2020, pp. 122–3.

[6] Wang Yangzhi, *Zhongguo dizhi xue jianshi,* p. 7. Wang cited an article in *Zhongguo dizhi bao* from 9 April 1984, which discusses the term.

[7] Shellen Wu, 'The Search for Coal in the Age of Empires: Ferdinand von Richthofen's Odyssey in China, 1860–1920'. *The American Historical Review,* Vol. 119, no. 2 (April 2014): pp. 339–62.

[8] Zhang Xinglang, 'Xiyang zhangdun yuju shi nianpu yijuan', *Dixue zazhi,* no. 2 (1933), pp. 1–51.

the Republican period. Zhang penned the preface to the first issue of *Geoscience Magazine*. The bulk of the introductory remarks in the first issue conjured up a social Darwinian vision of violent conflicts between nations and the survival of the fittest.[9] Zhang aligned geographical and geological research with the effort to protect China's borders and secure its frontier territories, as well as the means to exploit the country's subterranean riches.

Zhang's writing on the eve of the Republic employed a trope that would become ubiquitous in geological and geographical writings in the subsequent decades: that China was both large and rich in natural resources (*dida wubo*). These resources would be essential for building the nation (*jianguo*). Writings on the frontiers frequently used a variation of the phrase, *diguang renxi,* to describe the sprawling but sparsely populated borderlands in a distillation of the perceived disparities in population distribution across the Chinese geo-body. During the Republican period, both popular writers and geographers of the 'New Geography' school, who boasted overseas educational backgrounds and scientific outlooks, commonly wrote about the disparity between the density of population in coastal provinces and the allegedly open and wholly undeveloped lands on the frontiers, making these areas ripe for Han Chinese migration. Few considered the ethnic minorities who lived in these regions and their resistance to Han settlers encroaching upon their lands.[10] The German historian Robert Nelson describes this paradox of such writings this way, 'the realization that land was both empty and full at the same time, empty for colonizers, but full of "problem" populations, was at the heart of inner colonization'.[11]

China was not the only country in the world where geographers struggled to define the professional boundaries of the discipline. American geographers, for example, eagerly provided their services to the American government to gain institutional legitimacy. Both the academically oriented American Association of Geographers and its popular counterpart the National Geographic Society planned popular lecture series on an eclectic range of topics such as the exploration of Alaska, Venezuelan boundary disputes, conflicts in the Transvaal and Manchuria, progress in the Philippines, Arctic exploration, and the prospect of building a canal in Nicaragua. This programming openly sought to bolster support for American

[9] Zhang Xiangwen, 'Zhongguo Dixue Hui Qi', *Dixue zazhi* Vol. 1, No.1, (1910), p. 1.
[10] James Leibold, *Reconfiguring Chinese Nationalism: How the Qing frontier and its Indigenes became Chinese*. New York: Palgrave Macmillan, 2007; Ge Zhaoguang, 'Absorbing the "Four Borderlands" into 'China': Chinese Academic Discussions of "China" in the First Half of the Twentieth Century', *Chinese Studies in History*, 48:4 (2015), pp. 331–65. This theme carried over from the Republican into the PRC period. See Benno Weiner, *The Chinese Revolution on the Tibetan Frontier*. Ithaca: Cornell University Press, 2020, p. 4; Andres Rodriguez, 'Building the Nation, Serving the Frontier: Mobilizing and Reconstructing China's Borderlands during the War of Resistance (1937–1945)', *Modern Asian Studies* 45, no. 2 (2011): pp. 345–76.
[11] Robert Nelson (ed.) *Germans, Poland, and Colonial Expansion to the East, 1850 to the Present*. London: Palgrave Macmillan, 2009, chapter 3, 'The Archive for Inner Colonization, The German East, and World War I', p. 74.

expansion overseas by educating the public about these far-flung locales.[12] Meiji-era Japanese reformers promoted the study of geography as a key subject for building a new nation and what separated them from the backwards countries in Asia.[13]

Aside from a common interest in promoting nationalism, geographers in the Republican period belonged to a large and unwieldy discipline that splintered in multiple directions, incorporating both traditional historical geography and newly imported geosciences. Geologists dominated the scientifically oriented branch of physical geography, which emphasised fieldwork and empirically based analysis. In 1933, this group of scientifically oriented geographers founded the flagship journal of Chinese geography, *Acta Geographica Sinica* (*Dili Xuebao*). The meteorologist Zhu Kezhen, geologists Ding Wenjiang, Li Siguang, and geographer Hu Huanyong were among the founding members, as was historian Gu Jiegang. At the same time, various figures who were holdovers from the late Qing, including Zhang Xiangwen, continued to publish and advocate for a geography based in history and take part in compiling local gazetteers (*difangzhi*), a traditional genre that continued to flourish in the twentieth century.

Geographical Knowledge emerged out of this entangled and multi-faceted history. The first issue in 1950 provided a synopsis of the history of the flagship journal of geography in the Republican period, the *Acta Geographica Sinica*, and the affiliated Geographical Society. According to this synopsis, the Society incorporated various sub-fields of geography, including geology and meteorology. Membership in the organisation grew over the 1930s and 1940s from 60 members at the time of its founding in 1933 to 350 members in 1950, and an additional 50 permanent members.[14] Over the course of these decades, geographers from both the humanities and the sciences lobbied for the inclusion of geography in Academia Sinica, the national academy of sciences established in 1928. The decentralised nature of political power in the period meant that multiple geographical societies had sprung up around the country, along with departments of geography at universities, both at private institutions funded by missionary groups and government-backed universities. Intellectuals and political figures from across partisan divides agreed on the importance of geography to instilling nationalism but not on the actual content and theoretical foundations of the field.

The work of geographer Hu Huanyong, a frequent contributor to scientific publications and chair of the geography department at Central University (*Zhongyang daxue*), demonstrates some of the tensions in the field. During the war, in a collaboration with the political department of the KMT Military Affairs Committee, Hu penned a work of military geography to be used in KMT party training schools. The work, *Geography of National Defense* (*Guofang dili*), explicitly connected geography, national defence, and the need for resource management.[15] Such works

[12] Susan Schulten, *The Geographical Imagination in America, 1880–1950*. Chicago: University of Chicago, 2001, p. 66.

[13] Stefan Tanaka, *Japan's Orient: Rendering Pasts into History*. Berkeley: University of California Press, 1993, p. 63.

[14] Song Jiatai, 'Zhongguo dili xuehui gaikuang', *Dili Zhishi,* Vol. 1, Issue 1 (1950), p. 2.

[15] Hu Huanyong, *Guofang Dili* (Guomin zhengfu jushi weiyuan hui zhengzhibu bianyin, 8.21.1938), p. 2.

were very popular during the war, but what distinguished Hu's work is the section on border defence. Hu catalogued Korea as a former *shuguo*, or dependency. Hu also listed both Burma and Vietnam as *fanshu*, or outer dependencies, before their occupation by the British and the French respectively.[16]

The Japanese invasion may have radicalised geographers, who saw the war as posing an existential risk to China, but Hu had also picked up a strain of geographical thought dating to the late Qing, when intellectuals like Liang Qichao wrote extensively on history and geography, linking the two fields together with the development of civilisation. In his writings, as in earlier imperial era Chinese compendia of knowledge, Liang placed geography under the general heading of history. He separated China into two geographical components: the 18 provinces of the traditional heartland and the dependent regions (*shubu*), including Manchuria, Mongolia, and Tibet.[17] For Liang, the 'people of Asia' with whom the Chinese had interaction included only those successfully incorporated into the Qing Empire (the Tibetans, Mongols, Tongus, Xiongnu, Manchu, and the Han), a concept that corresponded to the territorial extent and constituency of what he was configuring as the modern Chinese nation.[18] On this basis, Liang Qichao advocated for the retention of empire under the umbrella of Chinese nationalism. Half a century later, in the struggle for survival, geographers like Hu once again staked the claims of the Chinese empire to bordering states, as well as the ethnic minorities within what they considered the rightful borders of a Greater China.

In Hu one sees the conjunction of the two dominant threads in early twentieth-century Chinese geography: the adoption of modern science and the surprising resilience of the traditional spatial epistemology. Hu had first coined the Heihe-Tengchong line in 1935 as a 'geo-demographic demarcation line'. East of the line contained 36 per cent of the territory of the nation, but 96 per cent of the population; west of the line was 64 per cent of the land area with only 4 per cent of the population. The Hu Huanyong line became a startling visualisation of the Republican mantra of *diguang renxi*, open lands sparse populations, and as a call to resettle Han Chinese excess populations further west and in the borderlands. Hu's background in human geography and ties to the KMT opened him to political attacks during the PRC. Hu was labelled a rightist in the 1950s, although his ideas did not differ dramatically from other ideas popular among intellectuals from across the political spectrum in the Republican period.

Geographers were not the only ones in the academic community to take up the patriotic call for the defence of the frontiers and borderlands. Developmental arguments filtered into the country through newly created disciplines in the social sciences. The career agricultural economist Tang Qiyu illustrates one such path.

[16] Ibid., section five.

[17] Liang Qichao, 'Zhongguoshi xulun', in *Yinbingshi wenji dianxiao*, eds. Wu Song *et al.* Kunming: Yunnan Jiaoyu chuban she, 2001, Vol. 3, p. 1802.

[18] Rebecca Karl, *Staging the World: Chinese Nationalism at the Turn of the Twentieth Century.* Durham: Duke University Press, 2002, p. 152.

Tang had received his Ph.D. in agricultural economics from Cornell University in 1924 with a thesis on 'An Economic Study of Chinese Agriculture'.[19] Already in his dissertation one could clearly trace the influence of widely circulating ideas about excess population and land scarcity in China. Tang argued that inner colonisation provided the one legitimate solution to surplus populations in China's densely populated core regions. By encouraging large-scale migration to the sparsely populated peripheries: Gansu, Yunnan, Xinjiang, Mongolia, and Manchuria, the country could address multiple issues with one solution.[20] Such migration would be beneficial in multiple ways by increasing the food supply, raising the standard of living for everyone, and producing the raw materials for the nation's industrial progress.

Tang posited that racism has closed prospects for Chinese migrants overseas, as Australia, South Africa, and the US respectively closed their borders and erected various legal barriers to Chinese settlers. For Tang, inner colonisation of China's own frontiers would provide the next best alternative. Tang wrote, 'China needs her millions to develop her natural resources and to construct her transportation system rather than send them to work for wages in other countries.'[21] Tang could look to efforts dating to the late Qing for precedent. From the late Qing, a series of efforts to develop frontier lands, introduce new crops, open small industries and mines to tap into local natural resources, had met with limited success. Sichuan provincial officials who sought to develop strategically important regions in eastern Tibet had difficulty recruiting Han Chinese settlers willing to farm in harsh and unfamiliar terrain. The few recruits they scrounged up either never turned up in the first place, absconded immediately upon seeing the remoteness of the area and the harsh climate, or abandoned the land after some months of making little or no progress.[22] Other regions faced similar challenges in sustaining development. Private enterprises quickly failed in the absence of substantial state investment in infrastructure.[23]

Upon his return to China after his overseas education, Tang Qiyu found a ready audience for his ideas. In early 1925, the warlord Feng Yuxiang established a 'Northwest Reclamation Planning Society' to draw up large-scale plans for northwestern development. Ambitious plans were reduced to modest levels when promised funds failed to come through from the central government, but the society did arrange a visit from a survey team headed by Tang Qiyu to Inner Mongolia.[24] With his return to China, Tang returned to writing and publishing in Chinese. What he

[19] Chi Yu Tang (Tang Qiyu), 'An Economic Study of Chinese Agriculture' (Ph.D. dissertation, Cornell University, 1924).

[20] Ibid., p. 272.

[21] Tang Qiyu, 'An Economic Study of Chinese Agriculture', p. 274.

[22] Xiuyu Wang, *China's Last Imperial Frontier: Late Qing Expansion in Sichuan's Tibetan Borderlands*. Lanham: Lexington Books, 2013, pp. 211–12.

[23] Ding Changqing, 'The Development of Capitalism in Modern Chinese Agriculture', in *The Chinese Economy in the Early Twentieth Century*, ed. Tim Wright. New York: St. Martin's Press, 1992, pp. 134–51.

[24] Justin Tighe, *Constructing Suiyuan: The Politics of Northwestern Territory and Development in Early Twentieth Century China*. Leiden: Brill, 2005, pp. 122–23.

described in English as 'inner colonization', he translated as *tunken* in Chinese, resituating inner colonisation as part of an unbroken Chinese historical tradition dating back to the first empire.

Twentieth-century Chinese writers rarely ever referred to Han settlement efforts using the Japanese neologism for colonialism, *shokumin,* preferring the classical Chinese terms for settlement and land reclamation, *kaiken,* or a newly coined word for military colonies, *tunken.*[25] Related to the terms *tuntian* and *kaiken,* both of which appear frequently in historical records dating from the Han dynasty, *tunken* did not come into vogue until 1920.[26] The etymology of the term strongly suggests that the popular adoption of *tunken* was popularised as a response to China's perceived besiegement from imperialist powers.

During the Japanese invasion, Tang Qiyu continued to actively promote inner colonisation. In 1938 he authored a pamphlet on refugees and reclamation, which was published by the Jiangxi Provincial Reclamation Bureau. Tang listed the various historical examples of deploying refugees to cultivate and reclaim wastelands. Refugees would become a fount of labour and a lifeline for the survival of the Han nationality. The nation needed not only warriors who shed their blood on the battlefield, but also farmers who through their sweat and labour helped to transform the wilds into farmlands. Science and technology would aid the work of reclamation.[27] The florid language of the pamphlet contrasted vividly with conditions on the ground. Jiangxi did open a reclamation zone in June of 1938. However, its operation stalled from the start because of a lack of funding. Out of a planned budget of Ch$19,000, only Ch$6,500 was dispensed. As a result, the planned construction of irrigation, roads, and buildings failed to materialise.[28]

In 1943, Tang participated in one of the wartime Nationalist Government's signature scientific research programmes: the Northwest Expedition to Xinjiang. Sponsored by the wartime Chongqing government and headed by Academia Sinica, the Northwest Expedition sought to reconnoitre and lay the foundation for the future development of the region. In addition to exploration and survey of the natural resources in the northwest, various wartime government agencies came up

[25] The August 1912 document, *Mongol Treatment Provisions,* issued by President Yuan Shikai, specifically enjoined the Central Government from using terms like 'dependent' [lifan] or 'colonial' [zhimin], to refer to Mongolia. Justin Tighe, *Constructing Suiyuan: The Politics of Northwestern Territory and Development in Early Twentieth Century China.* Leiden: Brill, 2005, p. 193.

[26] Qing documents referred to *tuntian* for military agricultural colonies on the frontiers, for example, in the northeast. James Rearon-Anderson, *Reluctant Pioneers: China's Expansion Northward, 1644–1937.* Stanford: Stanford University Press, 2005, pp. 78–9; James Millward, *Beyond the Pass: Economy, Ethnicity, and Empire in Qing Central China, 1759–1864.* Stanford: Stanford University Press, 1998, pp. 40–1; Peter Perdue, *China Marches West: Qing Conquest of Central Eurasia.* Cambridge, MA: Belknap Press, 2005, p. 345; Yuxin Peng, *Qingdai tudi kaiken shi.* Beijing: Nongye chuban she, 1990. On the use of *tunken* to refer to land reclamation bureaus in Xinjiang, see Peter Lavelle, *The Profits of Nature: Colonial Development and the Quest for Resources in Nineteenth-Century China.* New York: Columbia University Press, 2020, p. 119.

[27] SHAC, 23-3150.

[28] SHAC, 23-3092.

with blueprints for the post-war development of the Northwest. Among the proposals, the Ministry of Forestry department overseeing reclamation would move to the Northwest, as would a national agricultural experimental station. The government would situate in the region the industrial experimental department of the Ministry of Finance and a branch of the national geological survey. Finally, the Social Science Institute of Academia Sinica would be moved to Jiuquan in Gansu Province in the Northwest.

Tang continued to advocate for the inner colonisation of the Chinese frontier in the subsequent decades, including after the Communists came to power in 1949. Unlike many geographers who promoted similar ideas about reinforcing the frontiers through state-sponsored Han Chinese settlement, Tang quickly found his place in the new regime. Despite serving in various positions in the KMT wartime government, including as the deputy director of Department of Reclamation in the Ministry of Agriculture and Forestry, Tang managed the Republican-PRC transition with minimal political exposure, in part because, unlike the geographers, he was not part of the Chinese Academy of Sciences. Tang returned to Shanghai, where he was in the Agricultural Department's cadre school when a stroke forced him to retire. The health crisis turned out to be a blessing in disguise. Tang availed himself of the various resources in the Shanghai Library to work on a massive agricultural history book and only stopped writing with the outbreak of the Cultural Revolution.[29] He passed away in 1978 in Shanghai after suffering another stroke.

Tang's ideas about inner colonisation, the settlement of Han Chinese excess population in the Northwest, were also popular amongst geographers. The disciplinary demarcations turned out to be crucial, however. As an agronomist, Tang continued working well into the 1950s. His poor health, rather than past service in the KMT regime, led to his retirement. Geographers, on the other hand, quickly became embroiled in political attacks. The politicisation of geography began during the Republican period. Escalating tensions in the northeast between the Japanese and the Chinese military galvanised intellectuals in the rest of China, who saw the defence of the frontier as a patriotic duty, regardless of their political affiliation. Geographers launched new research programmes in support of this duty to the nation. The war brought about the geographers' long sought after inclusion in Academia Sinica and, in turn, their downfall.

Geographers during the crucible of war

Lobbying efforts to include geography among the disciplines at Academia Sinica reached fruition during the Japanese invasion in the 1940s. The wartime iteration of the Chinese Geographical Association was formed in the wake of the Japanese

[29] Tang Qiyu, *Zhongguo nongshi gao*. Beijing: nongye chubanshe, 1985; Tang Qiyu, *Zhongguo zuowu caipei shigao*. Beijing: nongye chubanshe, 1986.

invasion and retreat of the Nanjing government to the interior. The Association, along with a new Institute of Geography, were finally brought under the umbrella organisation of Academia Sinica. The new Institute of Geography (*Zhongguo dili yanjiu suo*) formally launched with four subgroups, in physical geography, human geography, geodesy, and oceanography, each with around 40–50 employees. The war exacerbated the difficulties of coordination between different units within the Institute. At the same time, all units emphasised fieldwork to the extent possible given the wartime constraints. Each research group sought out collaborations with different universities, most of which had relocated to the interior from major cities in coastal areas during the war. Oceanography was based on the Fujian coast at Xiamen University, but its work was largely dormant because of the lack of funding. The geodesy unit collaborated with the Tongji University geodesy group and, in 1943, moved to a village in Nanxi in western Sichuan. The rest of the geography group carried on their work in Beibei outside Chongqing, where the rest of Academia Sinica had also decamped. Within three months of its founding, the geographers formed two fieldwork groups. Led by Li Chengsan, Lin Chao, and six other geographers, the first group spent nine months in the field surveying the upper reaches of the Jialing River. A second group spent nine months in the field researching the valley between the Da Ba mountains in northeastern Sichuan and Hubei.

One of the leading geographers in the country, Huang Guozhang (1896—1966), contributed the preface to the inaugural issue of the *Journal of Geography* (*Dili*), the Geographical Association's wartime publication. In 1909, Zhang Xiangwen conjured up a social Darwinian vision of nations around the world vying for survival. Writing during an existential war for the survival of an independent China, Huang echoed these themes. A Hunan native, Huang had received his graduate degree in geography from the University of Chicago and returned to China in 1928 to take up a teaching post in Nanjing. Like many leading intellectuals, during the war he retreated with the KMT army into the interior and helped to establish the Geographical Association in December 1939 in Beibei, a village outside Chongqing, where Academia Sinica set up its wartime offices.[30] For Huang, geography could serve as a compass both for China's diplomatic efforts and its domestic policy. All organisms compete to survive and must adapt to their environment, according to Huang, but mankind goes a step further, to not only adapt to the environment but exploit its value.[31] Even with the forced move to Sichuan, an incomplete library of reference materials and transportation difficulties, geographers continued their research efforts in the field and in the pages of the publication. The *Journal of Geography*, Huang stated, amplified those efforts by broadcasting them to a broader reading public, as well as university students, middle school geography teachers, and all those with an interest in geographical research.[32]

[30] Huang Guozhang, *Dili,* Vol. 1 1(1941), p. 3.
[31] Ibid., p. 2.
[32] Ibid., p. 3.

Given these stated goals and the ongoing war effort, content in the subsequent issues of the *Journal of Geography* unsurprisingly focused largely on practical issues, including resource surveys. In the second issue, Huang contributed another article on the mission of secondary school geography teachers.[33] A nation is like a family, Huang opines, and just as a family would need to know how much land it possessed, so would a nation need to ascertain its boundaries. Other articles introduced German military geography and examined the impact of geography on the war effort in China, as well as the geographical foundation of World War II (WWII). Zhou Lisan penned an article on wartime migration patterns.[34]

Since geographers in the Association could only continue fieldwork in regions not under Japanese occupation, the bulk of the research articles dealt with areas traditionally considered borderlands, from Sichuan to Tibet and the Northwestern regions. Many of the articles examined the mapping of natural resources, transportation, and routes of future surveys. The content of the journal focused on issues of geopolitical concern, including the search for resources. Lin Chao contributed an article on the coal industry in Sichuan, in the vicinity of Jialing River.[35] The combined third and fourth issues in 1942 concentrated on Sichuan and the Tibetan and Qinghai plateaus. In 1944, yet another combined issue focused on the Northwest, including maps for a planned route to conduct a geological survey of the Qinghai Gansu region, reports on an expedition to Xinjiang, and gold mining in Altai.[36]

Some of the same figures who published in the *Journal of Geography* also spearheaded field expeditions. Geographers in the newly established Institute of Geography devoted considerable time and effort to fieldwork and conducting resource surveys. In 1946, the director of the Geographical Institute, Li Chengsan, reported on the institute's work in the last six years and strategic planning for the future.[37] Again, echoing the words of Zhang Xiangwen almost three decades earlier, Li spoke of the vastness of Chinese territory, the large population, diversity of ethnicities, and abundance of resources in the country. Yet, he pointed out, on questions of the frontiers, they need to consult foreign geographers and scientists, a matter that should cause embarrassment for the country. Li regarded national defence as the most important rationale for the founding of the Institute.

Despite the patriotic fervour with which geographers threw themselves into their research and their eagerness to contribute to the war effort, ominous signs already appeared at the end of the war that the state viewed geography as essential but geographers as expendable. In 1946, then president of the Chinese Geographical Association Li Chengsan wrote a report on the current state of geography and the prospects for the field after almost a decade of war.[38] Under the severe constraints

[33] Huang Guozhang, 'Zhongxue dili jiaoshi de lianzhong shimin', in *Dili,* Vol. 1, 2 (1941), pp. 121–6.
[34] Zhou Lisan, 'Zhanshi yimin dili yanjiu zhiyilie', *Dili,* Vol. 3, 1 and 2 (1943), pp. 1–4.
[35] Lin Chao, 'Jialing jiang sanxia meiye dili', in *Dili,* Vol. 2, 1 and 2 (1942), pp. 45–59.
[36] *Dili,* Vol. 4, 1 and 2 (1944).
[37] Second Historical Archives of China (SHAC), Nanjing: 393–2102.
[38] SHAC. 393–2102.

of wartime conditions, the society continued its research programme, carrying out surveys and participating in the long-term project on the Northwest. Yet, the end of the war brought no reprieve. Instead, the geographers found themselves isolated from the rest of the staff and members of the Academia Sinica when their building in Beibei was sold and they were moved to a separate building several kilometres away. With the end of the war, severe financial and material shortages continued. Staff morale further plummeted when Academia Sinica moved back to Nanjing but the geographers were left behind in Sichuan with no sense of when they might rejoin the larger organisation.

Noting all the challenges, Li Chengsan nevertheless ended his report on an optimistic note. Geography in China, he noted, was still in its infancy. Many people still did not see geography as a science. Li exhorted geographers to redouble their efforts to prove their place and their value to the larger national scientific organisation. The Institute continued to propose research projects in the post-war period, including research already begun during the war on the agricultural development of Qinghai and new research on silkworms, part of a reconstruction effort to rebuild the silk industry.[39] These post-war research projects proceeded under enormous financial pressures. The absurd budget requests for this research, including the Ch$18,750,497 budget request for four months of support for a silkworm study, exposed the hyperinflation plaguing the post-war economy and deterioration of the political situation for the KMT regime.[40] On 6 June 1949, the Geographical Institute completely ran out of funds to pay employees and had to disband.[41] Even as they ran out of funding, on the eve of the Communist takeover, Academia Sinica continued to plan fieldwork for 1950.[42] In 1949, the geological institute sent a team to Yunnan to investigate the copper and tin deposits in the vicinity of the Dayi mountains. The geologists planned to maintain 15 researchers working on the northeastern Sichuan/Subei area, with the expectation of spending at least a third of their time in the field.

Communist victory and the return of the country's leading geography organisation to Nanjing appeared to herald a new era. However, the issues that Li Chengsan alluded to in 1946, including the marginalisation of the geographers by Academia Sinica, were never resolved. Ideological debate about the place of geography among the sciences continued to haunt the discipline into the 1950s. Articles in the first year of *Geographical Knowledge*'s publication singled out for criticism the geo- and environmental determinism of geopolitical writings, particularly popular during the war. The Japanese invasion had spurred interest in the development of mining industries in China's peripheries and in the geographical study of vulnerable borderlands. China's desperate War of Resistance against Japanese invasion

[39] SHAC, 393–1476.
[40] SHAC, 393–2616.
[41] SHAC, 5.6793.
[42] SHAC, 393–2729.

motivated geographers, who threw their full support to the war effort and embraced the science of geopolitics.[43]

Geographers and geologists all sought to contribute to the war effort through their expertise. Various geographers wrote popular works illustrating the geography of the country and proposing ways for the country to meet its wartime industrial needs in the retreat to Chongqing. The geologist Weng Wenhao joined the wartime government by becoming director of the Industrial and Mining Adjustment Administration and put into action his views on the relationship between science and industry by overseeing the evacuation and relocation of Chinese industries.[44] KMT official Zhu Jiahua (1893–1963), a geologist by training who entered politics in the 1930s, encouraged and supported the research emphasis on the frontier regions of the Institute of Geography and the Geological Survey of China, which had relocated to Sichuan with the retreating Nationalist Government. Zhu not only funded the scientific study of the frontiers but also attempted to foster the building of a Nationalist Party structure in the Northwest, including in Xinjiang.[45] Into the 1950s, Zhu continued to advocate for a United Front against Communism from Taiwan and to inveigh against separatist movements in the borderlands.

Building New China from the outside in

From 1949, the PRC's consolidation of power finally established stability for social scientists to safely conduct research in the frontier regions. A grade school textbook of geography, written within the first months of Beijing's fall to Communist forces in 1949, clearly demonstrates the impact of the political change.[46] In the textbook all mentions of the KMT included the label of reactionaries, while introducing the Marxist historical narrative of China's liberation from feudal oppression:

> In the past the Han people represented the landlords. The capitalist political operatives promoted the "Greater Han Ethnic Principle", and oppressed the Mongolians, the Hui, the Tibetans, the Miao, the Yao and other minorities. Henceforth under the leadership of the Communist Party and the people, all ethnic minorities within the country could enjoy true freedom and equality.[47]

The rousing rhetoric of self-determination, however, retains certain similarities to the pre-1949 tone of paternalism: 'The Tibet question should be left for the Tibetan

[43] The war created severe shortages of raw materials for industries, in addition to millions of refugees who were displaced from their homes. See Micah Muscolino, 'Refugees, Land Reclamation, and Militarized Landscapes in Wartime China: Huanglongshan, Shaanxi, 1937–45' *The Journal of Asian Studies,* Vol. 69, Issue 2 (May 2010), pp. 453–78.

[44] Grace Shen, *Unearthing the Nation: Modern Geology and Nationalism in Republican China.* Chicago: University of Chicago Press, 2014, p. 155.

[45] Academia Sinica archives, Zhu Jiahua papers, 301-01-15-018; 301-01-15-019.

[46] Xiang Ruoyu (ed.), *Xinbian gaoji xiaoxue dili keben,* 4 Vol. Beiping: Haubei lianhe chubanshe, 1949.

[47] Ibid., Vol. 1 of 4, p. 2.

people themselves to decide. Obviously, we should help them [in their decision.]'[48] Even more ominously, 'We should follow the principle of ethnic equality, and offer them [the minorities] forceful help in the political, economic, and cultural arenas, bringing about a true unity of the ethnicities.'[49] Push came to shove, it seems, the new Chinese Communist Party (CCP) government was not quite willing to allow the ethnic minorities along the frontier the full right of self-determination.

For scientists and social scientists, the end of the Civil War and Communist victory finally allowed them the political stability and safety to pursue their research. The case of Li Anzhai illustrates the powerful appeal of the new regime. During the war, a group of Christian missionaries operated in the frontier areas of Gansu and Qinghai and issued a journal devoted to their religious cause entitled *Service on the Frontiers* (*Bianjiang Fuwu*). During the war, the number of missionaries increased from three to forty-eight by 1943.[50] Alongside reports from missionaries in the field and inspirational biographies, for example, a story translated from the English on 'Booker T. Washington, Apostle of Good Will', the journal also regularly featured articles by the prominent Tibetologist and religious study scholar Li Anzhai. Li endorsed the study of frontier regions not as the subject of colonisation but as equals.[51] During the war, Li and his wife conducted work based at the Tibetan frontier monastery of Labrang in present-day Gansu. During the war, Li had to collaborate with the small missionary group because the wartime KMT state failed to guarantee safety for researchers, particularly in areas where they had tenuous control.

The missionaries' work on the frontiers – providing medical assistance, promoting hygiene, and overseeing agricultural reclamation projects – overlapped to a considerable degree with the wartime Chongqing government's initiatives to reinforce national defence by promoting the development of the borderlands. After a temporary two-year hiatus from 1947, *Service on the Frontiers* returned to publication in March 1950 with an essay urging an honest self-examination among the missionaries on the true motives for their service through learning from the Communist Party 'the tool of self-criticism'. Li Anzhai was once again prominently featured in the journal, which included a brief note that Li and his wife followed the People's Liberation Army (PLA) army to Tibet.[52] In return for the army's protection and the ability to finally conduct research in Tibet, the tenor of Li's writings changed dramatically: Li declared Tibetan independence 'the result of imperialist manipulation'.[53] The last issue of *Service on the Frontiers* appeared two months later, in May 1950. The parting message of the journal repeated a *People's Daily* article from 7 September 1949 that 'Tibet is an inseparable part of Chinese territory'.[54]

48 Ibid., Vol. 2 of 4, p. 40.
49 Ibid., Vol. 3 of 4, p. 21.
50 Zhang Bohuai, 'Juantou yu', (preface) *Bianjiang Fuwu*, 1943 (1), p. 1.
51 Li Anzhai, 'Lun bianjiang fuwu', *Bianjiang Fuwu*, 1943 (1), p. 3.
52 *Bianjiang Fuwu*, 1950 (March), p. 32.
53 Li Anzhai, 'Xin shidaizhong de bianjiang', *Bianjaing Fuwu*, 1950 (March), p. 3.
54 'Xizang jieshao', *Bianjiang Fuwu*, 1950 (May), p. 13.

Li Anzhai was hardly the only one to toe the party line in the early 1950s. In Grace Shen's work on Chinese geology, she discusses the determination of the majority of the geological community to remain on the mainland in 1949.[55] At issue was not only the geosciences' commitment to fieldwork, developed during the Republican period, but also the sense of these scientists that they could contribute to the building of the nation in a new era of peace after over a decade of war. Moreover, in the initial phases of conquest, the Communists appeared much more open to the efforts of scholars to promote better relations between Han Chinese and minority groups on the frontiers.

Under the PRC, Academia Sinica was renamed the Chinese Academy of Sciences (CAS) and top Chinese scientists like geologist Weng Wenhao and meteorologist Zhu Kezhen were invited to return to the mainland, whatever their previous political affiliations.[56] Weng was one of the highest-level KMT officials to return to the mainland and his repatriation was a major coup for the new state. The considerable personnel overlap into the late 1950s also underscores similarities between the wartime research agendas of scientific institutions under the KMT and that of newly established academies in the PRC in the 1950s. The last years of the Civil War had proved exceptionally damaging to the scientific infrastructure in China and further decimated the scientific community's confidence in the KMT state. As we have seen with the example of Li Anzhai, the communist victory and, more importantly, the accompaniment of the communist army, allowed scientists to return to work and in stances that involved extensive fieldwork, to get back to the field.

The newly established PRC immediately sought to turn the troops to productive labour along the frontiers. A mere two months after the founding of the PRC and Mao's iconic moment atop Tiananmen Square on 1 October 1949, Mao issued a proclamation on troop participation in labour and production. The first order of business was land reclamation in the borderlands. The Reclaim Wastelands and Plant Crops Campaign (*kaihuang zhongdi*) launched with the target of bringing 600,000 *mou* of wasteland under cultivation for 1950. The PLA exceeded this target and reclaimed over 960,000 *mou* in 1950.

As with wartime reclamation projects under the previous regime, the statistics of progress concealed ugly conditions on the ground. The presence of nearly 200,000 soldiers in Xinjiang created an enormous gender disparity. Among the troops, men outnumbered women 160:1; while among those who were 30 years old and above and unmarried, the ratio was 300:1. Efforts to recruit a youth corps to Xinjiang from May to October in 1949 netted over 10,000 recruits but, of these recruits, only 1,127 or roughly 11 per cent were women.[57] Unlike some Nationalist commanders

[55] Grace Shen, *Unearthing the Nation: Modern Geology and Nationalism in Republican China*. Chicago: University of Chicago Press, 2014, pp. 185–9.

[56] Ibid., p. 140.

[57] Yong Yao, *Xiang lu nu bing zai Xinjiang*. Beijing, Guang ming ribao chuban she, 2012, p. 22.

during the war, who condoned and even encouraged relations with local women, PLA general Wang Zhen was adamant that ethnic Han troops not be allowed to marry there. Wang believed that such relations could potentially inflame tensions with the local population in Xinjiang. Such skewed gender ratio was obviously untenable for long-term occupation, however, and Wang Zhen personally requested permission from Marshal Peng Dehuai to recruit women settlers to Xinjiang.[58] As the consolidation of control over the borderlands proceeded on the ground, in the pages of *Geographical Knowledge* the discipline was brought to heel under the ideological tutelage of the Party.

The October 1950 issue of *Geographical Knowledge* recorded the harsh rebuke of the discipline delivered by Chinese Academy of Sciences vice president Zhu Kezhen from the May meeting of the CAS. Zhu informed the scientists that the creation of an Institute of Geography under the umbrella of the Academy was postponed because of the weak scientific nature of their field. Because of geography's broad coverage of research areas, Zhu argued, its scientific focus suffered. At a time when the People's Government strived to develop the natural sciences to better serve industry, agriculture, and national defence, geographers lacked a practical focus in their work. Zhu then informed the scientists that the geography group would be divided into three groups – general geography, geodesy, and cartography. At this point in 1950, Huang Guozhang was still an active member of the committee on reorganising geographical research and listed as one of the members of committees discussing the reorganisation of the discipline. Meetings among geographers took place both in Beijing and Nanjing as the Academy of Sciences moved from the former to the current capital.

As the months passed, it became clear that the centre of disciplinary power had shifted, along with the nation's capital, to Beijing. By 1951, issues of *Geographical Knowledge* increasingly featured translations of Russian geographers as criticism of the field mounted from the top down. From December 1951, publication of the journal was moved to Beijing from Nanjing. In that issue, the editorial committee conducted a scathing self-criticism, pointing out three fundamental errors in its conduct: 1) taking a vague political stance; 2) unintentionally adopting a bourgeois perspective; and 3) working against the scientific truth. The published self-criticism within the pages of *Geographical Knowledge* located the root cause of these major mistakes within the editorial committee, whose members were deemed to have had insufficient political training.[59]

Three months later, the journal published Zhu Kezhen's keynote address at the Chinese Geographical Association's first national congress. In the address, Zhu pointed to the chequered history of the field from the founding in 1909 by Zhang Xiangwen of the China Geography Association in Beijing, the 1933 organisation in Nanjing, to the 1940 founding of a China Geographical Research Institute in

[58] Yong Yao, *Xiang lu nu bing zai xinjiang*, p. 52.
[59] *Dili Zhishi,* 12 (1951), pp. 275–6.

Chongqing under the KMT government. Zhu cited these multiple efforts over the previous five decades as evidence of the lack of leadership and insufficient ideological soundness for the discipline of geography. Different organisations served the north and the south; the lack of cohesion forestalled large-scale fieldwork. The field, Zhu stated, needed significant reorganising. He left out his own participation in the establishment of a geographical organisation in 1933 and the publication of *Acta Geographica Sinica.*

As these debates over the fate of geography played out in the pages of *Geographical Knowledge*, translations from the Soviet Union further reinforced the idea that geography as a field had unduly fallen under imperialist influence. In 1952, Tian Meng translated a volume edited by the Soviet Academy of Science entitled, *Geography in the Service of American Imperialism.* The work roundly excoriated geopolitics as a tool of American chauvinism, a cause of the recent worldwide conflicts, and an intellectual cover for economic exploitation.[60] The suspect nature of geography was made clear by a series of essays on the various shortcomings of geography as practiced in Western Europe and the US. Over the next decade, the CAS and the Chinese Geographical Society published a series of searing critiques against imperialist components within geography. A 1955 translated volume critiquing geography, again of Soviet scientists, identified three aspects of bourgeois geography: geopolitics, Malthusian population discourse, and environmental determinism.[61] By the 1960s, the acceptable parameters of geography had narrowed to the study of soil composition and other practically oriented fields of physical geography (*ziran dili*) and cartography.

The substance of geographical research, however, differed little from wartime projects. The determined practical focus of research at the CAS meant that the pre-war trend of emphasising geology and the survey of natural resources continued. The pressing demands of the war reinforced the importance of defending the frontiers but also the need to further exploit the country's vertical frontiers to provide fuel for the wartime government's industries. This emphasis continued after the end of the war and the communist victory. In 1956, the CAS formed an Interdisciplinary Committee for Exploration to create the institutional structure for scientists across various disciplines to collaborate on fieldwork and research. Along similar lines as the CCP's economic planning, the CAS interdisciplinary committee established five-year plans and divided their research into macro regions. The research agenda largely continued the wartime focus on border regions in the northeast, the Northwest, and the southwest.[62] The interdisciplinary committee for the

[60] Soviet Academy of Sciences Geographical Research Institute (ed.), Tian Meng trans., *Wei mei di fuwu de zichan jieji dilixue*. Beijing: Bourgeois Geography in the Service of American Imperialism, 1952.

[61] Zhongguo dili xuehui zhishi bianji weiyuanhui (Chinese Geographical Society geographical knowledge editorial committee), *Guanyu zichan jieji dili sixiang de pipan* (Critique of Bourgeois Geographical Thinking) (Shanghai, 1955).

[62] Sun Honglie *et al.* (eds.), *Zhongguo ziran ziyuan zonghe kexue kaocha yu yanjiu*. Beijing: Shangwu yinshu guan, 2007; Zhang Jiuchen, *Ziran ziyuan zonghe kaocha weiyuan hui yanjiu*. Beijing: Kexue chubanshe, 2013.

exploration of natural resources focused on the development of practical science on the frontiers, including the survey and development of the coal and oil industries and other natural resources essential to the national economy.

The establishment of a scientific infrastructure in frontier territories preceded economic development and the building of industries. The CAS formed a Xinjiang research team in 1956–60; a Qinghai/Gansu research team in 1958–60; and an Inner Mongolia / Ningxia research team in 1961–6. Furthermore, the CAS's 12-year long-term planning called for the establishment of a chain of 14 regional research institutes in the provincial capitals of Urumqi in Xinjiang and Hohhot in Inner Mongolia. These institutes of geographical, geological, soil science, animal, and plant sciences, along with institutes covering economic and historical research, formally launched 1962–5 with anywhere from 22 to 60 researchers at each location.[63] In 1957, CAS president Guo Moruo lead a contingent of 120 CAS academicians to consult Soviet experts on their long-term research plans. The Soviets formed a 16-member panel of experts, which critiqued the CAS long-term plan as not setting sufficiently clear goals but otherwise endorsed the Chinese research agenda. As a result, CAS redoubled their efforts in places like Xinjiang with practically oriented research that would bear fruit for the development of agriculture, mining, and industry.[64] Under CAS auspices, large teams of geographers and geologists focused on the development of practical science on the frontiers, including surveys for developing the coal and oil industries and other natural resources essential to the national economy. The PRC's first major oil strike, on 26 September 1959, in the Songliao Basin in the northeast, resulted from one of these surveys and resulted in the development of an oil industry at Daqing.[65]

From 1952 onward, earlier leaders of the field like Huang Guozhang had disappeared from the pages of the journal either as authors or listed committee members. Each issue featured a photospread of different regions of the country and the great transformation of nature accomplished by the people, from terracing in Gansu, great hydroelectric projects around the country, to the accomplishments of the first Five-Year Plan. Ten years after the editorial committee stated that a tsunami was washing away Old China and liberating geography, the geographers held a national conference in Beijing in January 1960. Two hundred and ninety-three delegates from around the country met for 11 days, delivering 320 plus papers, totalling over three million characters. The largest number of papers, 197, covered topics in physical geography, followed by 75 papers on economic geography, and 41 papers on cartography. The triumphant tone of this post-conference synopsis, with its celebration of exact numbers – of papers, delegates, attendees, and surveys conducted – all reinforced the message that geography had been made anew and finally reached its potential as a liberated discipline serving the masses.

[63] Zhang Jiuchen, p. 88.

[64] Ibid., pp. 89–91.

[65] Li Hou, *Building for Oil: Daqing and the Formation of the Chinese Socialist State*. Cambridge, MA: Harvard University Asia Center, 2018, p. 33.

The reality was far more complicated. The work that CAS, with Soviet approval, proposed in its five years plans in the 1950s and 1960s continued the wartime focus of geographers from the 1940s. Nor was the PRC's interest in resource exploration new. Military and political leaders from the Republican period similarly saw mining as essential to delivering economic development and growth in isolated frontier areas. In a 1932 publication by the Society for the Study of the Northwest, for example, National Tongji University director of education and Nationalist Party cadre Guo Weiping had made clear that the call to develop the Northwest was the direct result of imperialist encroachment on the northeast, which left only the far west open for development.[66] He argued that the Northwest, which he broadly defined as including Shanxi, Gansu, Ningxia, Qinghai, Xinjiang, and Mongolia, contained the greatest resources in the country, including coal and oil. Guo called for greater control over the Northwest to bolster border defences and protect the country from the imperialist threat from both the Soviet Union and the UK. In Suiyuan, local notables began compiling a gazetteer in 1931 to document the development of the region. A first draft of the gazetteer was completed in 1937, just in time for the Japanese invasion.[67] Chinese ideas about the development of the mineral industries on the frontiers responded to both perceived and real threats of foreign encroachment.

The Republican history of these earlier development plans was erased at the same time that the PRC regime continued to claim coal and other natural resources across the expanse of Chinese territory as the essential fuel for the construction of a New China. The 1954 work, *Our Country's Coal*, begins with a dramatic exclamation, 'Coal is such a useful treasure! Where does this treasure come from? How much does our country have? As we go about building our nation, we need to learn more about coal.'[68] The book then goes on to explain that coal provides the source of energy that makes it possible for trains to run, lightbulbs to turn on, and factories to operate; coal provided the fuel for daily life in modern China. As such, coal played a crucial role as China embarked on its first Five-Year Plan. The author points out that even in the Soviet Union, where dams and hydraulic power generation had developed rapidly in recent decades, coal continued to generate two-thirds of the energy used in the country. A quote from Lenin, 'Coal leads various industries, is the food for industries', affirmed coal's revolutionary credentials.[69] A subsequent section placed China fourth in the world, behind the Soviet Union, the US, and Canada, in terms of coal reserves but explained that, as geological surveys continued and new and sizeable deposits are discovered in provinces like Xinjiang,

[66] Guo Weiping, 'Kaifa xibei tan', in *Kaifa Xibei zhi xian jue wenti* No.3–4 (1932) Xibei xue hui zonghui, p. 9.
[67] Suiyuan tong zhi guan, *Suiyuan tong zhi gao,* 100 juan (Huhehaote Shi, Nei Menggu ren min chu ban she, 2007).
[68] Zhou Jian (ed.), *Zuguo de mei*. Shanghai: Shanghai Ditu chuban she, 1954, p. 1.
[69] Ibid., p. 3.

Sichuan, and Xikang, China was bound to move up in the rankings.[70] The optimistic account depicted coal not only as the primary energy source for Chinese efforts to industrialise but also as essential to the make-up of the Chinese nation. Coal deposits of varying sizes dotted the country in every province and region.

A 1963 children's rhyming book reinforced the notion that the abundance of Chinese mineral resources was one of the country's defining features. Accompanying the map of China goes the catchy ditty:

> Our motherland, rich in mineral resources
>
> petroleum, coal, iron, and copper
>
> ...
>
> deposits spread throughout the provinces
>
> for hundreds and thousands of years, never to be exhausted.[71]

Both the words and the administrative map of China made clear that the country's abundant natural resources were a point of pride, found spread throughout the nation in all the provinces and autonomous regions. The presence of these resources formed the foundation for the building of the nation and naturalised its boundaries as the new regime set forth to conquer nature.[72]

Conclusion

In the twentieth century, areas in Inner Mongolia and Xinjiang have transformed into resource hinterlands, which provide essential mineral resources for Chinese and global industries in return for large-scale environmental damage as a result of strip mining and other polluting practices.[73] These changes took place over the course of the twentieth century as scientists and political and military leaders shaped the ideology of the nation-state to accommodate the territorial extent of the Qing Empire by emphasising the importance of mineral and natural resources to the construction of the nation. Geographers played an important role in the transformation. One could also trace the intellectual lineage of ideas about development and frontier defence across political divides from the late-Qing to the PRC period. Geographers and writers from across the political spectrum emphasised the unequal population and resource distribution across the expanse of Chinese territory and used this argument to promote Han Chinese migration to the borderlands.

This continuity in the conceptualisation of territory across political watersheds contrasts with the tragically personal points of rupture. Huang Guozhang, one of

[70] Ibid., pp. 14–15.

[71] *Zhongguo dili san zi jing* (Beijing, Renmin jiaoyu chuban she, 1963), p.4.

[72] Judith Shapiro, *Mao's War Against Nature: Politics and the Environment in Revolutionary China*. Cambridge: Cambridge University Press, 2001.

[73] On the issue of rare earth mining and the way these mines have polluted areas in Inner Mongolia occupied by ethnic minorities, see Julie Klinger. 'Mining Frontiers' (UC Berkeley dissertation, 2015).

the founders of the wartime Institute of Geography, committed suicide with his wife in 1966; noted human geographers Hu Huanyong and Li Xudan were labelled counterrevolutionaries in the 1950s and spent years in jail before being rehabilitated in the 1980s. More 'fortunate' members of that cohort of social scientists lived out the rest of their lives in exile in Taiwan and the US respectively. The contrast between the geographers' political persecution with the experience of someone like agronomist Tang Qiyu is deeply revealing of the contingent nature of the political and scientific disciplinary reorganisation in the 1950s.

The numerous examples of personal tragedies among geographers reflected less the controversial nature of their views than a series of unfortunate historical contingencies. The wartime inclusion of geography in Academia Sinica, the fruition of years of lobbying efforts, turned out to be a pyric victory. In turn, this decision led geography to be headquartered in Beijing with the CAS after 1949, which further exposed the discipline to scrutiny and political attack. During the war geographers worked hard to prove their utility to the state. They made the case for the importance of geography. The new regime absorbed the knowledge they created and used their expertise but deemed geographers dispensable for the construction of New China. Geographers won the battle and lost the war.

12

‘Marvelling at a World so Changed’: The Three Gorges Project in Mao’s China

COVELL F. MEYSKENS

SWIMMING ACROSS THE Yangtze River near China’s Three Gorges in 1956, Mao Zedong mused about ‘the great plans … afoot’ in the People’s Republic of China (PRC).[1] One grand idea that occupied his mind was building the Three Gorges Dam’s (TGD) ‘walls of stones’. By containing the Yangtze’s waters, the dam’s gargantuan edifice would usher in a sea change in Chinese history. Catastrophic floodwaters would no longer devastate Central China, and Yangtze rapids would be flattened into a ‘smooth lake’. These alterations to China’s largest waterway would be so monumental that it would inspire ‘the mountain goddess’, which custom held inhabited the Gorges, to ‘marvel at a world so changed’.

Mao was not the first leader to gaze in awe at the epoch-making contributions that the TGD could make for China’s development – that milestone goes to Sun Yat-sen in 1919. Sun’s party – the Kuomintang (KMT) – subsequently conducted geological surveys in the 1930s and partnered in the 1940s with the US and its top hydraulic engineer John Savage to draw up plans for a dam that would be at modern engineering’s forefront.[2] When the Chinese Communist Party (CCP) came to power in 1949, it heavily criticised the KMT. The CCP, however, shared its precursor’s view that a central mission of modern Chinese statecraft was ‘building (*jianguo*)’, and that what the Chinese state had most urgently to construct was the infrastructural sinews of the Chinese nation. For China to metamorphosise from a weak agrarian country, whose government could not provide for its people and defend them from geopolitical threats, into an industrial powerhouse, whose government could not only protect its citizenry but create better living conditions and

[1] Quotes in this paragraph are from Mao Zedong, ‘Swimming’, *Marxists Internet Archive*, June 1956, www.marxists.org/reference/archive/mao/selected-works/poems/poems23.htm.

[2] Covell Meyskens, ‘Dreaming of a Three Gorges Dam amid the Troubles of Republican China’. *Journal of Modern Chinese History* 15 (2022): 176–94.

Proceedings of the British Academy, **267**, 232–253, © The British Academy 2025.

command international respect and prestige, Chinese leaders had to oversee the construction of the nation's critical infrastructure, from roads, railroads, and powerlines to mines, factories, and dams.[3] As part of this process of building the infrastructural foundations of the Chinese nation, CCP leaders thought, like the KMT, that it was their political duty to catalogue the country's resources, make them serve economic growth, and domesticate resources that harmed the nation, like the Yangtze and its floods.[4] As Shellen Wu shows in her chapter on geology in this volume, surveying China's mineral resources was one major prong of this developmentalist pursuit. In parallel, government agents worked to gather and instrumentalise hydrological knowledge about China's rivers to make them into engines of national development.

Mao's interest in the TGD set the CCP's party-state working on the project in the early 1950s. Like Chiang Kai-shek, Mao favoured the TGD's erection as part of Chinese government efforts to engineer a new energy regime that would make China's rivers into state-directed engines of national development. Rivers would no longer cause floods and complicate transportation but instead power industrialisation and the circulation of people and materials nationwide.[5] With these promethean goals achieved, CCP leaders foresaw that China's global power and prestige would rise as the nation's riverine energies became state instruments of economic improvement. During Mao's time in power, the CCP undertook several projects to realise this titanic feat of environmental engineering. Party efforts primarily focused on mobilising huge regiments of manual labour to build roughly 90,000 small reservoirs and 2,000 medium and large reservoirs.[6]

The central pillar in the CCP's envisioned developmental disciplining of the Yangtze was the TGD. The CCP's hopes of realising this dream, however, came up against the same conundrum that the KMT had encountered in the 1940s, namely China did not have the technological, administrative, or fiscal means to undertake such a mammoth project. What the central government had adequate resources for was constructing small reservoirs for water storage and power generation, which it produced extensively. When the CCP did pursue large hydro-endeavours, it channelled resources mainly to projects on the Hai, Huai, and Yellow Rivers, such as the Sanmenxia Dam.[7]

While the CCP did not devote as much attention to the Three Gorges project as other hydro-ventures, it did not let the dam fall completely by the wayside and engaged in preparatory work. To make up for domestic resource constraints, the party employed the same developmental tactic the KMT used in the 1940s: it sought

[3] William Kirby, 'Engineering China: The Origins of the Chinese Developmental State', in *Becoming Chinese: Passages to Modernity*, ed. Wen-hsin Yeh. Berkeley: University of California Press, 2000, 137–60.

[4] Timothy Mitchell, 'Fixing the Economy', *Cultural Studies* 12 (1998): 82–101.

[5] Meyskens, 'Dreaming'.

[6] Vaclav Smil, 'China's Water Resources', *Current History* 77 (1979): 59.

[7] Smil, 'China's Water', 59.

foreign support. But where the KMT called on the US for assistance, the CCP by contrast appealed to Moscow in the 1950s. In both instances, it was a national crisis that stimulated an international partnership to construct this hydro-technical wonder.[8] In the KMT's case, it was the Sino-American struggle against Japan during World War II (WWII) that motivated locking-in substantial American backing for China's post-war development. In the CCP's case, it was the 1954 Yangtze floods that triggered soliciting Soviet patronage.[9]

Like Chinese and American experts in the 1940s, PRC and Soviet specialists held in high regard American John Savage's plan for the TGD, and like Savage, they envisaged the dam as part of the comprehensive development of China's river basins.[10] Constructing the dam was particularly important because it promised to transform the Yangtze River Basin's socioeconomic and environmental landscape. To achieve this, CCP leaders centralised technical personnel in the Yangtze River Water Conservancy Commission (YRC) and tasked its researchers with making legible the river's geological patterns. Based on this data, Chinese and Soviet experts could then design a dam that made the Yangtze into a predictable and productive instrument of national development, a techno-socialist dream that both Chinese and Soviet participants strove to materialise

Not everyone involved in the project supported its construction. Critics argued that small southwest dams could achieve comparable benefits, while not tying up so much of China's limited resources in a scheme with a long completion date and whose electrical output far exceeded national demands.[11] Mao's backing overrode criticisms and kept the project on the CCP's developmental agenda. Yet it was also Mao and Zhou Enlai that maintained technocratic-minded officials' influence over the dam until the late-1960s when Mao initiated the Cultural Revolution's assault on bureaucratic and technical elites and subsequently greenlighted a proposal to build a dam in the Three Gorges region, which sidelined technocrats, during the 1969 Sino-Soviet border crisis.

Advocated by Hubei officials, this hydro-political venture sought to simultaneously realise Mao's dream of a TGD and accelerate expanding China's military-industrial defences. This developmental push sought, like earlier efforts, to ensure the project operated at global technoscience's frontiers. Party leaders, however,

[8] Judd Kinzley, 'Crisis and the Development of China's Southwestern Periphery: The Transformation of Panzhihua, 1936–1969', *Modern China* 38 (2012): 559–84.

[9] Chris Courtney, 'At War with Water: The Maoist state and the 1954 Yangtze floods', *Modern Asian Studies* 52 (2018): 1807–36.

[10] Christopher Sneddon, *Concrete Revolution: Large Dams, Cold War Geopolitics, and the US Bureau of Reclamation*. Chicago: University of Chicago Press, 2015, 28–51. David Pietz, *The Yellow River: The Problem of Water in Modern China*. Cambridge, MA: Harvard University Press, 2015.

[11] Lin Yishan, 'Guanyu Changjiang liuyu guihua ruogan wenti de shantao', May 1956, in '*Zhongguo changjiang sanxia gongcheng lishi wenxian huibian' bianweihui, Zhongguo changjiang sanxia gongcheng lishi wenxian huibian (1949–1992)* (Beijing: Zhongguo sanxia chubanshe, 2010), 33–53. Sun Ronggang, *Mengxiang yu xianshi: Sanxia gongcheng jishi*. Beijing: Zhongyang wenxian chubanshe, 2005, 31.

promoted a knowledge regime that imparted technological practice's cutting edge with decidedly national roots while the guiding principle behind construction became Mao's idea that mass mobilisation, labour enthusiasm, and native techniques were more powerful contributors to industrialisation than technical aptitude and industrial equipment.[12] In the end, the CCP's employment of the Maoist developmental way produced a bumbling attempt to establish a TGD. To rectify the project, Zhou Enlai halted construction in 1972 and returned control to administrative and technical experts.

Making the Yangtze legible and controllable

As part of the CCP's drive to establish a new energy regime that transformed China's rivers into engines of national development, the CCP ordered the Water Conservancy Ministry in 1950 to gather all documentation on China's rivers and chart out their basic geological features. Attributes evaluated included annual height difference, waterflow volume, and electricity generation capacity. According to Ministry calculations, China's potential annual hydropower was estimated at 149,000 megawatts with roughly two-thirds from the southwest, coming mainly from the Yangtze River Basin. In 1954–5, the Ministry conducted another assessment of 1,598 major rivers that revised upward China's hydropower to 544,000 megawatts. Despite this more than threefold increase, the southwest and the Yangtze River Basin remained the key sources of national hydro-energy.[13]

To better comprehend and control the Yangtze's water resources, the CCP formed the YRC in 1949. Appointed as its director was Lin Yishan. Possessing a college degree from Beijing Normal University, Lin was better educated than the average party member.[14] His degree was in history, and he was made YRC director because of his leadership in the Shandong base area and contacts with high-ranking officials. It might thus appear that Lin's appointment is an indication that the CCP did not technocratically manage the Yangtze, but, while Lin did start out as a revolutionary, his party work made him into a technocrat.

Like many revolutionaries converted into bureaucrats, Lin attended science and technology classes in the 1950s to gain the basic skills and knowledge requisite for his position.[15] Lin also emphasised the YRC's technical competence, collecting documents from KMT agencies with managerial, technical, and geological information about the Yangtze and hiring around 2,600 experts formerly working on KMT

[12] Carl Riskin, *China's Political Economy*. Oxford: Oxford University Press, 1987, 201–22.

[13] Guojia nengyuan ju, *Zhongguo shuidian 100 nian: 1910–2010*. Beijing: Zhongguo dianli chubanshe, 2010, 44. For more on CCP efforts to control China's rivers in the 1950s and 1960s, Micah S. Muscolino, 'The Contradictions of Conservation: Fighting Erosion in Mao-Era China, 1953–66', *Environmental History*, 25(2): 237–62.

[14] Lin Yishan, *Lin Yishan huiyi lu*. Wuhan: Changjiang chubanshe, 2019, 13–14.

[15] Lin, *Lin*, 118–20.

Yangtze projects. The Commission also set up short courses in administration and technical skills for cadres, and it recruited educated urban youth and professionals proficient in water conservancy, mechanics, and accounting. Through these efforts, the Commission's rolls increased about 80 per cent to 4,700 employees by 1951.[16]

In the early 1950s, the YRC concentrated on flood control. Of particular concern was western Hubei near Jingzhou, where the Yangtze had repeatedly inundated southern Hubei and northern Hunan. A big 1949 flood in the area pressed home to CCP leaders the urgency of restraining the Yangtze. In 1950, the YRC proposed the Yangtze River Flood Diversion Project. Mao and Zhou Enlai knew that the project would not eliminate flooding. Yet they were also aware that China lacked the technical, financial, and material capability to achieve this objective. They thus endorsed the YRC's plan as an interim method of protecting Hubei and Hunan's 'millions of people and property' by reinforcing local dikes and building a flood diversion area.[17] Leading Soviet hydraulic expert Burkov[18] questioned whether China could execute this task and was only persuaded when YRC experiments modelling the flood diversion area displayed their technical know-how.[19]

The project's construction did not solely rely on China's advanced technological capabilities. It also depended on party power to quickly assemble a huge work force. In 1952, provincial leaders mobilised 300,000 people and placed them under a centralised command centre with members from the Ministries of Water Conservancy, Transportation, and Agriculture and Forestry as well as the Soviet Union.[20] Hydraulic engineer Zheng Hui brimmed with satisfaction upon seeing the final product in 1952. When he had worked in the US as an engineer during the Republican era, Americans had told him that the CCP could engage in war but not development. The CCP proved American doubters wrong, and the *New York Times* admitted as much, publishing a laudatory article, which caused Zheng to well up with tears of nationalist joy.[21]

China's press also expressed gratification about China regaining sovereignty over its waterways when Mao toured the Yangtze in 1953.[22] During his visit, Mao

[16] Zhonggong zhongyang dangshi yanjiu shi zhonggong hubei sheng Yichang shi weiyuanhui zhonggong hubei shengwei ganshi yanjiushi, *Zhongguo gongchan dang yu changjiang sanxia gongcheng*. Beijing: Zhonggong dangshi chubanshe, 2007, 6. Lin, *Lin*, 121–2.

[17] Sun, *Mengxiang*, 26.

[18] The original source text does not provide a Russian name. All that is given is a Chinese transliteration of a Russian name. I have not been able to locate the original Russian name, so I have transliterated the Chinese name into English.

[19] Lin Yishan, 'Changjiang sanxia goncheng juece guihua huigu', in Sanxia wenshi congheng (Beijing: Zhongguo sanxia chubanshe, 1997), 2.

[20] Sun, *Mengxiang*, 26–7.

[21] Lai Cenglin, 'Qing man changjiang', in *Sanxian wenshi congheng di er ji*, ed. Zhongguo renmin zhengxie shanghui Yichang shi weiyuanhui wenshi ziliao wenyuanhui. Yichang: Zhengxie yichang shi weiyuanhui wenshi ziliao hui, 2000, 20–1.

[22] Wang Zhousheng, '1953 nian Mao Zedong: Fandui diguo zhuyi qinlue yao jianli qiangda haijun', *Fenghuang wang*, 16 July 2007, http://news.ifeng.com/history/zhongguoxiandaishi/detail_2011_07/16/7735122_0.shtml.

met with Lin Yishan, questioned him about the Yangtze's weather patterns and floods, and told Lin that 'Taming this big river requires careful study', and that realising this goal was 'a scientific matter'. Tipping his hat to the importance of empirical evidence, Mao asked about collected data on Yangtze hydrology. Lin assured Mao that the YRC had lots from Republican-era studies and had conducted surveys to fill in gaps in existing records, an endeavour that Mao praised for its 'scientific' approach.[23]

After this opening discussion, Mao turned his attention to the flood diversion project and questioned whether it could terminate regional floods. When Lin replied no, Mao probed about what would.[24] Lin discussed how the Commission planned to handle this matter in three stages. First, it would bolster dikes to control the Yangtze and its tributaries. Then, polders would be built to check floods. Last, reservoirs would be constructed on the Yangtze's tributaries, which would completely remove the flood threat, provide hydroelectricity, improve transport, and expand irrigation. Mao had another idea, 'Do you think that building all these reservoirs on tributaries ... would be equivalent to a big Three Gorges reservoir?'[25] Lin replied that the YRC really hopes 'to build the TGD, but we dare not think about it now' due to the lack of fiscal, technical, and administrative resources for such an enormous enterprise.[26] Mao pushed back, retorting, 'But didn't you say that all these reservoirs ... would not be equivalent to a Three Gorges reservoir?' Lin responded by going over American engineer John Savage's TGD plan, drafted for the KMT in the 1940s. Mao took particular interest in its high cost – $1.3 billion – which would surely concern party elites. Mindful of the technical difficulties involved in such a complex project, Mao was pleased to hear that the YRC had 270 engineers and over 1,000 technicians thanks to its persistent efforts to expand its technical staff. Despite the Commission's significant concentration of technical personnel, Mao understood that it did not have a sufficient technical or empirical base to begin the Three Gorges project, and so he directed Lin to conduct a preliminary investigation.[27]

The YRC's Engineering Department then studied whether damming the Min, Wu, Jinsha, and Jialing Rivers in the southwest would settle the flood problem better than the TGD.[28] As part of this research venture, surveyors investigated potential reservoir sites, various flood control plans, and the different factors behind Middle and Lower Yangtze floods. The Commission also reviewed Savage's plan and uncovered that the location Savage chose for the TGD – Nanjing'guan – was riddled with caves, leaving Lin Yishan wondering whether Savage was as brilliant as most believed. To verify Savage's plan, Lin inspected possible dam sites

[23] Lin, *Lin*, 137–9. Quotes are on page 139.
[24] Zhonggong, *Zhongguo*, 8.
[25] Lin, *Lin*, 141.
[26] Zhonggong, *Zhongguo*, 7–8.
[27] Lin, *Lin*, 142, 146.
[28] Zhonggong, *Zhongguo*, 7–8

with former KMT hydraulic engineer Li Zhennan, concluding that the TGD's best location was not Nanjing'guan but further upstream at Sandouping's thick granite foundation. Survey participants understood that China lacked the technical, administrative, and fiscal capacity to build the TGD at this better location. But they took pride in showing that Chinese science could best American science in at least this instance.[29]

Investigating and comparing proposals

Efforts to tame the Yangtze gained intensity when it flooded in 1954, killing 33,000 and damaging 4.37 million homes. This hydrological crisis demonstrated the Yangtze River Flood Diversion Project's insufficiency to safeguard the nation. After the flood, Mao toured southern China with Zhou Enlai, Liu Shaoqi, and Vice Premier Chen Yun. While in Hubei, Mao met with Lin Yishan and asked whether China could independently build the TGD. Lin replied that the country was not ready, but it could be done with 'Soviet expert help'.[30] Otherwise, construction would have to wait until after China and the Soviets built the Danjiangkou Dam on the Han River.

From this experience, Chinese engineers could obtain world-class capabilities since Soviet experts were as good as American, and the US had made designs for the TGD. With this technical base established, China could pursue the TGD on its own. Another problem that the party had to address was shortages of the survey equipment needed to devise an accurate geological picture of the Three Gorges. Despite this national lack, Lin Yishan reported that Chinese surveyors had already surpassed the US in one respect. They had shown that Savage's plan to place the dam at Nanjing'guan was flawed, and that the dam should instead be built at Sandouping.[31] Like many CCP leaders, Mao was interested in finding the optimal place for the TGD because of the great boost it would give to forming a new riverine energy regime that would 'win the battle against' Yangtze floods.[32] Mao understood that Soviet assistance was necessary due to scarce domestic machinery and technical experts, and so he requested Soviet help with designs for the Yangtze River Basin and TGD. These activities were not part of the first Five-Year Plan. Nevertheless, Moscow still agreed and promptly dispatched 12 hydraulic experts. Lin Yishan integrated them into a 143-person group, which contained a who's-who of China's energy and hydrology sectors and surveyed the Yangtze River Basin for two months.[33]

[29] Lin, 'Changjiang', 2–3. Lin, *Lin*, 147.
[30] Zhonggong, *Zhongguo*, 9.
[31] Lin, *Lin*, 147.
[32] Zhonggong, *Zhongguo*, 9.
[33] Wu Shaojing, *Zhou Enlai yu zhishui*. Beijing: Zhongyang wenxian chubanshe, 1991, 35–6. Zhonggong, *Zhongguo*, 9.

Figure 12.1 Lin Yishan and Soviet experts at the Three Gorges in 1955

In the end, Soviet and Chinese surveyors came to different conclusions about the best way forward. The Soviets prioritised augmenting China's energy base and recommended building one dam near Chongqing and four more on the Min, Jinsha, Jialing, and Wu Rivers. Some Chinese participants agreed that the YRC should 'prioritize the Yangtze's tributaries and then work on the Yangtze'.[34] Lin Yishan disagreed and pointed out that the Soviet plan would not control the waters of the Wu and Jialing Rivers south of Chongqing, which had caused one of Central China's largest floods in 1870. Lin argued that the TGD could fix this hydrological problem and doubted that Sichuanese officials would support the Soviet plan because of how much of Sichuan's farmland would be lost.[35] In December 1955, Lin and the Soviet group's leader presented their proposals to Zhou Enlai, who endorsed the Chinese proposal for two reasons. The Soviet plan would submerge a huge chunk of arable land and not remedy flooding generated by the Wu and Jialing Rivers while the TGD could store so much water that Yangtze floods would become non-existent. The Soviets ultimately supported the Chinese proposal, but they were not willing to cast aside Savage's recommendation of Nanjing'guan as the TGD's location. Savage was a world-class engineer and further geological studies were

[34] Lin, 'Changjiang', 3.

[35] Wu, *Zhou*, 36–7. Li Zhennan, 'Huiyi sanxia gongcheng bazhi de zuanze', in *Sanxian wenshi congheng di yi ji*, 12.

necessary to disprove him. On the other hand, the Soviets were confident that China possessed enough engineering talent for the dam's construction.[36]

When Mao heard that the Three Gorges project was progressing, he was so delighted that regular speech was inadequate. Only poetry could convey his vision of the TGD's epochal significance. Mao encapsulated his techno-historic dream in the poem 'Swimming'. Composed after wading across the Yangtze in 1956, Mao contemplated a not-too-distant future when 'walls of stone will stand upstream … to hold back … the clouds and rain' of southwest China and bring about 'a smooth lake … in the narrow gorges', a technological triumph that would inspire the legendary goddess of the Three Gorges 'to marvel at a world so changed'.[37] Mao's words here were not just empty poetic rhetoric. From this point on, the lines of Mao's poem would reverberate in the halls of power of the Chinese state, serving as both inspiration and justification for future efforts to construct the TGD.

Preparing and debating

By March 1956, over 1,000 surveyors from water conservancy departments around China were gathering data for the TGD and other Yangtze River Basin projects. Research teams performed geological surveys, investigated possible dam locations, and studied socioeconomic conditions.[38] By October 1957, geological teams had assembled information from 17,448 hydrological stations about local rainfall, evaporation rates, and the sediment content, historic height, and waterflow volume of the Yangtze and its tributaries. Collected data included any available information from 1877 onwards.[39]

In May 1956, Lin Yishan published an article that summarised the state of Three Gorges research and endorsed building the TGD. He, nevertheless, acknowledged that China lacked the requisite construction machinery and hydropower equipment, that researchers had not determined the dam's location, and did not sufficiently understand silting trends and how to pour cement for such a large dam. The YRC had also not figured out the dam's socioeconomic impact. How the YRC would have made this evaluation is not apparent. What is clear is that Lin Yishan thought that only after addressing these many issues would the TGD be based on a solid empirical base, the most economical methods, and be as beneficial to China's development as scientifically possible. Yet even then it was still necessary to consider whether to allot sparse funds to the dam when a still mainly agrarian China could not consume so much electricity.[40]

[36] Lin, 'Changjiang', 3.
[37] Mao, 'Swimming'.
[38] 'Changjiang liuyu da guimo kance gongzuo quanmian zhankai', *Renmin ribao*, 25 March 1956.
[39] Li Min, 'Changjiang linian shuiwen ziliao zhengbian zhengli jieshu', *Renmin changjiang bao*, 19 October 1957.
[40] Lin, 'Guanyu', 20–32.

Although Lin admitted that China was not yet equipped to construct the TGD, he thought that the party should work towards acquiring the necessary knowledge, capabilities, and material resources because the dam would mark a major break with China's past. The party's top priority had to be eliminating flooding given that the Yangtze wreaked havoc every five-to-six-years; the TGD would solve this diluvian problem. Modelling experiments had shown that it could handle every single Yangtze flood in China's annals, and the dam would pacify the treacherous waters above the Three Gorges and make inland transport safer and more abundant. Irrigation networks emanating from the dam would additionally lift regional agriculture, and its power would massively stimulate industry. Lin emphasised that Soviet experts concurred with his views and affirmed that the dam could eventually be constructed with presently available technologies.[41]

Some Chinese experts did not agree with Lin. The most famous critic was the Director of the Energy Ministry's Hydroelectric Bureau Li Rui, who formerly served as Mao's secretary.[42] Li asserted that China should strive with Soviet support to reach the technological frontier that would make possible constructing a dam of such magnitude. China, however, should not attempt to build the dam soon. Li considered Lin Yishan's proposal too vague. Lin asserted that the dam would cease Yangtze floods and contribute to national development, but he had not conducted a detailed analysis of the dam's cost, socioeconomic consequences, and completion timeline. Comprehensive plans had to be drawn up that explained how the dam's water and energy would be funnelled into specific economic sectors. What is more, Lin had not adequately evaluated whether erecting the dam made sense.[43]

Before China could undertake such a colossal project, technical personnel needed a more comprehensive geological understanding of the Yangtze River Basin, and the party needed the capacity to produce the requisite survey and construction equipment. Even if China accomplished these objectives, Li Rui still questioned whether it would be wise to have a single project absorb so much of China's scarce fiscal and technological resources and thereby impede other more easily realisable projects. While Li supported the CCP's goal of building a new energy regime that made China's rivers serve the state's developmental objectives, Li contended that instead of constructing the TGD, the CCP should establish several small dams on the Yangtze and its tributaries whose economic fruits were obtainable with less money, resources, and time. These more modest enterprises would also avoid one huge issue that Lin Yishan ignored, namely the TGD required relocating about 2.15 million people. As other PRC dam projects had illustrated, people would likely

[41] Lin, 'Guanyu', 20–32.

[42] Zhonggong, *Zhongguo*, 9. Li, 'Huiyi', 12.

[43] Li Rui, 'Guanyu changjiang liuyu guihua de ji ge wenti', September 1956, in 'Zhongguo changjiang sanxia gongcheng lishi wenxian huibian' bianweihui, *Zhongguo changjiang sanxia gongcheng lishi wenxian huibian (1949–1992)*, 33–53.

be unwilling to move, and they should not be forced to do so. The party had to consult with them and arrive at a consensus on how to proceed.[44]

Stressing the importance of scientific debate in the Three Gorges project, Mao summoned Lin Yishan and Li Rui to present their competing views at a central party meeting in Nanning in Guangxi Province in January 1958.[45] In his talk, Lin drew on flood data since the Han dynasty to argue that the Three Gorges was the right location to finally resolve the Yangtze floods that had long afflicted Central China. As Lin talked, Mao's thoughts turned to party discussions about the dam's cost, leading him to inquire about John Savage's plan, its predicted expenses of around 780 million yuan, and decade-long construction timeline. Mao offered his annual expense account of 70–80 million to cover costs and questioned whether Savage's price estimate was achievable. Given that the CCP would have still covered Mao's expenses if he had followed through on this proposal, this idea might not seem to be very meaningful. However, by offering up his own personal expense account for the project, Mao was displaying his intense ardour for the dam.

Lin Yishan also took Mao's proposal seriously and replied that not only was Mao correct, but the Commission had already lowered anticipated costs from the Soviet's projected price tag of 20 billion yuan to 740 million yuan.[46] At this point, Zhou Enlai opined that the dam's cost was further reducible if its hydroelectric output was decreased from 25 million kW to 5 million kW. Lin agreed but retorted, 'For this sort of big high-dam, how can we just install 5 million kW of electric capacity?'[47] Lin proposed a different method of contracting costs – a temporary hydroelectric plant on the cofferdam built prior to the final dam. This stratagem would both lower costs and sooner stimulate economic growth.[48]

Mao then called on Li Rui to give his views on the TGD. Li argued that the dam would not completely end Yangtze floods. It would only address floods caused by waters above the Three Gorges, which only accounted for half the Yangtze's waters. The dam would do nothing to lessen downstream floods, which would still threaten Central China. Li also challenged Lin's positive portrayal of the dam's enormous electricity generation, questioning how an underdeveloped China could use so much power. Lin's proposed temporary hydroelectric plant found no favour with Li either, who condemned it as a huge waste of precious resources. The government also did not have enough geological data for the dam's construction. If the CCP were to move ahead, nonetheless, the project would be hamstrung by the fact that no engineer worldwide knew how to build the dam's hydroelectric powerplant or transport lock. It is not apparent what information, if any, Li Rui was drawing on to justify this claim. Given that John Savage had concluded in the 1940s that building the dam's power station and lock were feasible, there is reason to believe that the validity of Li

[44] Li, 'Guanyu', 33–53.
[45] Sun, *Mengxiang*, 30–1.
[46] Lin, *Lin*, 151–2.
[47] Zhonggong, *Zhongguo*, 10.
[48] Lin, *Lin*, 156–9.

Rui's assertion is questionable. Li, however, undoubtedly was opposed to the dam's construction and sought to discredit its supporters. Li thus also argued that the CCP should hold off on building the TGD because its massive concrete edifice would be a prime target for China's enemies. Given all these difficulties, Li suggested that, instead of the TGD, the party 'should first develop tributaries, starting with the small and then … the big', and focus on dikes for flood control.[49]

Mao praised Li Rui's comments in general terms and shared his concern about military safeguards. Nevertheless, he still sided with Lin Yishan, arguing that China should develop its hydroelectric sector and leave coal underground for future generations. What had to be done now was 'actively prepare' a 'completely reliable' regional development plan for the Yangtze centred on the TGD. As Mao was inclined to do, he made Zhou Enlai the supervisor of the Gorges project. Zhou initially genuflected, saying that 'Such a big affair would be better for the Chairman to handle.' Liu Shaoqi disagreed and asserted that 'Comrade Enlai is able to grasp the Chairman's intention, and so it would be better for him to be in command.'[50] Zhou yielded, and Mao ordered him to check on the project four times a year. With this decision, Mao again gave his imprimatur to the TGD, ensuring that critics did not prevail, and that machinery of the Chinese state continued to work towards the dam's realisation. By assigning his righthand man, Zhou Enlai, to oversee the project, Mao also not only illustrated how important the project was for him personally, but he guaranteed as well that the man at the project's helm had extensive power over the Chinese government apparatus.

Moving towards a Great Leap

Zhou immediately started work, leading a team to survey the Three Gorges in March 1958. The group included State Planning Commission Director Li Fuchun, Finance Minister Li Xiannian, Lin Yishan, and over 100 Chinese and Soviet technical experts and administrators. During their tour, they stopped at Nanjing'guan. Lin took Zhou there to show him local caves, disprove John Savage's suggestion to build the dam there, and thereby refute the widely held idea that China was scientifically behind the US. Lin argued for constructing the dam instead at Sandouping, which they proceeded to next. While there, Zhou was moved to reflect on the Three Gorges' sublimity and recited Mao's paean to their beauty and power – 'Swimming'. By publicly reading Mao's poem, Zhou reminded project participants that they were involved in no ordinary matter. Building the TGD was a historic endeavour that Mao thought to be of the utmost political importance as it would forever change China's landscape by domesticating the Yangtze River and channelling its hydraulic power into industrialising the nation.

[49] Sun, *Mengxiang*, 31.
[50] Wu, *Zhou Enlai*, 37.

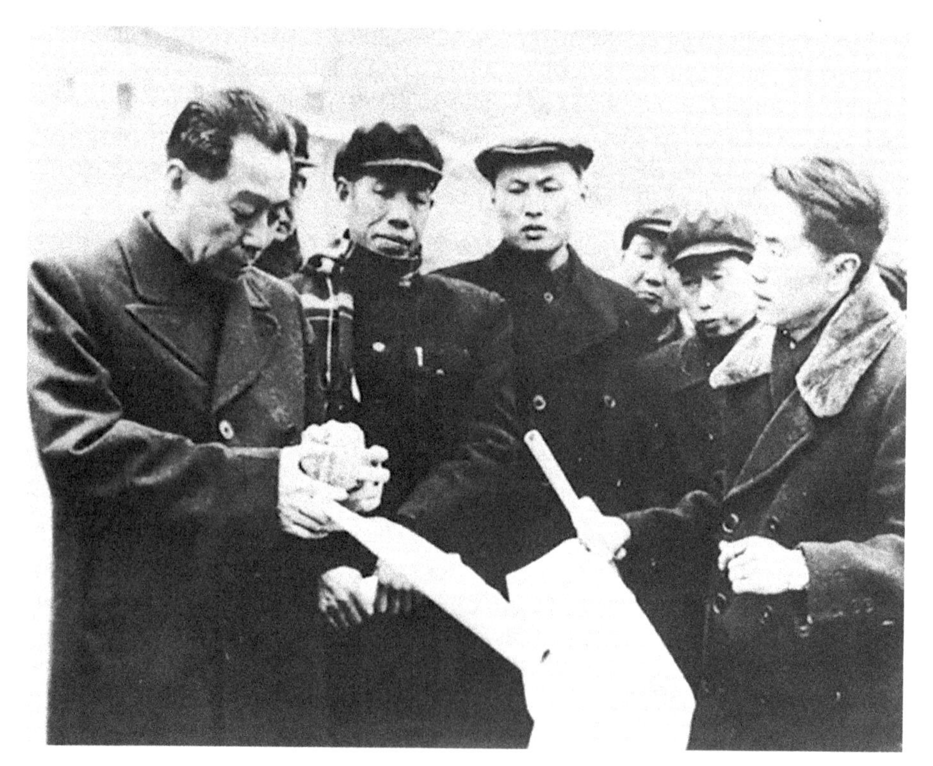

Figure 12.2 Zhou Enlai at the Three Gorges in 1958

After listening to Lin Yishan's report about the geological and technical advantages of building the TGD at Sandouping, Zhou acceded but informed Lin that he must convince Chinese and Soviet experts with further data and scientific experiments as many still viewed Savage's plan as the premier model.[51] Zhou encouraged project administrators to engage in debates of this sort because the TGD was a huge endeavour that could not be approached unilaterally. The Commission had also to formulate a detailed budget and ensure that TGD planning was coordinated with developmental plans for the entire Yangtze. At the end of his tour, Zhou declared that China could build the dam within 15–20 years with already available technologies. However, not wanting project leaders to be too forward-leaning, Zhou reiterated that Mao and himself must review and approve all plans and decisions.[52]

Zhou then headed to the Chengdu Conference where central and provincial leaders were conducting early discussions about the Great Leap Forward (GLF). At the meeting, CCP elites discussed building the TGD as one way to rapidly advance China's long-term economic goals. Chinese and Soviet experts' recent

[51] Du Xixiang, 'Zhou gong sanxia qing', in *Sanxian wenshi congheng di er ji*, 44–5.
[52] Zhonggong, *Zhongguo*, 10. Wu, *Zhou*, 40–1.

determination that it was technologically possible to construct the dam bolstered party support. Before construction began, leaders insisted on conducting more preparatory work, but the Leap's heated atmosphere led the central party to move more towards initiating the dam's construction, setting out a timeline for completing preparations for the first time in PRC history that mandated the YRC to finish the initial Yangtze development plan by 1959 and a preliminary TGD plan between 1959 and 1963.[53] After the Chengdu Conference, Mao and Lin Yishan steamed down the Yangtze from Chongqing to the Three Gorges. While onboard, Mao named Lin the 'King of the Yangtze' and asked him, 'Can you help me find someone to be (CCP) Chairman? Then, I'll be your assistant and help you build the TGD. How about it?'[54] Although Mao was joking, his assertion demonstrated how vital he considered the project to be for reigning in China's floods and furthering national development. At journey's end, Mao swam again across the Yangtze. In the late 1960s, project administrators would honour Mao by making that date – 30 March – into the codename for the Yangtze's first dam – Gezhouba – or Project 330. This event, however, was a decade away.

What was of the order of the day in March 1958 was the Great Leap, which sought to generate a Maoist way to speed up China's transformation into a premier socialist industrial nation. According to Mao, the knowledge regime behind Soviet economic methods was flawed. It produced economic advancements too slowly, and it accorded too much authority to bureaucrats and technical experts. To overcome these deficiencies, Mao advocated a different knowledge regime, which he proclaimed offered an alternate means of socialist industrialisation that was more suited to China's concrete conditions. Like Mao did with his poem 'Swimming', he aestheticised his strategy for rapidly reengineering China's economy, giving it the sobriquet of 'walking on two legs'. In this way, Mao made a top-down developmental vision seem to be an earthy everyday way of promoting economic progress that was as simple as walking.

According to Mao's two-leg developmental strategy, economic actors should modify how much labour and capital were employed in the industrialisation process depending on their availability. In China's case, it should offset its capital-poor economy by directing its large population into building infrastructure and relying on handicrafts, muscle power, and native techniques instead of industrial machinery, technical expertise, and foreign practices. Mao's developmental approach came from his military thought, which had not depended on technical experts, advanced weaponry, and foreign assistance to defeat the CCP's enemies in the 1930s and 1940s, but on mobilising all the people and resources at the party's disposal.[55]

Maoist developmental ideas are visible in Great Leap-era meetings about the TGD held in June and October 1958. During these meetings, attendees discussed

[53] Zhonggong, *Zhongguo*, 37–40.
[54] Lin, *Lin*, 151.
[55] Riskin, *China's*, 201–22.

possible dam locations and the best techniques to develop transport, produce electricity, control floods, and construct irrigation systems. Meeting participants, however, also echoed Leap talking points, stressing the national need to liberate the masses from oppressive foreign models that prized high technology and book knowledge instead of the masses' power to accomplish monumental feats of physical strength and create internationally trailblazing technological inventions. In this vein, attendees exercised their technical imagination and dreamed up a massive crane that could move mountains, and they envisioned a national future in which China's workers forged the frontiers of global technoscience, not through formal education and laboratory research, but through building the TGD.[56]

During the Leap, Maoist flights of scientific fantasy fuelled other hydraulic projects, most infamously the Yellow River's Sanmenxia Dam where, in line with a Maoist developmental approach, the party compensated for shortages of industrial capital with the mass mobilisation of labour. The result was a dam that both silted up and was unstable.[57] Less well-known is the Danjiangkou Dam in Hubei, which Lin Yishan had recommended as the TGD's training ground. Danjiangkou was not the model of technical proficiency that Lin envisioned. It was an object lesson in the problems that Maoist developmentalism can produce as it suffered from similar faults as Sanmenxia.[58]

As for the TGD, the YRC sought to ready its construction by establishing a test dam on a Yangtze tributary – the Lushui River – in 1958. The Commission took this course of action because it did not have experience constructing large dams. The test dam bore the Leap's imprint with technical personnel seeking techniques to accelerate the TGD's eventual construction. Methods examined leaned on both pillars of Mao's two-leg development strategy, employing capital-intensive approaches, such as increasing machine usage, as well as labour-intensive means such as stimulating labour enthusiasm and speeding up cement pouring and the building of the dam's different parts.[59] The party crossed another threshold in September 1959 when plans for the dam and Yangtze River Basin were finished, and the leading Soviet expert in China informed Zhou Enlai that construction preparations could commence. Another milestone was reached in April 1960 when the central party suggested investing 400 million Renminbi (RMB) in the TGD and starting construction in 1961.[60]

That never happened for a few reasons. First, leaders held divergent outlooks about the project. Although prominent critic Li Rui was imprisoned for two decades for opposing the Leap, some top officials still supported his view that the party

[56] Guowuyuan sanxia gongcheng jianshe weiyuanhui, *Bainian daji: Sanxia gongcheng 1919–1992 nian xinwen xuanji*. Wuhan: Changjiang chubanshe, 2005, 49–58.

[57] Pietz, *Yellow*, 225–9.

[58] Hubeisheng weiyuanhui xuexi wenshi ziliao weiyuanhui, *Hubei sheng wenshi ziliao, di sishi er*. Wuhan: Hubei sheng chubanshe, 1993, 1–7.

[59] Lin, *Lin*, 228–36.

[60] Zhonggong, *Zhongguo*, 11.

should first develop the upper Yangtze's tributaries before the TGD. Party Vice-Chairman Chen Yun disagreed with starting the Three Gorges project for another reason. The hydro-sector's resources were already stretched thin on the Sanmenxia Dam and other big hydroelectric plants. Sanmenxia also raised other issues, most importantly its silting up provoked concerns about the TGD clogging up China's main aqueous lifeline – the Yangtze.[61] Further stymieing support for the dam was Moscow's withdrawal of Soviet experts and the Leap's collapse into economic disorder and a massive famine.[62]

The YRC was not exempt from this catastrophic condition as most of its personnel shifted to farm work in 1961 to ward off starvation.[63] Some officials believed that the government should take more measures to address resource strains and disband the TGD project group. Mao and Zhou Enlai thought otherwise. They acknowledged that it was not time to construct the dam, as it would cost over 10 billion yuan, and China did not have the capacity to pursue such a vast venture, especially with Sino-Soviet relations deteriorating and Sino-American tensions intensifying.[64] The TGD project, however, could not completely shut down. Some personnel could be transferred to other ventures, but others had to continue to work towards fulfilling the techno-political dream of walls of stone rising in the Three Gorges and arousing the heaven's awe.[65]

The Cultural Revolution and the Gezhouba debacle

However, Zhou Enlai stipulated that before the dam's construction moved forward, the YRC must address a few problems. First, China's international situation had significantly worsened with the Sino-Soviet split, and so engineers had to find a way to defend the dam from air-raids.[66] The YRC also had to ensure that the TGD could last a long time as Mao had called during the Leap for a dam that 'could operate for at least 200 to 300 years'.[67] Lin Yishan and other Commission experts knew this was technologically impossible. Yet they also understood how important it was to ensure the TGD's longevity. The key puzzle was the Yangtze's heavy silt content. In 1964, the YRC stood up a special group to study silting patterns on all China's dams and techniques for managing them. The technical experts involved also gathered information about dams worldwide and checked that China's silt management

[61] Lin, 'Changjiang', pp. 5–6.
[62] Riskin, *China's*, 130.
[63] Zhonggong, *Zhongguo*, 11.
[64] Lorenz M. Lüthi, *The Sino-Soviet Split: Cold War in the Communist World.* Trenton: Princeton University Press, 2008, 246–72. Warren I. Cohen, *America's Response to China: A History of Sino-American Relations.* New York: Columbia University Press, 2010, 206–14.
[65] Lin, 'Changjiang', 6. Wu, *Zhou*, 43–4.
[66] Wu, *Zhou*, 44.
[67] Lin, *Lin*, 160–1.

methods accorded with international practices. In 1966, the group completed their study and submitted a report to Zhou Enlai, explaining the techniques they had uncovered for controlling silt accumulation and extending a dam's life span.

By the time the YRC submitted this report, it had over 10,000 employees, making it one of the PRC's larger technical institutes.[68] While this strength helped with TGD preparations, it became a liability when Mao launched the Cultural Revolution in 1966 and instituted a new knowledge regime that targeted leading administrators and technicians for supposedly being 'capitalist-roaders' who were more concerned about their own bureaucratic and technical privileges than about socialist China's revolutionary progress.[69] Like at many other government offices, the Commission's administrators and technicians were attacked in late 1966. When the Cultural Revolution began, Lin Yishan initially thought that Mao was right to have the masses rectify cadre errors. He maintained this stance even when big character posters denounced him. His views changed when rebel groups started in 1967 to physically struggle against him and his YRC colleagues roughly three times a week for about two years. At one point, he was imprisoned underground for a month, starved, and had bones broken. Eventually, he was allowed to return to work. However, tarred as a capitalist roader, he was forbidden to perform administrative or technical work and was instead engaged as a cafeteria worker.

Throughout this ordeal, Lin's desire to augment China's industrial infrastructure did not subside. To escape the shattered dreams of his present, Lin occasionally examined Chinese geological maps hidden in his personal belongings and elaborated his thoughts about different hydroelectric projects.[70] During the Cultural Revolution, rebel groups also castigated Lin's colleague Zheng Hui as an 'American special agent' for studying there in the Republican era. Zheng coped with these nativist assaults similarly to Lin Yishan and other technically minded party workaholics, finding solace in secretly working on plans for various developmental ventures.[71]

When Lin Yishan and other technocratically-predisposed officials were deposed throughout China's bureaucracy, CCP leaders did not lose interest in building a new energy regime that remade the nation's rivers into engines of national development. Oversight over the Three Gorges project, however, shifted from the YRC to the People's Liberation Army (PLA), which Mao deployed in 1968 to suppress Cultural Revolution factional conflicts and bring national order.[72] The key figures in this leadership transition in Hubei were provincial First Party Secretary Zeng Siyu and his second-in-command, Zhang Tixue. Their governmental push to revive the Three Gorges project was brought on by the CCP's initiation of the Third Front

[68] Lin, *Lin*, 162–3.

[69] Riskin, *China's*, 180–1.

[70] Lin, *Lin*, 245–8.

[71] Lai, 'Qing man changjiang', 34–7.

[72] Roderick MacFarquhar and Michael Schoenhals, *Mao's Last Revolution*. Cambridge, MA: Harvard University Press, 2006, 239–72.

campaign to ramp up military industrialisation in Central China in the wake of Sino-Soviet border clashes in March 1969. The Third Front's security focus resulted in a very different industrial geography than previous schemes, which were celebrated with public fanfare and directed energy to major cities. New plans were kept out of the public eye as they funnelled electricity to secret military industries hidden in China's mountains. Like the Great Leap, the party rooted the Third Front's drive to hasten industrialisation in the Maoist developmental strategy of mass mobilising labour, handicrafts, and native techniques to compensate for scant industrial equipment and technical expertise as well as non-existent foreign aid.[73]

In the same month that Sino-Soviet skirmishes erupted, Zhang Tixue invited YRC members to investigate the Three Gorges. Afterwards, he recommended building the dam to the vice heads of the Ministry of Water Conservancy and Electric Power Qian Zhengying and Wang Yingxian, who backed Zhang's proposal, much to his delight. In May, Mao met in Hubei with Zhang Tixue and Zeng Siyu, who took their meeting as an opportunity to 'work towards' the Chairman's longstanding dream of building the TGD.[74] Mao, however, stated that with Sino-Soviet tensions mounting, 'Right now is time to prepare for war. It is inadvisable to consider this … If a basin of water fell on your head, would you be scared?'[75] Unfazed, Zhang sought other options for a dam in the Three Gorges region. Qian Zhengying suggested constructing a dam at Gezhouba, which was a few kilometres downstream but could still generate copious energy. Also, as a low dam, if it were bombed, it would be less disastrous than the Three Gorges' high dam.[76]

Wanting different viewpoints represented in TGD discussions, Zhou Enlai ordered Lin Yishan to be reappointed as a YRC administrator in March 1970.[77] Given Zhou's power and prestige, it is not surprising that military leaders then directing the YRC approved Lin's return and authorised his travel to Beijing, where he learned to his great surprise that top Hubei officials – Zhang Tixue and Zeng Siyu – were pushing to build the TGD while Mao was still alive.[78] In October 1970, Zhang and Zeng reported their ideas to the State Council. During the meeting, Zhang swore to Zhou Enlai and other party elites present that, 'If there was a problem with the Gezhouba Dam, then take my head and hang it from Tiananmen.'[79]

Despite Zhang's apparent ardour for Gezhouba, he still preferred the TGD, and so he and Zeng Siyu went to see Mao in December. Mao began the meeting on the offensive, chastising Zhang for the Danjiangkou Dam's many problems. Mao quickly tempered his criticism and extolled Zhang's experience building dams and fighting battles. Yet Mao was still uncertain about whether Zhang and

[73] Covell Meyskens, *Mao's Third Front*. Cambridge: Cambridge University Press, 2020, 10–12.
[74] Wu, *Zhou*, 50. MacFarquhar and Schoenhals, *Mao's Last Revolution*, 48.
[75] Zhonggong, *Zhongguo*, 12.
[76] Zhonggong, ibid.
[77] Lin, *Lin*, 245–8.
[78] Hubeisheng, *Hubei*, 32.
[79] Zhonggong, *Zhongguo*, 12.

his colleagues were ready for the Yangtze's first dam. Well-versed in the flattery prevalent in what Frederick Teiwes has called Mao's court, Zhang replied with his own compliment, quoting Mao about the superiority of learning by doing.[80] Mao responded in kind, praising him for having guts. Zhang then talked about the past decade of preparations, and how the dam would realise Mao's poem 'Swimming'. Unconvinced, Mao rejected his proposal, citing inadequate funds, scientific information, designs, and construction plans. Also, if an atomic bomb hit the TGD, it could trigger flooding all the way to Shanghai. At that point, Zhang switched tactics and recommended building Gezhouba, stating that experts had conducted research and were ready to begin construction. Mao assented and enjoined him to coordinate with Zhou Enlai and relevant ministries.[81] Zhou then, possibly at Mao's behest, told Lin Yishan to write a position paper on the Three Gorges and Gezhouba Dam. Seeking to make up for lost time, Lin had, since being reinstated, gone on a whirlwind tour of Yangtze hydraulic projects and was eager to offer his expert opinion on Gezhouba.[82]

In November 1970, Zhou Enlai organised a State Council meeting to discuss the different options. Lin Yishan argued that the TGD should be built first because Gezhouba would raise water levels by 20 metres, making the TGD more complicated and expensive.[83] There were also not sufficient plans for Gezhouba. The CCP's intention had always been to build the TGD, and so engineers had made designs. Although the YRC had performed studies of Gezhouba since the 1950s, they were far from enough. Only since 1969 was there much work on Gezhouba, and these efforts had yet to yield formal designs. Despite Lin's criticisms, the State Council provisionally approved Gezhouba's construction.[84] Zhou, however, thought Lin's position paper was 'a powerful opinion', and so he submitted it to Mao along with his own letter and a report spearheaded by Zhang Tixue and Zeng Siyu, which depicted Gezhouba as accomplishing Mao's poetic ideal of a 'smooth lake rising in the narrow gorges'. In addition, Gezhouba would conquer Yangtze floods and dramatically enhance China's energy and river transport. As a low dam, it posed less of a security challenge, and Gezhouba planning was based on a decade of geological research and dam modelling experiments, so the PRC's past dam-building mistakes could be averted. What Zhang and Zeng completely glossed over was that the scientific research that they trumpeted was for the TGD, not Gezhouba.[85] When Mao read the Gezhouba file on his 77th birthday, 26 December 1970, he gave it his imprimatur with the caveat that revisions be made as appropriate.[86]

[80] Frederick Teiwes, 'Mao and His Lieutenants', *Australian Journal of Chinese Affairs* 19/20 (1988): 18.
[81] Hubeisheng, *Hubei*, 10–17.
[82] Lin, *Lin*, 250.
[83] Zhonggong, *Zhongguo*, 12.
[84] Hubeisheng, *Hubei*, 20, 33–4, 54.
[85] Hubeisheng, *Hubei*, 4.
[86] Mao Zedong, 'Guanyu Xingjian Gezhouba shuili shuniu gongcheng de pishi', 26 December 1970, in *Zhongguo gongchan dang yu changjiang sanxia gongcheng*, 59.

When Mao made this decision, he surely knew that Zhang and Zeng were stretching the truth with their claim that Gezhouba was the subject of extensive research. From the 1950s, Mao repeatedly consulted with Lin Yishan and Zhou Enlai about a dam in the Three Gorges region, but they had never focused on Gezhouba. Mao also had to understand that a mammoth hydraulic enterprise like Gezhouba would require managerial and technical experts to direct project planning and execution, and that although Lin Yishan and other technocratically-inclined personnel were involved in the project, their voices were not the most influential because in the late 1960s and early 1970s the CCP was loudly advocating nationwide the Maoist developmental way of quickening industrialisation by substituting a mass-mobilised activist labour force for industrial machinery and technical expertise.

A Maoist developmental path is exactly what Hubei leaders followed when they started Gezhouba's construction in late 1970, promptly assembling 100,000 militia members and implementing a Great Leap-era policy known as the three simultaneities in which geological surveys, project designs, and construction work were carried out concurrently. As the dam took shape, technicians alerted Gezhouba's military leadership about a range of issues, from the lack of anti-silting plans and qualified construction workers to labourers pouring concrete without stopping its expansion with ice. Critics also spoke out against project leaders not building a lock on Gezhouba and sacrificing Yangtze transport in order to increase the dam's electricity. Military leaders excoriated these errant views for betraying the Maoist developmental way.[87]

After a few months, Gezhouba rose above the Yangtze, but it was teetering on disaster's edge. One significant issue was Hubei leaders' decision that the YRC's big endowment of technical experts was not properly Maoist. Their proposed solution was to reduce its technical staff from over 8,000 people to only 450. Zhou Enlai vehemently rejected this idea, stating that 'the YRC cannot be changed'.[88] Gezhouba also had perilous structural problems. Its innards were like a honeycomb due to the constant application of cement without pauses for it to set in place, and the lack of a lock meant that upstream Yangtze traffic would soon stop to the dam's west. In late 1972, Zhou Enlai learned about these calamitous complications and called the project's leadership to Beijing where he lambasted them, yelling that, 'If the Yangtze River has a problem, it's not one person's problem, it's the entire country's problem.'[89] To prevent Gezhouba's collapse, Zhou demanded construction stop immediately and not resume until technical personnel under Lin Yishan's direction had plans that guaranteed the dam's stability. In 1974, building started up again. By that time, however, CCP

[87] Hubeisheng, *Hubei*, 34, 75–9.

[88] Wu, *Zhou*, 69.

[89] Zhonggong, *Zhongguo*, 12–13.

Figure 12.3 Building the Gezhouba Dam in 1970

leaders had begun moving China towards a more technocratic approach to manifesting Mao's poetic prophecy of a TGD technologically taming the Yangtze and making it into a hydraulic engine of national development.[90]

Conclusion

Although CCP leaders made significant changes to China's approach to building the TGD after Deng Xiaoping came to power in the late 1970s, there were also notable areas of overlap with the developmental techniques the party pursued in Mao's China. Under both Mao and Deng, China's supreme leaders' interest in the project kept it on the party's agenda. Mao and Deng both endorsed the dam as part of a nationwide drive to engineer a new energy regime that transformed China's rivers into flood-free, bustling thoroughfares whose tamed waters served as engines of industrialisation. China's pursuit of this developmental dream under Mao and Deng ran into the same conundrum. China did not have the technical, administrative, or fiscal means for such a gigantic undertaking. This material condition

[90] Zhonggong, ibid.

prompted the CCP to solicit Soviet aid in the 1950s and American, European, and Japanese assistance in the 1980s.

In all instances, Beijing's partners sought to make China into a strong industrial nation with the notable difference that Moscow aimed for the PRC to help replace an American-led capitalist world order with a new socialist world economy, while Washington strived to weaken the Soviets and bolster US power in Asia through engaging with China. Despite this geopolitical divergence, both Chinese and foreign specialists endorsed entrusting centralised experts in the YRC with making the Yangtze's geological patterns legible and turning the river into a motor of national industrial progress.

The Gorges project also had critics who repeatedly argued that small dams could achieve benefits quicker, for less. Mao overrode criticisms, yet also accorded significant influence to technically minded officials until the 1969 Sino-Soviet border crisis when Mao, amid Cultural Revolution attacks on bureaucrats and technical experts, backed Hubei leaders' proposal to realise his dream of a dam in the Three Gorges region by building Gezhouba, however this time without mass media attention since its power was to be directed to a covert military-industrial complex.

Despite this difference, project managers endeavoured, like in earlier efforts, to keep the project at the global technological frontier, though they imparted science's cutting edge with nativist roots, stressing Mao's developmental idea that mass mobilisation was a more powerful economic force than industrial aptitude and machinery, a decision that produced a botched dam. To remedy the project, Zhou Enlai authorised YRC experts to oversee its completion, and yet factional disputes still plagued the project until after Mao died and, as people said in China at the time, went to meet Marx.

After Mao, CCP leaders still prescribed accelerated development for the Chinese nation, but under a new knowledge regime where the Maoist developmental way was no longer conceived of as global science's forefront, but as a major cause of underdevelopment. The Chinese masses were also no longer viewed as a reservoir of native talent and productive power, but as technically backward compared to the world's leading nations, and whose manual labour was no longer glorified, but rather shunted back into the hidden abode of production, as CCP leaders exalted engineers and encouraged the Chinese people to be spectators to their technical marvels.

To overcome national backwardness, the CCP again looked abroad for partners to help with China's economic rise. The YRC took advantage of this new geopolitical situation and collaborated with foreign experts on building Gezhouba. Projects leaders, however, no longer prioritised huge labour regiments and using indigenous methods to construct a clandestine military-industrial base, but rather empowered bureaucrats and engineers to command a multitude of machines and draw on global knowledge networks to fulfil the Chinese state's hydraulic dream of harnessing the Yangtze's power and making it work for the rapid development of the nation's cities and coastline for all the world to see.

13

Publishing as Making: Defining Knowledge and Cultivating Organic Intellectuals, 1961–1965

ROBERT CULP

> Knowledge is power. A revolutionary cadre must have rich ancient and modern, Chinese and foreign knowledge to serve as a foundation for engaging in work and studying theory. The *Knowledge Series* was published in order to satisfy this need. The content encompasses knowledge about philosophy, social science, natural science, history, geography, international issues, literature, art, and daily life.
>
> The *Knowledge Series* Editorial Committee[1]

THUS BEGAN THE inscription of every volume of the *Knowledge Series* published by a consortium of Chinese publishers between 1961 and 1965.[2] Long-time publishing hand Hu Yuzhi initiated the series in 1961 to provide basic education to China's cadres in the aftermath of the Great Leap Forward (GLF), which was marked by disastrous policy decisions that led to the deaths of millions. Yet, as the inscription suggests, the series introduced cadres not only to functional technical information that could inform policy, but also to a world of knowledge and culture extending far beyond the practical. Framed in this way, the series raises a host of questions

[1] Portions of this chapter are drawn from *The Power of Print in Modern China*, by Robert J. Culp. Copyright © 2019 Columbia University Press. Reprinted with permission of Columbia University Press.
[2] Hu Yuzhi's protégé Chen Yuan borrowed Francis Bacon's phrase to introduce the series, believing 'once knowledge was grasped by the people, it would transform into material power'. 'Knowledge is Power' (*zhishi jiushi liliang*) was also the name of a Soviet science journal, and it was adopted as the name of a Chinese journal for science popularisation that was published between 1956 and 1963, reflecting the positivism of Soviet-style state socialism. See Song Yuanfang (ed.), *Zhongguo chuban shiliao: Xiandai bufen* (Historical materials for Chinese publishing: Modern section), Vol. 3.1. Jinan: Shandong jiaoyu chubanshe, 2000, 199–200; Marc Andre Matten, 'Coping with Invisible Threats: Nuclear Radiation and Science Dissemination in Maoist China', *East Asian Science, Technology and Society* 12, no. 3 (September 2018), 238.

Proceedings of the British Academy, **267**, 254–272, © The British Academy 2025.

for students of the early People's Republic of China (PRC). How was 'knowledge' (*zhishi*) defined during the Mao era? How did that definition configure or 'make' a domain of knowledge? Why were cadres singled out as a prime audience for that knowledge? What kind of power was knowledge supposed to provide them?

This chapter argues that the *Knowledge Series* marked a turn back to a comprehensive, global conception of knowledge that more closely resembled that associated with Republican China's projects of popular enlightenment than early-PRC forms of knowledge production. After a decade of privileging Soviet scholarship, Chinese publishers were now encouraged to draw widely from cosmopolitan intellectual sources that ranged far beyond the Warsaw Pact countries.[3] This resurgent cosmopolitanism, which harkened back to the Chinese Communist Party (CCP)'s May Fourth heritage, created a Chinese socialist knowledge system that was neither Soviet-centred nor narrowly culturalist, self-enclosed, and autochthonous, a rare combination in the Cold War socialist world. Moreover, the *Knowledge Series*' initiators differentiated knowledge from both ideology (*sixiang*, *yishi xingtai*) and theory (*lilun*). Instead, they embraced a multi-dimensional definition of knowledge that encompassed the scientific and technical as well as basic forms of ordinary knowledge (*changshi*), cultural literacy, foundational national culture, and understanding of world history, literature, and intellectual traditions. This approach to instilling comprehensive knowledge directly paralleled the large-scale series publications of the Republican period, such as Commercial Press's *Complete Library* (Wanyou wenku) and World Book Company's *ABC Series* that had aimed to provide a broad cross-section of knowledge to China's citizens.

As the inscription makes clear, however, cadres, not the national citizenry as a whole, were the primary intended audience of the *Knowledge Series*. Why target cadres in this way, and why expose them to some forms of knowledge that seem quite irrelevant to much of their daily work? I argue that the *Knowledge Series* was a central feature of the effort to cultivate cadres as 'organic intellectuals' or 'indigenous representatives of the proletariat'.[4] As such, they needed to be trained as both 'red' (politically committed) and 'expert' (technically competent),[5] while also embodying forms of cultural literacy associated with the late imperial literati and forms of cosmopolitan knowledge identified with Republican-period intellectuals immersed in modern academic scholarship from Europe, the US, and Japan.

[3] For the prevalence of Soviet influence during the early Mao era, see the chapters by Wu, Meyskens, and Volland in this volume.

[4] Eddy U, 'Intellectuals and Alternative Socialist Paths in the Early Mao Years', *The China Journal*, no. 70 (July 2013), 15–21. The formulation of the 'organic intellectual' comes from Antonio Gramsci's analysis of the social and political roles of intellectuals in his seminal essay 'The Intellectuals' in *Selections from the Prison Notebooks of Antonio Gramsci*, eds. Quintin Hoare and Geoffrey Nowell Smith. New York: International Publishers, 1971.

[5] Maurice Meisner, *Mao's China and After: A History of the People's Republic*, 3rd edition. New York: The Free Press, 1991, 118, 212. Sigrid Schmalzer further draws parallels between 'redness' (*hong*) and 'native' (*tu*), and 'expert' (*zhuan*) and 'foreign' (*yang*). See *Red Revolution, Green Revolution: Scientific Farming in Socialist China*. Chicago: University of Chicago Press, 2016, 37–9.

For cadres to be recognisable as organic intellectuals, they needed to embody the cultural authority historically associated with Chinese intellectuals in the past. Knowledge, in this way, empowered by providing both technical mastery and cultural authority. Instilling such cultural authority in cadres also reified a hierarchical distinction between governors and the governed.

Organising knowledge production

During the Republican period (1912–49), China's leading commercial publishing companies developed comprehensive series that they marketed to a broad reading public.[6] Two of the most prominent examples of these comprehensive series were Commercial Press's *Complete Library* (*Wanyou wenku*) and World Book Company's *ABC Series*. Each of these series included hundreds if not thousands of titles, and their initiators intended for them to provide an encyclopedic survey of modern subjects, ranging from history, literature, and philosophy to science and technology. Each publisher mobilised staff editors in their editing departments and expert authors from outside the company to translate foreign works and write synthetic monographs that introduced particular subjects or academic disciplines. The target audience for these series was what Commercial Press's Wang Yunwu characterised as the 'general reading public' (*yiban dushu jie*) and World's Xu Weinan described as the 'entire populace' (*quanti minzhong*). By making a wide spectrum of modern knowledge available to the broadest possible cross-section of readers, the commercial publishers sought to maximise profits while enlightening the Republican citizenry.

Private publishing companies like Commercial Press and Zhonghua Book Company continued to operate alongside the People's Publishing Company (*Renmin chubanshe*) and other state-owned publishers during the early 1950s. As late as 1953, roughly half of the titles published nationwide were produced by private companies.[7] In fact, Commercial Press continued to publish one of its signature comprehensive book series, the *University Series* (Daxue congshu), into the early 1950s.[8] In the changed political climate after 1949, however, publications from and related to the Soviet Union as well as series on science and technology began to dominate.[9] For example, in its initial post-1949 publication list distributed in the summer of 1950, Zhonghua Book Company offered two robust series of titles on Marxism-Leninism and the Soviet Union that were translated from Russian, the *New Era Small Series* (Xin shidai xiao congshu) and the *Soviet Knowledge*

[6] Robert Culp, *The Power of Print in Modern China: Intellectuals and Industrial Publishing from the End of Empire to Maoist State Socialism*. New York: Columbia University Press, 2019, chapter 5.

[7] Song Yuanfang, *Zhongguo chuban shiliao: Xiandai bufen*, 3.1:131.

[8] Culp, *The Power of Print*, 217.

[9] See Volland's chapter in this volume.

Small Series (Sulian zhishi xiao congshu).[10] Starting in 1951 Commercial Press also began to produce numerous new series on the themes of science and technology, several of which drew primarily on Soviet sources.[11]

The inaugural national publishing conference in 1950 introduced new forms of state oversight and restructured the publishing industry in fundamental ways. Most directly relevant for this discussion was a push toward coordinated division of labour among China's public and private publishers, with each company developing a specialised focus in its publications.[12] Thus, by 1958 the reconstituted Commercial Press came to focus on publishing translations of foreign scholarship and synthetic works about it, while Zhonghua Book Company also reorganised to emphasise republication of classical texts and scholarship about Chinese history and culture. In step with this move toward specialisation and division of labour, in 1954 private companies like Zhonghua and Commercial Press reorganised as public-private joint management (*gong si heying*) enterprises with increasing degrees of party and government involvement.[13] Within this new system of division of labour under government oversight, publishing companies continued to produce series publications, but most series became specialised to accord with the company's publishing mandate, and they aimed to satisfy demand from distinct groups of readers, depending on the content of the series.[14]

Running counter to this trend of specialisation, in May 1961 vice minister of the Ministry of Culture in charge of publishing Hu Yuzhi proposed publication of the *Knowledge Series* (Zhishi congshu) as a comprehensive series publication.[15] To spark the project, he hosted a lunch meeting at Beijing's Sichuan Restaurant. Joining Hu and his protégé Chen Yuan were key figures in the Mao-era cultural apparatus and publishing industry: the director of the CCP Propaganda Department's Publishing Office at the time Bao Zhijing; the director of the Publishing Bureau of

[10] *Zhonghua shuju jiefanghou chuban xinshu mulu* (Catalog of new books published by Zhonghua Book Company after Liberation) (N.p.: Zhonghua shuju, 1950), 4–12.

[11] Shangwu yinshuguan (ed.), *Shangwu yinshuguan jiushinian, 1897–1987—Wo he Shangwu yinshuguan* (Ninety years of the Commercial Press, 1897–1987—Me and the Commercial Press). Beijing: Shangwu yinshuguan, 1987, 443. Hereafter SYJ.

[12] Zhongguo chuban kexue yanjiusuo and Zhongyang dang'anguan, *Zhonghua renmin gongheguo chuban shiliao, di'er juan*: 1950 (Historical materials for publishing in the People's Republic of China), Vol. 2: 1950. Beijing: Zhongguo shuji chubanshe, 1996, 642–4. Hereafter PRCCS1950.

[13] For detailed discussion of Commercial Press's and Zhonghua Book Company's reorganisation, see Culp, *The Power of Print*, chapter 6.

[14] SYJ, 460–1; Zhonghua shuju bianjibu (ed.), *Huiyi Zhonghua shuju* (Remembering Zhonghua Book Company). Beijing: Zhonghua shuju, 2001 [1987], 2: 16, 124–7. Hereafter HZS.

[15] Hu had spent the better part of his adult life working in the publishing industry. After joining the Commercial Press in 1914 as a young adult, he worked there on and off for much of the next two decades. Hu later helped Zou Taofen launch *Life Weekly* (*Shenghuo zhoukan*) and managed a series of journals aimed at overseas Chinese in Southeast Asia during and after the War of Resistance, joining the CCP in 1933. After 1949 Hu served as the first director of the General Publishing Administration. Xu Youchun *et al.* (eds.), *Minguo renwu da cidian* (Biographical dictionary of Republican China). Shijiazhuang: Hebei renmin chubanshe, 1991, 578.

the Ministry of Culture Wang Yi; the People's Press's Wang Ziye; Commercial Press's Chen Hanbo; and Zhonghua Book Company's Jin Canran.[16] At the meal Hu Yuzhi floated the idea of publishing a series to raise the cultural level of China's cadres, with each volume introducing them to intermediate-level knowledge on a cultural, scientific, or practical topic. Hu asserted that the basic level of knowledge presented in the series meant that the authors need not be elite experts; rather, staff editors in the publishers could write most of the manuscripts, as they often had at Commercial Press, Zhonghua Book Company, and Kaiming Bookstore before the war. As models Chen Yuan cited Commercial Press's *Complete Library* but also the international examples of *Everyman's Library* and the Little Blue Book series. Hu's colleagues immediately supported the initiative and voiced suggestions for identifying series authors among 'compilers, teachers, and engineers and technicians'.

Hu further raised specific ideas about the content of the series, suggesting a focus on reading materials about science and technology and about 'problems that are currently of the greatest concern to the masses'.[17] In the aftermath of the GLF, when cadres' mismanagement had played a role in creating an economic disaster, the need for raising cadres' technical knowledge was quickly acknowledged by Hu's colleagues. In contrast, they balked at the idea of publishing volumes on questions like 'Why is the market in short supply?' or 'Why is there a lack of non-staple foods?' Insofar as these issues related to the state's economic policies, in the wake of the Lushan Plenum (July–August 1959) and Peng Dehuai's purge (23 July 1959), they were viewed as too politically sensitive.

Similarly, people raised the possibility of publishing texts that introduced basic elements of Marxist-Leninist theory.[18] However, the Central Propaganda Department's Bao Zhijing warned that publishing volumes on what were essentially contested concepts in Marxist theory could be difficult to do correctly and potentially politically sensitive. He suggested leaving this work to a group in the Central Propaganda Department that was writing books explaining the basic principles of Marxism. In these discussions, government and party leaders responsible for managing the publishing industry and representatives of several leading publishers negotiated how to address a clear market demand while limiting risk, in a context in which culture and learning were increasingly politicised. In doing so, they carefully differentiated the domain of knowledge from that of theory, leaving the latter to party theorists. They also steered away from issues of current government policy, which were considered politically sensitive.

Based on these discussions, on 15 May 1961, the Ministry of Culture's party group submitted a preliminary proposal to the Central Propaganda Department.[19]

[16] Song Yuanfang, *Zhongguo chuban shiliao: Xiandai bufen*, 3.1: 187–9.
[17] Song Yuanfang, *Zhongguo chuban shiliao: Xiandai bufen*, 3.1: 189–91.
[18] Song Yuanfang, *Zhongguo chuban shiliao: Xiandai bufen*, 3.1: 190–1.
[19] Zhongguo chuban kexue yanjiusuo and Zhongyang dang'anguan, *Zhonghua renmin gongheguo chuban shiliao* (Historical materials for publishing in the People's Republic of China), Vol. 11, 1961 nian juan. Beijing: Zhongguo shuji chubanshe, 2007, 152–5. Hereafter PRCCS1961.

They planned to produce some 2,000 volumes between 1961 and 1965, with each averaging 50,000 characters or about 100 pages. They proposed organising an editorial committee composed of 'leading comrades of cultural and academic circles, figures in academic circles, and relevant publishing companies' editors-in-chief' to oversee and support the project. This organisational structure followed the precedent of the editorial committee Wang Yunwu had organised to support the *University Series*, which Commercial Press had published from the mid-1930s into the early 1950s.[20] The thematic areas covered would include theoretical knowledge (but not Marxism-Leninism), historical knowledge, world knowledge, knowledge about production skills and daily life, and knowledge about literature and the arts. The range of subject areas generally paralleled that of the comprehensive series publications of the Republican period, such as Commercial Press's encyclopedic *Universal Library* (Baike xiao congshu) and World Book Company's *ABC Series*.

Given the division of labour among China's publishing companies after the industry's reorganisation during the 1950s, however, no single publisher was equipped to produce this kind of comprehensive series since each specialised in particular kinds of publications. Instead, six different publishers had to coordinate, with each to focus on a specific subject area.[21] People's Publishing Company would concentrate on philosophy and social science; People's Literature Publishing Company would cover literature and the arts; Zhonghua Book Company would take up Chinese history and cultural heritage; Commercial Press would engage foreign history and cultural heritage; the Science Publishing Company would specialise in science, technology, and productive life; while the World Knowledge Publishing Company would address international knowledge. Each company was then to formulate a list of titles for its subject area and solicit manuscripts to serve as models for its part of the list, which were to be published in the fourth quarter of 1961. Making knowledge in the Mao era required extensive coordination across enterprises.

On 18 May the Central Propaganda Department held a high-level meeting with the Ministry of Culture's Qi Yanming, Hu Yuzhi, and Chen Yuan to discuss the initial proposal.[22] In the meeting Propaganda Department director Lu Dingyi identified Commercial Press's *Complete Library* as a model and noted the way it incorporated ancient and modern, Chinese and foreign knowledge. Assistant director Yao Zhen reiterated cadres' demand for information about foreign countries and saw the value of individual volumes introducing specific countries. Zhang Ziyi asked whether 2,000 volumes was too many, and suggested fewer, shorter volumes of 10,000–20,000 characters each. Assistant Director Zhou Yang, who chaired the meeting, concluded it by calling for an

[20] Culp, *The Power of Print*, 171–2.
[21] PRCCS1961, 155.
[22] Song Yuanfang, *Zhongguo chuban shiliao: Xiandai bufen*, 3.1: 192–4.

emphasis on richness of material rather than insisting that each volume embody a Marxist-Leninist point of view, a statement that clearly differentiated knowledge from ideology and theory. And he proposed that Hu Yuzhi serve as the project's executive editor.

On 23 May, the Ministry of Culture submitted a slightly revised plan that was approved by the Central Propaganda Department on 8 June.[23] The most significant change was moderation of the scale, with a goal of publishing 300–400 books in two to three years. In this approval process, Ministry of Culture officials collaborated with publishing company directors to formulate a plan that was acceptable to cadres in the CCP's Propaganda Department. These leading cultural cadres all had a hand in making this element of the Mao-era knowledge system by defining the parameters of knowledge and supporting its material production through a coordinated publishing process for the *Knowledge Series*. In doing so they built on publishing strategies that had long histories prior to the founding of the PRC in 1949 and that were rooted in the cosmopolitan knowledge culture of the 1910s and 1920s.

Work on the series began in August 1961 with a meeting of the expanded editorial committee. Hu Yuzhi quickly realised that the editorial committee, composed as it was of leading cultural and political figures that were widely dispersed, could not serve as an immediate managing body for the series.[24] Instead, he organised a standing editorial committee (*changwu bianwei*) composed of Chen Yuan, Wang Cheng, and representatives of each of the participating publishers, who could meet every month or so to manage it, with the Central Propaganda Department's Xu Liyi also frequently participating.[25] To brainstorm titles for each component subject area of the series, Hu initiated a series of forums between outside experts and representatives of the publishers to discuss possible topics of focus. Forums were held that year on economics, international issues, and philosophy. Building on all these discussions, the Ministry of Culture's Chen Yuan, Wang Cheng, and Han Zhongmin formulated a list of about 1,000 topics, on the model of an encyclopedia, echoing Republican-era projects like the *Complete Library* and *ABC Series*. The standing editorial committee then worked with the publishers to identify volumes that fit the series. For instance, at Zhonghua Jin Canran engaged the linguist Wang Li to write a 50,000–60,000 character volume about poetic metre. Written in an accessible language, the committee eagerly adopted it as a model, also feeling that its focus on Chinese cultural heritage was relatively safe politically.[26] At this level, the 'making' of knowledge entailed cultural bureaucrats' mobilisation of intellectuals to write individual manuscripts.

[23] PRCCS1961, 156–8.
[24] Song Yuanfang, *Zhongguo chuban shiliao: Xiandai bufen*, 3.1: 194–8.
[25] Song Yuanfang, *Zhongguo chuban shiliao: Xiandai bufen*, 3.1: 195.
[26] Song Yuanfang, *Zhongguo chuban shiliao: Xiandai bufen*, 3.1: 197–8.

Cosmopolitan knowledge and cadre power

The *Knowledge Series* demarcated and organised knowledge in ways that departed from the Soviet-centred series publications of the early 1950s and revived an earlier late-Qing and Republican form of cosmopolitanism that looked to Europe, Japan, and the US for models.[27] This form of cosmopolitanism, which embraced intellectual and cultural influences from beyond the socialist world, stood in contrast to the socialist cosmopolitanism that Nicolai Volland has identified as a central feature of early Mao-era cultural dynamics.[28] The broad outlines of that way of defining knowledge are already clear from our review of the series' organisational process. Significantly, in the aftermath of the Sino-Soviet split, no explicit reference was made to Soviet academic models, disavowing the emphasis on scholarship from the Soviet Union that characterised most of the 1950s. Instead, earlier Chinese and foreign publication series, especially Commercial Press's *Complete Library*, were referenced, despite its associations with Republican Chinese publishing's arch villain, Wang Yunwu, who became a leading Nationalist Party official during and after the War of Resistance and fled to Taiwan after 1949.[29]

A fuller picture of the definition of knowledge associated with this series comes from a talk by then assistant director of the CCP's Central Propaganda Department Zhou Yang at the expanded conference on the *Knowledge Series* held in August 1961. Zhou began by making the case for knowledge as a part of the revolutionary project:

> Now we urgently need knowledge, just as we need grain and non-staple foods. Can we say that we've carried out these years of revolution without knowledge? Without knowledge could the revolution have been victorious? In the past people said, the CCP has skills but not learning; the democratic personages have learning but no skills. This saying is mistaken. Actually, this is not endorsing carrying out revolutionary activities. One should say that the entire CCP and progressive circles have learning (*xuewen*), with a level that should be said to not be very low, encompassing scientific circles and literary and art circles.[30]

[27] Leo Lee has been most explicit about characterising the late Qing and especially Republican foreign influences as formulated by cultural producers in Shanghai as a form of 'cosmopolitanism'. See *Shanghai Modern: The Flowering of a New Urban Culture in China, 1930–1945*. Cambridge, MA: Harvard University Press, 1999, 313–15.

[28] Nicolai Volland, *Socialist Cosmopolitanism: The Chinese Literary Universe, 1945–1965*. New York: Columbia University Press, 2020. Joseph Levenson described as cosmopolitan Mao-era translations of realist foreign plays from multiple different countries, embracing a broad cultural field that extended beyond the Soviet Union, the ideologically socialist, or the technically practical. Joseph R. Levenson, *Revolution and Cosmopolitanism: The Western Stage and the Chinese Stages*. Berkeley: University of California Press, 1971, 11–18. Levenson, too, sees this cosmopolitanism, which echoed that of an earlier period, to have been under threat from the iconoclasm and provincialism of the Cultural Revolution (p. 51).

[29] In other fields, too, pre-1949 paradigms and approaches persisted during the Mao era. See the chapters by Wu, Meyskens, and Brazelton in this volume.

[30] PRCCS1961, 220.

Zhou's statement pointed to the value of a cross-section of academic and cultural knowledge to the revolutionary process. He went on to stress the need to extend that knowledge to a broad public of workers, peasants, and cadres, stating that 'we must raise the level of knowledge of the worker and peasant masses and cadres'.[31]

Zhou Yang justified a broad-based approach to knowledge production by reference to both Lenin and Mao. At one point he argued:

> We certainly must grasp cultural knowledge. Lenin said, 'Only if one uses all the intellectual wealth created by humankind to enrich one's mind can one become a communist.' These words from Lenin are very deep. Without cultural knowledge and without experts it is impossible to construct socialism.[32]

Lenin's identification of 'all the intellectual wealth created by humankind' as essential to the revolutionary project opened the door to a diverse array of learning from both foreign and Chinese sources being collected and disseminated. Indeed, Zhou went on to cite Mao's definition of meaningful forms of knowledge as 'knowledge about the struggle for production and knowledge about social struggle' as pointing to the value of natural science and social science, thus embracing a wide spectrum of academic learning.[33]

In this talk to editors, publishers, and scholars who had faced the full brunt of the Anti-Rightist Campaign just a few years before, Zhou Yang also directly addressed concerns about ideological criticism for academic and publishing work. In doing so, he effectively segregated the domain of knowledge from that of ideology and theory.

> With fear [about writing incorrectly and being criticised], ideological misgivings, we must resolve it … . The requirements for the *Knowledge Series* are that as long as the material is substantial, there are not any reactionary viewpoints, and it does not violate the six political criteria (*liutiao zhengzhi biaozhun*)[34] that is acceptable. Isn't it always rare that natural science and history books violate the six criteria? As for theoretical errors (*lilun shang de cuowu*) or errors in academic viewpoint (*xueshu guandian*), they are unavoidable. Everyone can be at ease and write boldly; if there are errors, everyone can discuss them. We must separate political problems, ideological problems, and academic problems, must draw a boundary line between them and not lightly cross it.[35]

[31] PRCCS1961, 221.
[32] PRCCS1961, 221–2.
[33] PRCCS1961, 222.
[34] Mao had laid out six criteria for differentiating so-called 'fragrant flowers' from 'poisonous weeds' in the context of the 100 Flowers Movement. See 'On the Correct Handling of Contradictions among the People' (published 19 June 1957), accessed 7 November 2023, www.marxists.org/reference/archive/mao/selected-works/volume-5/mswv5_58.htm. For the formulation of the six criteria and their publication in the revised version of this speech, see Roderick MacFarquhar, *The Origins of the Cultural Revolution, Vol. 1, Contradictions among the People, 1956–1957*. New York: Columbia University Press, 1974, 263–7.
[35] PRCCS1961, 222–3.

In stressing the need for a clear line between ideological and theoretical problems and academic problems, Zhou essentially distinguished a sphere of academic learning, which was specialised and technical, from the concerns of politics, ideology, and theory.[36] In doing so, he was not suggesting that academic work could never be tainted by theoretical or ideological errors, but he did suggest that knowledge claims could, and should, be evaluated according to academic standards, apart from their political valence, creating a significant degree of autonomy for a domain of knowledge. This formulation created a basis for producing and circulating knowledge on its own terms.

According to Chen Yuan, by 1965 at least 83 titles had been published in the *Knowledge Series*, making a significant start toward the goal of 300–400 books overall.[37] Of the 78 extant titles I have been able to identify,[38] the publishing distribution was as follows: 40 from the Popular Science Publishing Company (*Kexue puji chubanshe*); 17 from Commercial Press; 15 from Zhonghua Book Company; 4 from World Knowledge Publishing Company; and 2 from Sanlian Publishing Company (which had not been listed among the original participants). World Knowledge Publishing Company also produced a *World Knowledge Series* (Shijie zhishi congshu), focused on the socialist and developing worlds, which can be seen as a companion list and included at least an additional 20 titles by 1965. Noticeably absent were contributions from People's Publishing Company and People's Literature Publishing Company, which had been included in the original publishing plans.

Reviewing the subject areas covered by these books provides a concrete way for us to delineate the domain of knowledge from that of theory or ideology as cultural cadres like Zhou Yang and Hu Yuzhi articulated it during this period. Most of the published titles produced by Popular Science Publishing Company addressed scientific and technical knowledge, such as television, film, electronic computing, wireless, cells, phenology, soybeans, genes and heredity, as well as human origins and development. Many of these subjects tackled the practical concerns about helping cadres govern more rationally that were raised in the aftermath of the GLF. At the same time, Commercial Press and Zhonghua Book Company published books for the series that ranged more widely in introducing general knowledge about world history and thought along with Chinese history and the classics. For instance, Commercial Press contributed titles on Adam Smith and England's classical political economy, the Italian Renaissance, Malthusianism, Keynesianism, physiocratic political economy, mercantilism, and other Western thinkers and schools of thought. Zhonghua offered titles on subjects like the *Book of Songs*

[36] Of course, any separation between knowledge and politics was never absolute. As the papers by Brazelton and Wu show in this volume, the political nature of knowledge, in terms of nationalism and class, had been a feature of academic discourse from the start of the Mao era.

[37] Song Yuanfang, *Zhongguo chuban shiliao: Xiandai bufen*, 3.1: 200.

[38] This overview of titles is based on a survey of OCLC and extant holdings at the Shanghai Municipal Library.

(Shijing), Qu Yuan, Wen Tianxiang, Song-Yuan storytelling scripts, and the 1911 Revolution. This more diverse group of topics replicated the expansive approach to spreading ordinary knowledge that had characterised the comprehensive series of the Republican period.

Analysing the forms of knowledge encompassed in this list reveals that it ranges far beyond scientific or technical knowledge that would have allowed cadres to make better decisions about farming techniques, steel production, or allocation of labour that had contributed to the rural economic crisis during the GLF.[39] Also given prominent place here are basic forms of cultural literacy in the Chinese cultural tradition, both Chinese and foreign history, as well as a wide spectrum of foreign philosophy and thought, ranging across disciplines. In sum, the series promised to provide cadres with a broad-based academic and cultural education in the humanities, sciences, and social sciences (especially political economy, economics, and demography), along with exposure to various forms of ordinary knowledge. Through their inclusion, the series legitimised various forms of Western scholarship and Chinese history, literature, and culture as valid forms of knowledge.

Why the state's cultural managers thought cadres needed these kinds of knowledge was not explicitly articulated. As noted above, the overt justification expressed in publication meetings and meetings with the Ministry of Culture and Propaganda Department had to do with more rational and efficient governance. It is not immediately clear, though, how an understanding of the Italian Renaissance or Chinese poetic metre could contribute to that goal. Rather, this comprehensive approach to knowledge for cadres becomes comprehensible only when they are identified as 'organic intellectuals' of state socialism who could embody both the qualities of 'red' and 'expert' and serve as 'indigenous representatives of the proletariat'.[40] Viewed in that light, the forms of theoretical learning and cultural literacy embedded in the *Knowledge Series* served to educate the cadre so that he could embody enough cultural capital to operate as an intellectual. Two vital sources of cultural capital inscribed in this list were Euro-American academic theories from the social sciences and the humanities along with knowledge of key practices, figures, and texts in China's classical heritage. These domains of knowledge corresponded, respectively, to those mastered by the foreign-trained modern intellectual and the late imperial literati, who were the main kinds of intellectuals that cadres as

[39] More technical and specialised publications along these lines were also published for internal distribution to cadres during the early 1960s. For examples related to economic management, see Zhongguo banben tushuguan comp, *Quanguo neibu faxing tushu zongmu (1949–1986)*. Beijing: Zhonghua shuju, 1988, 152–219 passim.

[40] U, 'Intellectuals and Alternative Socialist Paths', 19–21. See also Joel Andreas, *Rise of the Red Engineers: The Cultural Revolution and the Origins of China's New Class*. Stanford: Stanford University Press, 2009, chapters 1–3. Andreas sees efforts to augment cultural capital as focused mostly on providing technical education to students of the post-revolutionary generation. Meyskens in this volume also describes how cadres were trained in technical skills to support the Yangzi River Water Conservancy Commission during the early 1950s. By contrast, the *Knowledge Series* aimed to reach adult cadres with lessons about both technical subjects and a wide range of other cultural knowledge.

'organic intellectuals' were intended to replace. Thus, arming cadres with a basic understanding of these two domains of knowledge was essential to providing them with the cultural authority to legitimise them as intellectuals and by extension political leaders. Correspondingly, the opening inscription's statement 'knowledge is power' could be seen to speak to the status competition between cadres and the remnant intellectuals of the late-Qing and Republican periods, for which CCP cultural officials sought to empower cadres culturally and intellectually.

The relationship between knowledge and power also comes into play in terms of the intended readership for the comprehensive knowledge presented in this series. During the Republican period, publishers had sought to spread similar kinds of comprehensive knowledge to a broad cross-section of Chinese citizens, the 'general reading public', to prepare them to be active members of the nation and civic community.[41] Although in practice a much smaller readership of students and intellectuals with a secondary level of education and higher actually encountered those materials, more universal access was the goal.[42] By making cadres the primary focus of publishing comprehensive knowledge, cultural cadres and publishers tacitly signalled that they were the ones who needed knowledge to act politically. The 'entire populace' might be mobilised for political action, as it was many times during the Mao era, but those who needed knowledge to guide that action were cadres, not necessarily average citizens. Thus, cultural cadres and publishers functionally segmented, and vertically stratified, the general reading public that Republican publishers had sought to create, fashioning discrete reading communities that they targeted with specific kinds of published knowledge.

In terms of the *Knowledge Series*, though, this stratification was not strict and absolute. In their publication plans, the Ministry of Culture and Propaganda Department also acknowledged that the *Knowledge Series* could serve as supplemental reading materials for youths with secondary-level educational backgrounds and higher.[43] Just as importantly, the series was not designated for 'internal distribution' (*neibu faxing*), which would have allowed the publishers to systematically limit access to cadres alone. By the early 1960s, the mechanism of internal distribution, which originally had been developed to protect state and party internal communication, was increasingly deployed to demarcate readerships for different kinds of publications.[44] By allowing the *Knowledge Series* to be 'openly distributed'

[41] Culp, *The Power of Print*, chapter 5.

[42] I have estimated that readership during the 1920s and 1930s to be in the range of 12 to 24 million. See Culp, *The Power of Print*, 16.

[43] PRCCS1961, 152, 153, 157.

[44] See Culp, 'Market Configuration: Internal Distribution's Transformation of Texts, Reading Communities, and Reading Practices' (paper presented at the Biennial Conference of the European Association for Chinese Studies, Olomouc, Czech Republic, 26 August 2022). For an example of how publishers and cultural cadres designated different kinds of books for specific groups of readers, and chose distribution methods accordingly, see Commercial Press's proposed methods for distributing works of foreign philosophy and social science in 1961, contemporaneous with the launch of the *Knowledge Series*. PRCCS1961, 193–200.

(*gongkai faxing*) through New China Bookstores with fairly robust print runs[45], the publishers and the state's and party's cultural cadres left open the possibility that this series that had been produced with cadres as the intended audience might be 'overheard', to use Michael Warner's evocative formulation, by other kinds of readers.[46] In addition, the cultural cadres did not script the reading process, as they did with many internal circulation materials. Internal publications were often explicitly designed for 'study' (*xuexi*) or 'reference' (*cankao*).[47] The former emphasised collective reading to discern government policy and party line; the latter constituted a form of critical reading of material that could be politically suspect for the purpose of political criticism or building state socialism. Cultural cadres' discussion of the *Knowledge Series*, by contrast, suggested a flexible, selective reading process on the part of individual cadres ('read some volumes among them according to one's own needs' (*an ziji xuyao kan qizhong ruogan ben*).[48]

Framing knowledge

Despite Zhou Yang's careful separation of knowledge and ideology, a decade into the Mao era it was clear that topics like Malthusianism, utopian socialism, and genetics could become quite politicised.[49] Thus, as they published titles for the *Knowledge Series* that presented a broad spectrum of information and ideas, some of which was clearly marked as 'feudal' or 'bourgeois', publishers and authors used different strategies to frame their subjects and forestall political or ideological criticism. In his inaugural volume for the *Knowledge Series*, for example, linguist Wang Li astutely aligned the project of understanding and appreciating rhymed poetry with Mao's calls to rescue the democratic essence in the large number of 'brilliant and immortal works (*guanghui canlan de buxiu zuopin*)' in the classical tradition while eliminating 'feudal dregs' (*fengjianxing de zaopo*).[50] In addition, Wang placed Mao's own poetry in a continuous line with rhymed poems of previous centuries, showing how they built on particular formal techniques and suggesting that

[45] Print runs for the *Knowledge Series* ranged from 5,000 to 30,500 in the publications I have seen. These print runs were somewhat smaller but similar in scale to those of contemporaneous popular series for ordinary readers, like Zhonghua's *History Small Series* (*Lishi xiao congshu*), which ranged from 19,700 to 50,000 copies in the titles I have reviewed.

[46] Michael Warner, *Publics and Counterpublics*. New York: Zone Books, 2002, 76–87.

[47] Culp, 'Market Configuration'. For more on the Mao-era dynamics of 'study', see Aaron Moore's chapter in this volume.

[48] Song Yuanfang, *Zhongguo chuban shiliao: Xiandai bufen*, 3.1: 193. Who actually read these books, and how they read them, is a separate question, extending beyond the scope of this chapter.

[49] E.g., Judith Shapiro, *Mao's War Against Nature: Politics and the Environment in Revolutionary China*. Cambridge: Cambridge University Press, 2001, chapter 1. See also Brazelton's chapter in this volume.

[50] Wang Li, *Shici gelu* (Poetic Meter) (Beijing: Zhonghua shuju, 1962), 1–2. For discussion of how officials in the Ministry of Culture drew on Mao's 'On New Democracy' to justify renewed publishing of classical texts, see Culp, *The Power of Print*, 204–6.

basic knowledge of poetic form was necessary to truly understand and appreciate Mao's writings.[51] In fact, by illustrating how Mao was fully conversant with formal conventions but adapted them for his own purposes, Wang made an implicit case for continuities from classical culture to revolutionary culture, suggesting that classical literary learning was in line with socialist modernity. For cadres to be able to understand Mao fully, in other words, they needed enough classical learning to be able to intelligently read his poetry by grasping poetic form.

Commercial Press's 1962 publication on Keynesianism by Yang Xuezhang, by contrast, took a polemical approach to framing the text, explicitly countering Keynes' theoretical arguments by invoking Marxist economic theory. In various sections of his discussion of Keynesianism, Yang provided a corresponding Marxist critique. While presenting Keynes' theory of rates of employment driven by effective demand, for example, Yang began with a step-by-step explanation of the logic of the argument.[52] Then he went on to offer a Marxist critique of each aspect of the theory, as one can see with the following section, where Yang offers a class-based refutation of Keynes' analysis of the relationship between income and consumption.

> Based on the 'basic psychological law of consumer trends' that he concocted, although consumption increases with the increase in income, it is always less than the latter. Keynes completely failed to mention the class character of consumption in capitalist society and concealed the true reasons for the contradiction between social production and consumption. For all of society he put forward a supra-class, supra-historical, unified law of consumption. So he ignored that the innate character between the bourgeoisie's consumption and the working class' consumption differed.[53]

Through these critical insertions, the text politically or ideologically framed Keynes' capitalist economics, identifying some of its social implications. Still, interwoven with the critical text was a clear synthetic account of Keynes' theories. Consequently, although the overt political intentionality of the text would have been clear, the discerning reader could literally read between the lines to piece together a reasonable understanding of some of the most current economic theory from the capitalist world. When we extrapolate this contrast between framing and content over the series as a whole, we can posit that the *Knowledge Series* made widely available to literate readers a broad cross-section of ideas that at the time would have otherwise been characterised as 'feudal', 'bourgeois', 'revisionist', or 'reactionary'.

Similar forms of framing and qualification mark Wu Yifeng's[54] volume on utopian socialist economic theory published by Commercial Press in 1964. Wu

[51] Wang Li, *Shici gelu*, 2, 144–5.

[52] Yang Xuezhang, *Kai'ensi zhuyi* (Keynesianism). Beijing: Shangwu yinshuguan, 1962, e.g., 45–6.

[53] Yang Xuezhang, *Kai'ensi zhuyi*, 48.

[54] Wu (1932–) graduated from Yangzhou Normal School in 1950 and taught in several schools in Jiangsu before studying political economy at People's University. After graduation in 1959, he started teaching in the Economics Department at Renda, where he remained. His published work during the Mao era largely focused on Western political economy. https://baike.baidu.com/item/吴易风, accessed 4 October 2019.

characterised utopian socialist economic theory as a precursor to Marxism that was less developed than classical political economy, which provided the basis for Marx's analysis. Moreover, Wu directly criticised the utopian socialists for being idealist, not recognising the privileged historical role of the proletariat, and for rejecting class struggle and revolutionary action.[55] However, the book's account of individual theorists offered generally accurate portrayals of their views, if always framed in teleological terms of how they built toward but differed from Marxist-Leninist theory. For example, Wu praised Saint Simon for being historicist, in contrast to the ahistorical classical political economists, and explained how he described a historical trajectory toward the full development of industrial society. He also described Saint Simon's formulation of the need for centralised economic planning in industrial society and his celebration of labour over leisure.[56] As with the previous volume on Keynesianism, *Knowledge Series* titles offered basic introductions to diverse forms of cosmopolitan thought that were interspersed with Marxist criticism to orient the reader politically.

By contrast, other texts, especially those in scientific or technical areas, could be completely, or at least largely, depoliticised in how they were framed. For instance, the Popular Science Publishing Company in 1962 published a volume titled *Genes and Heredity* written by the Genetics Research Institute (*Yichuanxue yanjiusuo*) at Fudan University.[57] Established by China's leading geneticist, C.C. Tan, Fudan's Genetics Institute was the country's premier genetics research centre, and in the early 1960s it published a number of pieces aimed at a general audience to raise public awareness about genetics, of which the volume in the *Knowledge Series* was a prominent example.[58] Despite the fact that for the better part of the 1950s Mendel-Morgan genetics had been displaced by Lysenkoism,[59] the Genetics Institute offered a completely unframed, scientific account of the history and theory of genetics that bracketed off any explicit mention of politics.

Instead, the text presented Mendel-Morgan genetics as providing bio-chemical explanations for patterns of heredity that had long been recognised in Chinese popular culture. The book began by stating:

> Humankind from early on noticed the phenomena of biological heredity and variation. Our country's ancient sayings were 'if you plant melons, you get melons, if you plant

[55] Wu Yifeng, *Kongxiang shehui zhuyizhe de jingji xueshuo* (Utopian Socialist economic theory). Beijing: Shangwu yinshuguan, 1964, 5–9.
[56] Wu Yifeng, *Kongxiang shehui zhuyizhe*, 61–75.
[57] Shanghai Fudan daxue yichuanxue yanjiusuo, *Jiyin he yichuan* (Genes and heredity). Beijing: Kexue puji chubanshe, 1962.
[58] L.A. Schneider, 'Learning from Russia: Lysenkoism and the Fate of Genetics in China, 1950–1986', in *Science and Technology in Post-Mao China*, eds. Denis Fred Simon and Merle Goldman. Cambridge: Harvard University Council on East Asian Studies, 1989, 60–1.
[59] See Schneider, 'Learning from Russia', 45–61; and James F. Crow, 'Genetics in Postwar China', in *Science and Medicine in Twentieth-Century China: Research and Education*, eds. John Z. Bowers, J. William Hess, Nathan Sivin. Ann Arbor: Center for Chinese Studies, University of Michigan, 1988, 155–62.

> beans, you get beans'; 'if one mother bears nine sons, the nine sons will each differ.' What they were discussing was the phenomena of heredity and variation.[60]

The authors then went on to explain how the scientific discovery of genes explained the mechanism by which that process of heredity occurred. Thus, the text harmonised the relationship between folk wisdom from the masses, which was privileged during the Mao era, and Western biomedical science, in the form of Euro-American genetic theory, resigning Soviet pseudoscience to the margins.[61]

The efforts by the initiators and publishers of the *Knowledge Series* to separate knowledge from politics, however, stood in tension with Mao's emphasis on class struggle, ideological indoctrination, and mass line politics. Publication of the *Knowledge Series* coincided with the introduction and implementation of the Socialist Education Movement (SEM) between 1962 and 1965. Basic-level cadres were the main target of the SEM, but, instead of knowledge building, the campaign focused on ideological and political rectification.[62] In the initial phase of the movement in 1962 and 1963, Mao called for renewed class struggle in the countryside and viewed basic-level cadres' lack of socialist consciousness as key, requiring ideological struggle through criticism and self-criticism. After work team investigations by central-level cadres, including Liu Shaoqi's wife Wang Guangmei, the party centre launched an intensive rectification of rural cadre mismanagement and malfeasance in the form of the 'Big Four Cleanups' (*Da siqing*). But in 1965 Mao was critical of the work teams' autocratic work style and abandonment of mass line political methods as well as its focus on cadre corruption instead of 'powerholders within the Party who take the capitalist road'. His frustration with being unable to implement rectification on his terms through party institutions was one motivation for launching the Cultural Revolution the following year. In general, the SEM's emphasis on ideological rigour, political struggle, and cadre rectification ran counter to the focus on broader knowledge, administrative effectiveness, and building cultural authority reflected in the *Knowledge Series* initiative. Indeed, the effort by cultural cadres like Hu Yuzhi and Zhou Yang to focus on cultivating knowledge among cadres can be seen as offering an alternative to Mao's efforts to use class struggle to ensure political rectification.

Consequently, it is perhaps not surprising that, despite Zhou Yang's efforts to insulate the dissemination of knowledge from ideological struggle, the SEM and Four Cleanups impacted the development of the *Knowledge Series*. For instance,

[60] Shanghai Fudan daxue yichuanxue yanjiusuo, *Jiyin he yichuan*, 1.

[61] Schmalzer offers other examples of Mao-era scientific and technical synthesis and enhancement of collective popular techniques and knowledge. See Sigrid Schmalzer, 'Layer upon Layer: Mao-Era History and the Construction of China's Agricultural Heritage', *East Asian Science, Technology and Society* (2019), 13: 413–41.

[62] Richard Baum, *Prelude to Revolution: Mao, The Party, and the Peasant Question 1962–66*. New York: Columbia University Press, 1975 and Richard Baum and Frederick C. Teiwes, *Ssu-Ch'ing: The Socialist Education Movement of 1962–1966*. Berkeley: Center for Chinese Studies, UC Berkeley, 1968.

after Mao's call at the Central Committee's Tenth Plenum to 'never forget class struggle' (*qianwan buyao wangji jieji douzheng*) in 1962, Chen Yuan claims that Wang Li's model volume for the series, *Poetic Metre*, was singled out for having 'opposed' or at the very least 'weakened' class struggle.[63] This criticism made clear that in a Maoist framework, the very effort to depoliticise knowledge could be viewed as a negative political act for failing to advance the political struggle. When that perspective became ascendant during the Cultural Revolution, the slogan 'knowledge is power' itself was criticised for 'denying class struggle' and being 'counterrevolutionary', with the series' promoters being accused of 'stressing knowledge to assault politics'. With the start of the Cultural Revolution, the *Knowledge Series* ceased publication.[64] During the subsequent decade, an almost completely politicised approach to knowledge displaced the relatively depoliticised cosmopolitanism of the early 1960s.

Conclusion: Red and expert after Mao

In the aftermath of the GLF, the *Knowledge Series* presented a global, comprehensive approach to knowledge that was insulated from political ideology and theory. This approach marked a shift away from the socialist cosmopolitanism of the early Mao era back to a mode of cosmopolitanism oriented more to Europe, Japan, and North America. Hu Yuzhi and his colleagues abandoned the dependence on Soviet and Warsaw Pact academic discourse of the early 1950s and returned to strategies of knowledge production from China's Republican period, when comprehensive series publications flourished. Because of the reorganisation of Chinese publishers into more specialised companies during the 1950s, however, only a consortium of publishers could produce a collection like the *Knowledge Series*, whose subjects spanned numerous fields, disciplines, and thematic areas. As a result, knowledge was made during this portion of the Mao era through close collaborations between intellectuals, representatives of these publishing companies, and cultural cadres in the party and state. Also distinct from Republican period series publishing, the target audience of the *Knowledge Series* was not the 'general reading public' but primarily cadres. By exposing cadres to a wide spectrum of learning in the sciences, social sciences, arts, and humanities, the PRC's cultural bureaucrats sought to cultivate cadres as organic intellectuals for state socialism by making them both 'red' and 'expert'. Vital to that project was providing them with the cultural capital to compete with late-Qing literati and foreign-trained intellectuals to convincingly claim recognition as intellectuals and government officials.

From one perspective, the period from 1961 to 1965 when the *Knowledge Series* was published can appear to be nothing more than a brief interlude between the

[63] Song Yuanfang, *Zhongguo chuban shiliao: Xiandai bufen*, 3.1: 198.
[64] Song Yuanfang, *Zhongguo chuban shiliao: Xiandai bufen*, 3.1: 200.

GLF and the Cultural Revolution, when thought was highly politicised and efforts to separate knowledge from politics were characterised as reactionary. Moreover, as we have seen, cultural leaders' efforts to enlighten basic-level cadres still coincided with intensive efforts at cadre rectification and ideological purification in the form of the SEM. By taking a somewhat longer view, however, we can see ways in which Deng Xiaoping in the Reform era reintroduced views of knowledge and the cadre that echoed this period of the early 1960s. In his 'Speech at the Opening Ceremony of the National Conference on Science' in March 1978, for example, Deng criticised the Gang of Four for the way it 'could wantonly sabotage the cause of science and persecute intellectuals'.[65] Instead, Deng identified science and technology as a key part of the productive forces in society that the PRC needed to develop, and he argued that such development in the domain of knowledge had no necessary implications for the sphere of politics. Intellectuals, he affirmed, could be both 'red' and 'expert', evoking this Mao-era ideal. Moreover, he asserted that even if there was 'a group of scientists and technicians whose bourgeois world outlook has not fundamentally changed … as long as they are not opposed to the Party and socialism, we should unite with them and educate them, promote their special skills, respect their work, take an interest in their progress and give them a warm helping hand.'[66] 'Scientists and technicians', he argued, 'should concentrate their energies on their professional work.' Through these arguments formulated soon after the overthrow of the Gang of Four, Deng made the case for allowing the development of a domain of knowledge segregated from ideology and politics that would help drive socialist development.

At the same time, he called for cultivation of expertise in cadres so that they could better embody the 'red and expert' ideal. In his programmatic lecture 'The Present Situation and the Tasks Before Us', Deng promoted development of cadres who had both a socialist orientation and also professional knowledge and competence. 'Being "expert" ', he said, 'does not necessarily mean one is "red", but being "red" means one must strive to be "expert". No matter what one's line of work, if he does not possess expertise, if he does not know his own job but issues arbitrary orders, harming the interests of the people and holding up production and construction, he cannot be considered "red".'[67] Echoing Hu Yuzhi and his colleagues in 1961, Deng viewed technical knowledge to be essential to good governance, and by advocating for cadres who were both 'red' and 'expert', he returned to a model of the cadre as an organic intellectual. Deng's overwhelming emphasis in this discussion of cadre training was cultivation of knowledge related to science and technology. Work by Richard Kraus and Elizabeth Perry, though, makes clear that political

[65] Deng Xiaoping, *Selected Works of Deng Xiaoping (1975–1982)*, trans. The Bureau for the Compilation and Translation of the Works of Marx, Engels, Lenin and Stalin under the Central Committee of the Communist Party of China. Beijing: Foreign Languages Press, 1984, 101.
[66] Deng Xiaoping, *Selected Works*, 109.
[67] Deng Xiaoping, *Selected Works*, 247–8.

leaders' embodiment of certain forms of cultural capital has been equally important for building the cultural authority to govern and claim status vis-a-vis other cultural elites.[68] In sum, Reform era calls for cadres to master technical expertise and acquire certain forms of cultural capital seem to mark a return to the ideal of the cadre as organic intellectual that flourished during the early 1960s. I raise these parallels to suggest that the categorisation of knowledge and formulation of the cadre as organic intellectual established in the early 1960s might not have been just a short interlude in a dominant trend of Maoism's politicisation of knowledge but rather a viable alternative system for conceptualising knowledge and cadres during the Mao era, as well as a harbinger of things to come.

[68] Richard Curt Kraus, *Brushes with Power: Modern Politics and the Chinese Art of Calligraphy*. Berkeley: University of California Press, 1991; Elizabeth J. Perry, *Anyuan: Mining China's Revolutionary Tradition*. Berkeley: University of California Press, 2012, Conclusion.

Part V

Socialist Internationalism at Home

14

Plagues from the Skies: Bacteriological Expertise in the 1952 Germ Warfare Allegations

MARY AUGUSTA BRAZELTON

IN 1951, SHORTLY after the Chinese People's Liberation Army (PLA) entered the 'war to resist America and aid Korea', the Chinese Communist Party (CCP) and the Communist Workers' Party of Korea started publishing reports that the American military was dropping strange bombs in northern Korea and northeast China. The munitions were filled with a panoply of horrors: dead flies and spiders infected with anthrax, plague, and other infectious diseases. These allegations of germ warfare gained the world's attention in 1952 when several American pilots, captured as prisoners of war, confessed to dropping germ-filled bombs on enemy territory. Chinese and North Korean media outlets broadcast these admissions alongside the revelation that local bacteriologists had proven these items carried smallpox and other infectious diseases.[1]

Outside transnational bodies of governance like the United Nations or World Health Organization, the People's Republic of China (PRC) agitated for global recognition of American bacteriological warfare as a war crime. In June 1952, the Chinese state established an 'International Scientific Commission [ISC] for the Facts Concerning Bacterial Warfare in China and Korea' through the Soviet-aligned World Peace Council; it brought in European experts to verify the findings of the Chinese researchers in a lengthy report. Chinese and Korean media organisations

[1] Albert Cowdrey, '"Germ Warfare" and Public Health in the Korean Conflict', *Journal of the History of Medicine and Allied Sciences* 39, no. 2 (April 1984): 153–72 at 157; and 'Wo xijunxue zhuanjia Wei Xi, Liu Weitong qinshen zhengshi meiguo xijunzhan zuixing fennu zhi huan meiguo zhengfu guoqu suo shouyu de xunzhang jiangzhuang' [Our bacteriology experts Wei Xi and Liu Weitong personally confirm the American crime of germ warfare: Angrily throwing away the Medal of Commendation once received by the American government], *Renmin ribao*, 19 May 1952.

Proceedings of the British Academy, **267**, 275–292, © The British Academy 2025.

also released films supporting the allegations. At a crucial moment in the post-war making of a new geopolitical order, these materials constructed moral narratives about global scientific expertise suggesting that, in contrast to American researchers who used their knowledge to develop weapons of mass destruction, Chinese bacteriologists deployed the same expert professional knowledge to contribute to public health. In doing so, the latter suggested the successful integration of their work into an emerging Maoist ideal: that of an anti-imperialist, class-conscious science that served the masses.

Controversy over the accusations erupted immediately and continues today. At the time, Secretary of State Dean Acheson and other American officials denied the allegations. Since 1952, some historians have suggested that the accusations were plausible, pointing to the well-documented interest and competence of the US military in bacteriological warfare research.[2] Others have highlighted the claims' weaknesses, calling attention to their lack of corroboration and their potential identification in Russian archives as a disinformation campaign.[3] Still others have demonstrated that regardless of the accusations' veracity, Chinese military and political leaders *acted* as if they were true; subsequent nationwide public health programmes offered large-scale hygienic interventions and science education.[4] In China, military histories have generally reinforced official narratives.[5] In 1991, Qi Dexue suggested that unless and until the US declassified relevant documents, the truth would remain elusive.[6] In 2013, a military doctor involved in the accusations,

[2] Bruce Cumings and Jon Halliday, *Korea: The Unknown War*. London: Penguin Books, 1990, pp. 182–4; Gavan McCormack, 'Korea: Wilfred Burchett's Thirty Year's War', in *Burchett: Reporting the Other Side of the World, 1939–1983*, ed. Ben Kiernan. London: Quartet Books, 1986, p. 204; Stephen Endicott and Edward Hagerman, *The United States and Biological Warfare: Secrets the Early Cold War and Korea*. Bloomington: Indiana University Press, 1998.

[3] Zhang Shuguang, *Mao's Military Romanticism: China and the Korean War, 1950–53*. Lawrence: University Press of Kansas, 1995, p. 186; Kathryn Weathersby, 'Deceiving the Deceivers: Beijing, Moscow, Pyongyang, and the Allegations of Bacteriological Weapons Use in Korea', *Cold War International History Project Bulletin* 11 (Winter 1998): 176–85; Milton Leitenberg, 'New Russian Evidence on the Korean War Biological Warfare Allegations: Background and Analysis', *Cold War International History Project Bulletin* 11 (Winter 1998): 185–99.

[4] Chen Jian, *Mao's China and the Cold War*. Chapel Hill: University of North Carolina, 2010, p. 110; Cowdrey, '"Germ Warfare,"' 153–72; Andrew Kuech, 'Cultivating, Cleansing, and Performing the American Germ Invasion: The Anatomy of a Chinese Korean War Propaganda Campaign', *Modern China* (2019): 1–30.

[5] Zhu Kewen, Gao Enxian, and Gong Chun (eds.), *Zhongguo junshi yixue shi* [A history of military medicine in China]. Beijing: Renmin junyi chubanshe, 1996, p. 481; Qu Aiguo, 'Shi meijun de zuixing haishi zhong chao fangmian de "huangyan": guanyu kangmei yuanchao zhanzheng fan xijunzhan douzheng de lishi kaocha' [Is it American crime or Chinese-Korean lies? A historical survey of the conflict over bacteriological warfare in the War to Aid Korea and Resist America], *Junshi lishi*, no. 2 (2008): 1–8.

[6] Qi Dexue, *Chaoxian zhanzheng juece neimu* [Inside stories of decision-making in the Korean War] (Shenyang: Liaoning daxue chubanshe, 1991), pp. 286–7. Cited in Zhang, *Mao's Military Romanticism*, p. 186n98.

Wu Zhili, wrote, 'The matter of bacteriological warfare was settled by being left unsettled' (*buliao liaozhi*).[7]

I am not interested in judging the veracity of the allegations. As Ruth Rogaski has suggested, it is more productive to analyse the significance of this controversy – in her case, how it articulated political narratives of environmental disaster and national victimisation.[8] In this chapter, I apply lessons from the sociology of scientific knowledge to consider what kinds of bacteriological truth counted and were constructed as scientific and authoritative in this dispute, and to whom. Typically, controversy studies focus on disagreements within expert communities over claims relating to research.[9] The case at hand, however, involved complex configurations of expertise and engaged multiple communities. In this fraught scenario, the work of Bruno Latour and Steve Woolgar on literary inscriptions in scientific controversies provides a helpful means of navigating the various perspectives involved, insofar as it interrogates the transformation of labour and materials into documentary outputs as means of building credibility in scientific communities.[10] Chinese bacteriologists and entomologists sought to provide believable narratives of biological warfare; scientists from the Soviet Union and Europe who participated in the ISC sought to analyse and favourably interpret these accounts; and news of these efforts was communicated to global audiences. Investigation of these processes suggests that what was on trial in 1952 was not only the American military, but also the validity and authority of Chinese science.

Chinese accounts and propaganda, as well as the report of the Commission and the diaries of prominent Commission member Joseph Needham, suggest a need to reconsider the perspectives of the Chinese researchers who provided the evidence

[7] Wu Zhili, *Yi ming jun yi de zishu* [A military doctor's account in his own words]. Beijing: Huaxia chubanshe, 2004, pp. 40–1. See also Milton Leitenberg, 'China's False Allegations of the Use of Biological Weapons by the United States during the Korean War', *Cold War International History Project*, Working Paper 78, March 2016.

[8] Ruth Rogaski, 'Nature, Annihilation, and Modernity: China's Korean War Germ-Warfare Experience Reconsidered', *Journal of Asian Studies* 61, no. 2 (May 2002): 381–415 at 382.

[9] David Bloor, *Knowledge and Social Imagery*. Chicago: University of Chicago Press, 1991, 2nd edition); Steve Shapin and Simon Schaffer, *Leviathan and the Air-Pump*. Princeton: Princeton University Press, 1985; H.M. Collins and Trevor Pinch, 'The Construction of the Paranormal: Nothing Unscientific is Happening', in *On the Margins of Science: The Social Construction of Rejected Knowledge*, ed. R. Wallis. Keele: University of Keele Sociological Review Monograph No. 27, 1979, pp. 237–70; H.M. Collins, 'Knowledge and Controversy: Studies in Modern Natural Science', *Social Studies of Science* 11, no. 1 (1981): 3–158; Michel Callon, 'Some elements of a sociology of translation: Domestication of the scallops and fishermen of St Brieuc Bay', in *Power, Action and Belief: A New Sociology of Knowledge?* ed. John Law. London: Routledge, 1986, pp. 196–229; Bruno Latour, *Science in Action* (Cambridge, MA: Harvard University Press, 1987).

[10] Bruno Latour and Steve Woolgar, *Laboratory Life*. London: Sage Books, 1979. See also G. Nigel Gilbert and Michael Mulkay, *Opening Pandora's Box: A Sociological Analysis of Scientists' Discourse*. Cambridge: Cambridge University Press, 1984, and Greg Myers, *Writing Biology: Texts and the Social Construction of Scientific Knowledge*. Madison: University of Wisconsin Press, 1990.

and witnesses that the report discussed. They included leading bacteriologists, immunologists, and microbiologists of their time, such as Xie Shaowen and Tang Feifan. Rogaski nevertheless notes that 'the nature of the participation of these scientists remains the most elusive question of the germ warfare allegations'.[11] Gordon Barrett has shown that scientists in the early PRC were not totally isolated from global research networks.[12] For microbiologists, the ISC furnished opportunities for further international engagement and even prominence. Its sponsorship by the Communist-dominated World Peace Council provided an example of the increasing significance of international socialist networks for PRC participants in the 1950s.

This chapter argues that the allegations took on different meanings for Chinese researchers in domestic and international contexts; both contributed to a broader project to incorporate science into the making of a new socialist order. The first section focuses on the domestic sphere, describing how the accusations gave bacteriologists key roles in the creation of a Maoist political order and publicised Chinese microbiologists as figures whose scientific credentials lent authority to the accusations. For these researchers, confirming the allegations remade the construction of scientific knowledge as a project embedded in class politics and political conflict – a process also explored by Robert Culp in his contribution to this volume. The second section considers the ways in which the engagement of researchers with the ISC represented an effort to be recognised as legitimate, competent participants in a global research community, even as the aforementioned newly established definitions of scientific knowledge posed challenges for acceptance into that community. The Commission's final report claimed scientific authority based on independent, objective analysis and sought to confer this authority to Chinese researchers, yet its arguments ultimately relied on sensational stories rendered plausible by experimental inquiry. The depiction of the claims in photographs and films, discussed in the final section, adopted and underscored this approach to the construction of scientific facts, which celebrated its differences from foreign scientific norms even as it also sought commensurability with them.

Domestic narratives of germ warfare and the ascent of bacteriology

In 1952, the CCP relied on the expertise of Chinese researchers to lend credence to allegations of germ warfare. In doing so, they incentivised scientists to adopt an approach to their work consonant with the construction of a new socialist order.

[11] Rogaski, 'Nature, Annihilation, and Modernity', 402.
[12] Gordon Barrett, 'Between Sovereignty and Legitimacy: China and UNESCO, 1946–1953', *Modern Asian Studies* 53, no. 5 (2019): 1516–42. See also William C. Kirby, 'China's Internationalization in the Early People's Republic: Dreams of a Socialist World Economy', *The China Quarterly*, no. 188 (Dec 2006): 870–90.

Bacteriologists and immunologists were involved from an early stage; for instance, the autobiography of Wu Zhili, then Director of the Chinese People's Volunteer Army Health Division, stated that soon after the allegations were publicised in February, a researcher named Wei Xi provided testimony suggesting that they were false. Then the head of bacteriology at Dalian Medical College, Wei was one of the first microbiologists called in to verify the accusations. Wei and colleague He Qi initially asserted that the samples they inspected were not bacteriological weapons, and the concern was therefore merely a 'false alarm'. After Wu reported this finding to Peng Dehuai, the general leading Chinese military operations, Peng claimed that Wu was parroting the speech of American imperialism and convened a meeting in which colleagues accused Wu of rightist sympathies.[13] Wu and Wei subsequently changed their stories. In a 1952 edition of the *Chinese Medical Journal*, Wei co-authored an article that supported the accusations of biological warfare, saying that 'like an octopus, [the US] was stretching its tentacles far to the rear of the battle front'.[14] Wei's retraction of his previous denial suggested a choice between continuing to deny the allegations because the evidence was unconvincing, or supporting them for the sake of political survival.

As Wei's recanting indicated, the accusations took place in a rapidly changing political environment for Chinese scientists, especially those trained in the West. The CCP conducted nationwide campaigns to purge 'bourgeois elements' throughout the 1950s.[15] Against the backdrop of anti-American sentiment in the Korean War, intellectuals with American educations – which described many Chinese microbiologists in research positions at the beginning of the PRC – were targeted.[16] To support accusations of germ warfare was therefore not just an opportunity to declare one's support for the party, but also a means of avoiding its persecution.

Throughout 1952, microbiologists published pieces supporting the claims of American germ warfare and joined military medical teams at the Korean front. Rogaski has noted, 'Individual reasons for such actions may have been diverse, but [scientists'] actions as a group were decidedly orchestrated by the government for purposes of political mobilization.'[17] Wei Xi, the bacteriologist mentioned above who retracted his original denial of the accusations, provides a case in point. Because he had previously worked with American colleagues, his U-turn to support the allegations provided a useful example of Chinese rejection of American

[13] Wu, *Yi ming jun yi*, pp. 40–1; and Leitenberg, 'China's False Allegations'.

[14] Wei Xi (Hsi) and Zhong Huilan (Chung Huei-lan), 'Peace and Pestilence at War', *Chinese Medical Journal* 70, supplement (1952): 8–19 at 8.

[15] Julia C. Strauss, 'Morality, Coercion and State Building by Campaign in the Early PRC: Regime Consolidation and after, 1949–1956', *The China Quarterly*, no. 188 (Dec 2006): 891–912 at 897; and Theodore Hsi-en Chen and Wen-Hui C. Chen, 'The "Three-Anti" and "Five-Anti" Movements in Communist China', *Pacific Affairs* 26, no. 1 (1953): 3–23 at 4–5.

[16] James Zheng Gao, *The Communist Takeover of Hangzhou*. Honolulu: University of Hawaii Press, 2004, pp. 159–60; and Kenneth G. Lieberthal, *Revolution and Tradition in Tientsin, 1949–1952*. Stanford: Stanford University Press, 1980, p. 125.

[17] Rogaski, 'Nature, Annihilation, and Modernity', 402.

power. For instance, a *People's Daily* article published on 19 May 1952 explained that Wei and a colleague, Liu, had received accolades for work with American military medical researchers on typhus prevention during World War II (WWII). The article described Wei's repudiation of these honours upon confirming the allegations of bacteriological warfare, representing him as supporting an anti-American political outlook.[18]

Where Rogaski asserts this episode's prioritisation of politics over science, I suggest that the allegations' publicisation helped shape their scientific meanings. Affirming charges of germ warfare moved the work of Chinese microbiologists increasingly into the public eye, making the construction of scientific knowledge about bacteriology a matter for discussion in daily newspapers as well as specialist journals. Just as professors at research institutes like the former Peking Union Medical College were undergoing criticism in the *People's Daily* during the Three-Antis Campaign to stamp out bureaucratism and corruption, the same newspaper was presenting bacteriologists like Tang Feifan (head of the National Vaccine and Serum Institute in Beijing and president of the Chinese Society for Microbiology), Wei Xi, and Xie Shaowen (head of immunology and bacteriology at Peking Union Medical College) as experts faithfully serving their country.[19] Between 1950 and 1952, the Ministry of Health sent researchers like Wei, Tang, Xie, and others to laboratories at the Korean front. Tang and Xie published articles in the *People's Daily* confirming and condemning American germ warfare.[20] Tang, Xie, and 26 other microbiologists from elite institutes signed a June 1952 editorial in the *People's Daily*. The authors wrote: 'We have already totally proven that the American army has dropped bombs on North Korea and northeast China that contained many worms and other poisonous organisms. Bacteriologists in every place, using all different kinds of experiments, have already extracted every kind of illness-inducing bacterium from the organisms.'[21] In an uncertain political environment, endorsing

[18] 'Wo xijunxue zhuanjia Wei Xi, Liu Weitong qinshen zhengshi meiguo xijunzhan zuixing fennu zhi huan meiguo zhengfu guoqu suo shouyu de xunzhang jiangzhuang' [Our bacteriology experts Wei Hsi and Liu Weitong personally confirm the American crime of germ warfare: Angrily throwing away the Medal of Commendation once received by the American government], *Renmin ribao*, 19 May 1952.

[19] Bai Sheng, 'Zhongguo xiehe yixueyuan jiji kaizhan fan tanwu yundong' [The China Union Medical College actively undertakes the anti-corruption movement], *Renmin ribao*, 12 March 1952.

[20] Tang Feifan, 'Women yao yong shiji xingdong lai kangyi mei diguozhuyi de baoxing' [We must take action to resist the atrocities of American imperialism], *Renmin ribao*, 9 March 1952; Tang Feifan, 'Guoji kexue weiyuanhui chedi jiechuan mei diguozhuyi jinxing xijunzhan de zuixing' [An international scientific committee thoroughly exposes the American imperialists' crime of conducting germ warfare], *Renmin ribao*, 19 June 1952; Tang Feifan, 'Xijunxue zhe yao xianchu yiqie liliang, baowei renmin, baowei zuguo' [Bacteriologists give everything in their power to defend the people and the motherland], *Renmin ribao*, 23 February 1952; Xie Shaowen, 'Women you kexue shang de tiezheng' [We have iron-clad scientific proof], *Renmin ribao*, 4 April 1952.

[21] 'Shoudu xijunxue gongzuozhe duiyu Koulan he du Bosi er shi fouren meiguo jinxing xijunzhan de yanlun de shengming' [The capital's bacteriological workers' voices against the claims of the two American scientists Coughlin and Du Bois denying that America conducted germ warfare], *Renmin ribao*, 9 June 1952.

allegations of bacteriological warfare provided means for immunologists to maintain political legitimacy and build public authority.

Allegations of germ warfare in the Korean conflict also justified broader state investments in preventative health – initiatives that microbiologists were well placed to lead. In a March 1952 memo to Zhou Enlai, Mao Zedong wrote, 'Please prepare to vaccinate all soldiers east and west of the Liao River. Central and East Hebei, as well as Beijing and Tianjin, should also make preparations.'[22] A military medical history notes that one of the army's primary responses to the allegations was 'to supply large quantities of plague, typhus, and other vaccines, carry out urgent immunisation among the troops, and simultaneously give vaccinations to residents near military encampments.'[23] Anti-germ warfare health campaigns quickly expanded in scope. In 1952 the Chinese Ministry of Health launched the Patriotic Hygiene Campaigns, mass hygienic movements that promoted a range of health measures, including immunisation against multiple illnesses, nationwide. Tang Feifan, head of the National Vaccine and Serum Institute in Beijing, directed the institute that took primary responsibility for manufacturing vaccines for the nation. The campaigns emphasised direct links between fighting bacteriological warfare and strengthening public health.[24]

The deployment of bacteriological expertise in the media and in public health reflected broader trends in the early PRC to incorporate science into social revolution, a dynamic that Sigrid Schmalzer and Miriam Gross have explored in their work.[25] Microbiologists' confirmation of germ warfare helped articulate a distinctively self-reliant, class-oriented approach to science, and asserted the authority of this form of knowledge production on a global scale. The domestic publicity afforded the allegations reveals the ways in which health administrators interpreted science within novel theoretical frameworks. For example, a 1952 account by Zheng Lingcai, head of the Epidemic Prevention Team of the Southwestern Hygiene Department, far from Manchuria, recapitulated the allegations and stated: 'The laws of history have also proved that this is the dying struggle of American imperialism. At the same time it has also showed that science has class.'[26]

The assertion of 'laws of history' placed the allegations within Marxist theories of the stages of social development, which would see imperialism and capitalism

[22] 'Guanyu Fushun', *Jianguo yi lai Mao*, p. 303.

[23] Zhu, Gao, and Gong, *Zhongguo junshi yixue shi*, p. 481.

[24] Cowdrey, '"Germ Warfare,"' 153–72; Rogaski, 'Nature, Annihilation, and Modernity', 381–415; and 'Zhengwu yuan', 258–9.

[25] Sigrid Schmalzer, *Red Revolution, Green Revolution: Scientific Farming in Socialist China*. Chicago: University of Chicago Press, 2016; Sigrid Schmalzer, 'Self-Reliant Science: The Impact of the Cold War on Science in Socialist China', in *Science and Technology in the Global Cold War*, eds. Naomi Oreskes and John Krige. Cambridge, MA: MIT Press, 2014, pp. 75–106; Miriam Gross, *Farewell to the God of Plague: Chairman Mao's Campaign to Deworm China*. Oakland: University of California Press, 2016.

[26] Zheng Lingcai, 'Zenyang fangyu "xijunzhan"' [How to stop 'germ warfare'], *Xinan weisheng* 5, no. 6 (1952): 8–12 at 8. Here 'class' refers to class identity.

destroyed in socialist revolution; yet the explicit statement that 'science has class' represented something slightly different. It suggested that, in the new Maoist order, science was not simply a means of producing new technologies to facilitate the industrialisation of society in the transitions to socialism and communism, nor was it simply the occupation of a subset of intellectuals; as something to which class identity could be attributed, science provided a space for a variety of policies and ideals associated with Maoism, including upholding the peasantry, rebelling against authority and imperialism, and adopting practice as 'the sole criterion of truth'.[27]

Zheng did not stop at affirming the class nature of science, but went on to assert its reorientation away from American influences: 'This is really a powerful warning to health workers who still hold purely technical ideas and supra-class thinking. At the same time it has only added to our hatred for American imperialism.'[28] The statement intimated that researchers who did not embrace class conflict would be politically recriminated as elitists. It supports Rogaski's point that 'the participation of scientists in the 1952 movements was a harbinger of a new sort of modernity for China' in which science could be separated from its 'capitalistic and imperialistic origins' and used to serve the people.[29] In the case of bacteriology, expert researchers sought to publicly repudiate their identification with elite, foreign institutions in mass media and medical publications, and they took on new identities and positions as Chinese researchers leading national mass movements to apply bacteriological techniques to further public health. In doing so, they stressed the value of a self-reliant, Chinese science, in which the pursuit of scientific knowledge was a semi-public enterprise subject to class politics. These narratives existed in tension with the ambition to demonstrate that self-reliance and accomplishment to foreign audiences. They were complicated by the arrival of an old friend from Cambridge.

The ISC and its 'mass of facts'

In the summer of 1952, Chinese politicians sought transnational traction for their allegations of American bacteriological warfare. They established the ISC 'for the Investigation of the Facts Concerning Biological Warfare in Korea and China' in coordination with the World Peace Council, a European group of left-wing intellectuals. The Council reflected the networks of socialist internationalism in which the PRC was enmeshed in the early 1950s, in which initiatives to 'learn from the Soviet Union' stressed exchanges with the Soviet Union and its allies in a variety of fields, including the sciences.[30] The Commission it assembled articulated

[27] Lovell, *Maoism,* p. 26.
[28] Zheng, 'Zenyang fangyu "xijunzhan"', p. 8.
[29] Rogaski, 'Nature, Annihilation, and Modernity', 405.
[30] Günter Wernicke, 'The Unity of Peace and Socialism? The World Peace Council on a Cold War Tightrope between the Peace Struggle and Intrasystemic Communist Conflicts', *Peace & Change* 26, no. 3 (July 2001): 332–51; Gao Xi, 'Learning from the Soviet Union: Pavlovian Influence on Chinese

a distinctively socialist approach to gathering, assessing, and disseminating evidence: one that took villagers' experiences seriously, valourised practice over theory, and reflexively acknowledged the social nature of scientific inquiry. The Commission met 40 times between 23 June and 31 August, mostly in Beijing with excursions to field sites in Manchuria and northern Korea. In its final 560-page report, the Commission affirmed the expertise of Chinese science, praising the work of local researchers and confirming their conclusions.[31] In undertaking the latter task, the Commission members sought to construct events of bacteriological warfare as objective truths, and thus confer scientific legitimacy upon the scientists who had supplied the evidence.

Reading the Commission's report alongside the correspondence of Joseph Needham, the Cambridge scientist and historian who was its most famous co-author, reveals new details about the processes by which this group sought to construct scientific authority. Needham was reluctant to join the Commission, but subsequently became its central figure.[32] The next most prominent member was Nicolai Nicolaievitch Zhukov-Verezhnikov, a Soviet bacteriologist. The other members – Swedish clinician Andrea Andreen, French physiologist Jean Malterre, Italian anatomist Oliviero Olivo, and Brazilian entomologist Samuel Pessoa – were fellow-travellers in the socialist cause. For Needham, a major motivation for going to China in 1952 was to assert publicly 'Chinese scientists are ok'.[33] He relished the opportunity to see acquaintances from his wartime travels, when he had supported researchers in unoccupied China as director of the Sino-British Science Cooperation Office. The Commission was hosted by a group of scientists led by Qian Sanqiang, a physicist who had trained under the Joliot-Curies and now led China's nuclear weapons programme. The Chinese 'committee of reception' included prominent health authorities. It was led by Li Dequan, then president of the Chinese Red Cross Society, and included He Cheng, president of the Chinese Medical Association; Wu Zaidong, a pathologist at Nanjing University Medical College; Zhong Huilan, director of the People's Hospital in Beijing and a professor of clinical medicine

Medicine, 1950s', in *Public Health and National Reconstruction in Post-War Asia*, eds. Liping Bu and Ka-che Yip. New York: Routledge, 2014, pp. 72–89; Izabella Goikhman, 'Soviet-Chinese Academic Interactions in the 1950s: Questioning the "Impact-Response" Approach', in *China Learns from the Soviet Union, 1949-Present*, ed. Hua-Yu Li. Lanham: Rowman & Littlefield, pp. 275–302.

[31] Cowdrey, '"Germ Warfare,"' 225; Chen Shiwei, 'History of Three Mobilizations: A Reexamination of the Chinese Biological Warfare Allegations Against the United States in the Korean War', *Journal of American-East Asian Relations* 16, no. 3 (2009): 222–8; and Andrea Andreen, Jean Malterre, Joseph Needham *et al.*, *Report of the International Scientific Commission for the Investigation of the Facts Concerning Bacterial Warfare in Korea and China (with appendices)*. Peking: ISC, 1952, pp. 55–62.

[32] Buchanan, 'Courage of Galileo', 506–11; and Joseph Needham, Beijing, to Dorothy Needham, Cambridge, 15 August 1952, Needham Correspondence, Needham Research Institute, Cambridge [hereafter NC]. Needham's personal diary suggests that Zhukov-Verezhnikov chaired the first meetings and often suggested directions for inquiry. Thanks to Stephanie Frow for this observation.

[33] Joseph Needham to Dorothy Needham, undated from Beijing, Imperial War Museum, JNP/12. Cited in Buchanan, 'Courage of Galileo', 507n17.

at the China Union Medical College; Fang Gang, a bacteriologist at the Central Research Institute of Health; and other professors of medicine, entomology, and pathology at universities in Beijing and Shanghai.[34]

As mentioned above, Needham's desire to rekindle old friendships – he had met Qian, Fang, and Wu during the war – may have presented problems in a context in which associations with the West brought scientists danger rather than prestige. Needham recalled being told by Qian that 'I might have thought that the scientists were not as forthcoming as in the old days, but I should realise that this was because of my current international character.'[35] Buchanan suggests that Needham's desire to endorse the work of these scientists was overpowering: 'ultimately, Needham's belief that bacteriological warfare had occurred owed much to his noble, but perhaps quixotic, willingness to accept the word of the Chinese scientists.'[36] Such an assertion emphasises the role that personal scientific networks contributed to narratives that affirmed Chinese expertise, even though the political context of that expertise threatened transnational relationships in those networks.

Certainly, the Commission's report confirmed and praised the capabilities of Chinese and Korean scientists: 'there could be no doubt whatsoever as to their high competence'.[37] The final appendix of the report gave a 'biographical register of Chinese and Korean scientists and medical men' listing their field of work, positions held, training, and publications: essentially, reinforcing the identities of these figures as scientists.[38] The report attempted to guard against accusations of bias, saying, 'Although there was no reason to doubt the competence and probity of the medical men and other scientists in China and Korea, the Commission left no precaution untaken. It never wearied in analysing the cases, and took the greatest pains to enter into direct contact with the original facts whenever this was at all possible.'[39]

In addition to confirming the expertise of Chinese researchers, the ISC sought to independently affirm their conclusions. As such, the authors described their report as the result of a 'mass of facts', insisting that inexorable logic had led them to agree that American forces had employed bacteriological warfare. They wrote that the Commission's 'members held themselves continually on guard against political, ethical or emotional influences, and its work was done in an atmosphere of calm and scientific objectivity.'[40] The use of the term 'objectivity' here is suggestive. The authors may have thought that they needed to stress the objective nature of their inquiry because they were making such a contested claim – or because they were making that claim in a space outside the laboratories and fields of twentieth-century Anglophone knowledge production.

[34] Andreen *et al.*, *Report of the International Scientific Commission*, pp. 4–5.
[35] Joseph Needham, Beijing, to Dorothy Needham, Cambridge, 6 July 1952, NC.
[36] Buchanan, 'Courage of Galileo', 513, 521.
[37] Andreen *et al.*, *Report of the International Scientific Commission*, p. 14.
[38] Andreen *et al.*, *Report of the International Scientific Commission*, pp. 635–65.
[39] Andreen *et al.*, *Report of the International Scientific Commission*, p. 8.
[40] Andreen *et al.*, *Report of the International Scientific Commission*, p. 8.

If, in the words of Woolgar and Latour, 'the elimination of alternative interpretations of scientific data and the rendering of these alternatives less plausible is a central characteristic of scientific activity', then in this case, the Commission's report sought to assert the fact of biological warfare by privileging such an 'alternative interpretation' to the views adopted by microbiologists in other parts of the world, and calling attention to the contingent and messy processes by which such alternatives were typically dismissed.[41] This attention is reflected in the report's extremely cautious argumentation, in contrast to the certainty of its conclusions. The document repeatedly asserted that the evidence was circumstantial and acknowledged the limits of current knowledge. For instance, a discussion of bacterial infection of insects noted, 'It is a difficult matter to isolate pathogenic micro-organisms from such material when no one knows exactly what should be looked for, all the more so when artificially selected bacteria and viruses are in question.'[42] In the terms of science studies, one might say that the report sought to render alternative interpretations of data more, rather than less, relevant.

Some ambiguity stemmed from the under-explored environment of northeast Asia. Commission authors could not say definitively whether the presence of insects not native to northeast China and Korea indicated the introduction of these organisms through bacteriological warfare, because no one really knew what insects were native to the region in the first place. 'Even after the work of half a century, the systematic classification of many groups of insects in the Chinese subcontinent remains imperfectly known', they wrote; 'It was therefore impossible to assert that all new introductions could be definitely recognised as such.'[43] This stipulation subtly undermined the logical structure that the Commission had articulated as a means of producing facts, as well as the authority of the Asian researchers it was attempting to bolster, because it implied the unreliable nature of any data from this region. The unknown entomology of China and Korea frustrated the construction of scientific authority about these places.

The document's claim to objective truth is fascinatingly self-contradictory, in part because the authors alternately defined and denied the limits of their knowledge, but also because the report was full of emotive declarations. Consider the final statement: 'The Commission reached these conclusions, passing from one logical step to another. It did so reluctantly because its members had not been disposed to believe that such an inhuman technique could have been put into execution in the face of its universal condemnation by the peoples of the nations.'[44] The narrative structure of the Commission's report explains both its emotional affect and its claims to truth. It sought to affirm allegations of bacteriological warfare by presenting a model of a chain of causal events to which reported events could be compared. The justification for this logic was that it was extremely difficult to point

[41] Latour and Woolgar, *Laboratory Life*, p. 36.

[42] Andreen *et al.*, *Report of the International Scientific Commission*, p. 57.

[43] Andreen *et al.*, *Report of the International Scientific Commission*, p. 14.

[44] Andreen *et al.*, *Report of the International Scientific Commission*, p. 60.

to one particular test or condition that could confirm the clear presence of bacteriological warfare. The authors therefore claimed it 'necessary to envisage a manner of grouping events into a coherent pattern so that they can throw light upon each other and perhaps build up a circumstantial case.'[45] Most Commission members were experimentalists accustomed to working in the controlled spaces of laboratories. The evidence presented to them, however, had originated outside laboratories and demanded expertise in field sciences like ecology and entomology. The final report, then, might be understood as the efforts of a group of laboratory scientists to translate events in an uncertain field environment into the experimental conditions of the laboratories to which they were accustomed.[46]

These conditions were specified in a chart (Figure 14.1), which constructed a model of bacteriological warfare that depended on the notion of 'anomaly'. For something to count as part of a causal chain in this model, it had to be demonstrably unusual. For instance, the report stressed the appearance of swarms of insects despite typical weather as one condition supporting claims of bacteriological weapons. Most of the document consisted of case studies – events that 'attained most nearly the demonstrative character of the ideal pattern'.[47] The narrative form that this representation assumed was not simply the result of translation, but also reflected the original format in which the evidence was presented to the Commission. It suggested the central significance of the laboratory and field investigations that contributed to the identification of biological warfare, befitting Maoist emphases on practice over theory.

Because this pattern was narrative and sequential, the authors selected case studies that also provided arresting stories. For instance, one case in which a Korean woman and her husband ate clams infected with cholera has a cinematic quality: 'A country girl picking herbs on the hillsides found a straw package containing a certain kind of clam. She took some of the clams home and she and her husband made a meal of them raw; on the evening of the same day both fell suddenly ill and by the evening of the following day both were dead.' The authors acknowledged the 'bizarre' improbability of cholera transmission via clams, but then cited Japanese literature on the cultivation of cholera vibrios in molluscs to suggest that it was at least within the realm of possibility that this incident could occur.[48] The authors concluded plaintively, 'The young couple who died, impoverished by war devastation, had the imprudence to eat some of the clams which had been intended as the vehicles of contamination.'[49]

The result of this narrative structure was a document that, in tone, resembled science fiction. Appeals to emotion and class identity reflected the conceptualisation

[45] Andreen *et al.*, *Report of the International Scientific Commission*, p. 13.
[46] On laboratory/field distinctions, see Jeremy Vetter (ed.), *Knowing Global Environments: New Historical Perspectives on the Field Sciences*. New Brunswick: Rutgers University Press, 2011.
[47] Andreen *et al.*, *Report of the International Scientific Commission*, p. 13.
[48] Andreen *et al.*, *Report of the International Scientific Commission*, p. 35; see also p. 447.
[49] Andreen *et al.*, *Report of the International Scientific Commission*, p. 36.

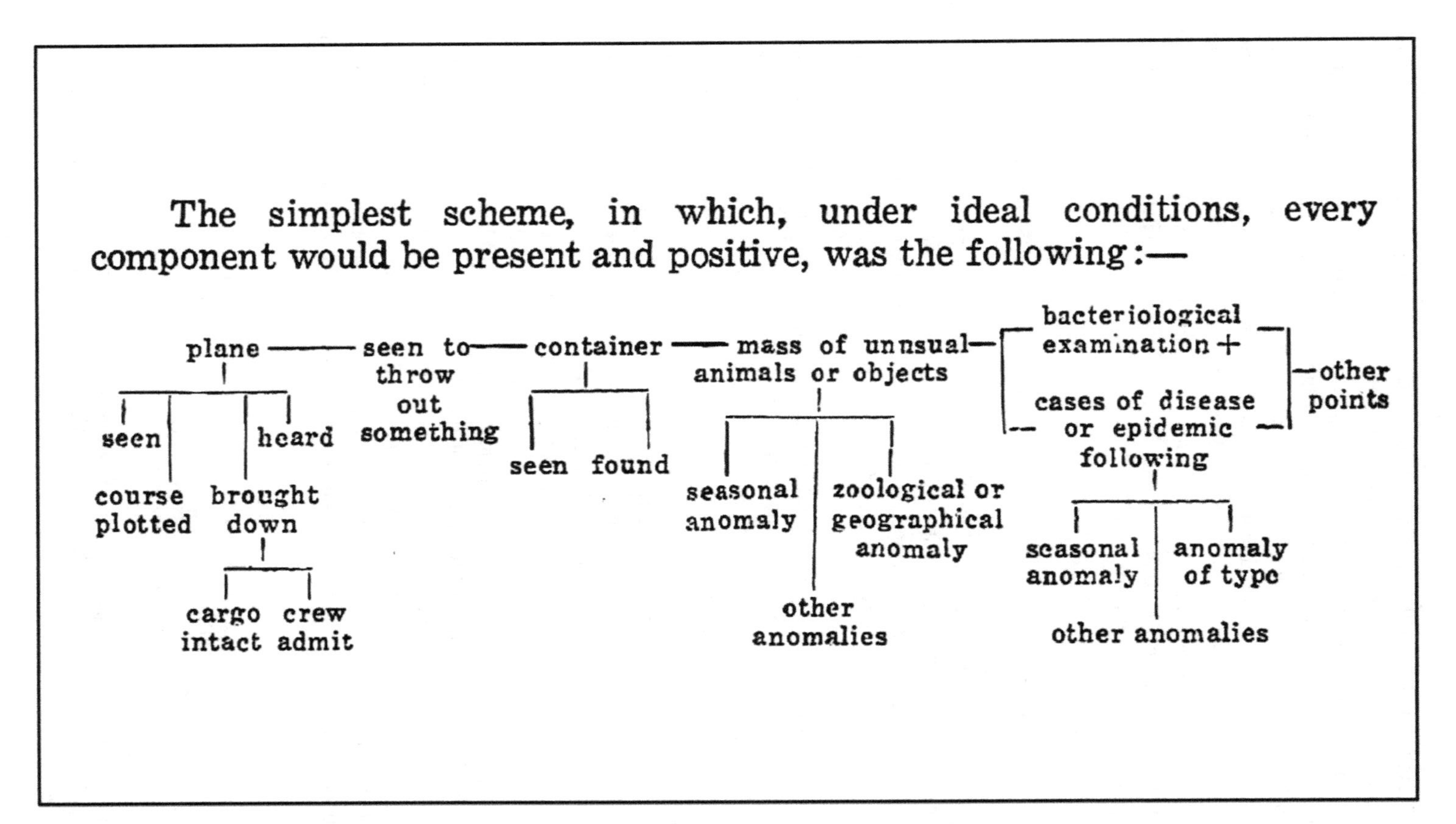

The simplest scheme, in which, under ideal conditions, every component would be present and positive, was the following:—

Figure 14.1 *Report of the International Scientific Commission*, p. 13

of Chinese science, with its attention to class and sociopolitical context, then emerging – yet they sat uneasily alongside notions of objective, value-free, apolitical science then gaining currency in other Cold War contexts.[50] In his personal correspondence, Needham gestured towards the self-contradictory nature of the Commission's work and reporting. In a letter to his wife Dorothy, he wrote, 'I am sorry to have to say, that no doubt whatever is left in any of our minds' about the veracity of the accusations. A few lines later, he confessed, 'there are things one simply wouldn't believe. A lot of the time this summer I have spent wondering whether I am dreaming or not, or whether I can really believe what I am seeing with my eyes and hearing with my ears.'[51]

Facts and fictions: Capturing the allegations on film

The cinematic nature of the Commission's report was supported by dramatic images depicting interviews with witnesses, close-ups of dead insects and animals, and cultures in petri dishes. In March 1952, Chinese delegates distributed Needham's collection of 25 photos from the trip at the World Peace Council Executive Committee meeting in Oslo.[52] And a propaganda film disseminated in the same year extensively discussed the allegations in promoting China's Patriotic Hygiene Campaigns. Christos Lynteris has argued that photographs taken after epidemics end can create 'a state of epistemic suspension' in which questions about disease transmission remain unresolved.[53] In 1952, the extensive documentation of epidemic events forcefully asserted the fact of bacteriological warfare – and yet the wealth of images it involved largely suggested such an epistemic suspension, calling into question the effectiveness of the approaches to science and knowledge production that Chinese researchers had constructed.

The report appendices began with general overviews of the fields and disciplines relevant to the allegations. Specific cases were then discussed, including historic precursors, like Japanese biological warfare in World War II (WWII), and the episodes that constituted the basis for the 1952 allegations. Images supported discussions of nearly every stage in the posited process by which germ bombs were distributed. Photographs of insects and seed pods believed to have been disseminated by American planes appeared alongside microscopic photos of fungi and bacteria cultured from samples taken from these suspicious flora and fauna. Yet

[50] Audra Wolfe, *Freedom's Laboratory: The Cold War Struggle for the Soul of Science*. Baltimore: Johns Hopkins University Press, 2018.

[51] Joseph Needham, Beijing, to Dorothy Needham, Cambridge, 15 August, 1952, NC.

[52] World Peace Council Original Photos, NRI2/10/1/31/7, Needham Research Institute. Thanks to John Moffett for this observation.

[53] Christos Lynteris, 'Photography, zoonosis and epistemic suspension after the end of epidemics', in *The Anthropology of Epidemics*, eds. Ann Kelly, Frédéric Keck, and Christos Lynteris. London: Routledge, 2019, pp. 84–101 at 84.

the overall effect was less than conclusive. Key stages in the posited attacks – the actual dropping of the bombs, for instance – remained off-camera, depicted instead by post-event photographs of shattered bomb shells and local witnesses.

In some cases, the images feel extraneous, as for instance in one appendix considering leaves from a type of oak tree, typically found in southern Korea, which had appeared in the north. A photograph of the leaves appeared above a photo of the species entry in a Chinese herbarium; the unnecessary inclusion of the latter suggested a desire to stress the authority of local compendia.[54] The inclusion of so many images also drew readers' attention away from the events in question by identifying certain kinds of plants or animals as native or not to northeast China and/or northern Korea – even though biologists had not traditionally drawn such regional distinctions.[55] Indeed, many of these images raised destabilising questions about the ecological bases on which processes of disease transmission were posited.

Film provided another medium for the allegations. Between 1950 and 1952, the Peking Film Studio and National Film Studio of Korea co-produced propaganda videos in multiple languages that animated the narratives later presented in the Commission's report. 'Oppose Bacteriological Warfare' depicted the Commission of the International Association of Democratic Lawyers (a Paris-based Communist organisation), conducting 'on-the-spot investigations' in March 1952.[56] Versions of the film exist in English, Spanish, and other European languages, as well as Chinese. It portrayed the discovery of germ warfare as a scientific process that started with wide shots of health workers in northeastern villages gathering insects or small animals that were deemed unnatural to the environment. Suspicious samples were then sent to laboratories, where the film focused on specialists preparing them for bacteriological scrutiny, and then researchers using microscopes to confirm the presence of pathogenic organisms in the samples. The filmmakers repeated this narrative over and over: health workers isolated unnatural specimens, laboratory assistants processed them into objects of bacteriological analysis, and researchers turned them into verified artefacts that provided evidence for the state's accusations.

In stressing mundane objects and processes, these images gave priority to the microscopic view. Sean Lei has shown that, during the Manchurian plague, the microscope became a powerful tool for establishing scientific facts. In 'Oppose Bacteriological Warfare', viewers themselves looked through the microscope, suggesting that this tool still dictated medical truth. The film focused on the materiality of scientific inquiry, lavishing attention on researchers' white clothing and masks, as well as the 'aseptic techniques' of laboratory workers and processes like

[54] Andreen *et al.*, *Report of the International Scientific Commission*, p. 194.

[55] Jaehwan Hyun, 'Brokering science, blaming culture: the US-South Korea ecological survey in the Demilitarized Zone, 1963–8', *History of Science* 59, no. 3 (2021): 315–43.

[56] Peking Film Studio of China and the National Film Studio of Korea, 'Oppose Bacteriological Warfare', 1952, National Archives and Records Administration, ARC 1630600 / LI 263.1006, also online at https://archive.org/details/gov.archives.arc.1630600, 35:15–45; 40:07–30.

sterilisation with DDT and microscopic slides.[57] The film featured 'eminent specialists' analysing specimens, signing reports confirming their discovery of pathogenic bacteria, and discussing these findings with each other.[58] Most of these objects and processes were routine. Why give them pride of place? Emphasising the material culture of bacteriological inquiry demonstrated Chinese scientists' abilities to conform to global standards of research, and therefore claimed authority for their findings, as well as the socialist values that had come to be associated with their practices. If new attention to class changed *who* participated in science, this visual rhetoric suggested that *what* happened in scientific knowledge production – laboratory inquiries using microscopes and prepared materials – remained commensurable to comparable protocols in the foreign contexts in which the film's claims competed.

Conclusion: Failures in fact-making

The work of the ISC and the scientists they championed remained illegible to Anglophone audiences, who ignored or rejected the allegations. Across the broader American public, an opinion poll in July 1952 suggested that about 64 per cent of respondents knew of the charges, which Steven Casey suggests was 'a far lower level of public awareness than on other major issues'.[59] The British government mobilised politicians, journalists, and scholars to undermine the allegations; Needham suffered significant reputational damage.[60]

The defamation of Needham laid bare his detractors' assumptions about science and China. On 17 October 1952, Sirs Henry Dale and Robert Robinson, writing as former Presidents of the Royal Society, published a letter in *The Times* claiming that it was inappropriate for Needham to have served on the Commission because the work of confirming bacteriological warfare was fundamentally unscientific.[61] 'The very conditions under which a modern man of science has to make his contributions to a common stock of progressive discovery are unfavourable to his efficiency as a detective', Dale wrote to Robinson. 'If he had to consider, in every case, the possibility that the results presented by another scientist

[57] Christos Lynteris, 'Plague Masks: The Visual Emergence of Anti-Epidemic Personal Protection Equipment', *Medical Anthropology* 37, no. 6 (2018): 442–57; and Sean Hsiang-lin Lei, 'Sovereignty and the Microscope: Constituting Notifiable Infectious Disease and Containing the Manchurian Plague', in *Health and Hygiene in Chinese East Asia*, eds. Angela Ki Che Leung and Charlotte Furth. Durham: Duke University Press, 2011, pp. 73–106.

[58] 'Oppose Bacteriological Warfare', 26:00–28:00.

[59] Steven Casey, *Selling the Korean War: Propaganda, Politics, and Public Opinion in the United States, 1950–53*. Oxford: Oxford University Press, 2008, p. 422n105; Jacob Darwin Hamblin, *Arming Mother Nature: The Birth of Catastrophic Environmentalism*. Oxford: Oxford University Press, 2013, p. 54.

[60] Buchanan, 'Courage of Galileo', 517–22.

[61] Robert Robinson and Henry H. Dale, 'Report on Germ Warfare: Need for Judicial Assessment', letter to the editor, *The Times*, 17 October 1952, issue 52445, p. 7.

might be, not merely erroneous, but deliberately fraudulent, his own work would become impossible.'[62] Such a statement contrasts with Latour and Woolgar's claim that practices of literary inscription and social networks in scientific laboratories are produced with the intent of eliminating alternative accounts.[63] Furthermore, in their assertion that trust in the scientific community was a necessary condition of research, Dale and Robinson assumed the exclusion of Chinese scientists from that community.

Needham responded with a missive of his own to *The Times*, in which he made the trustworthiness and reliability of Chinese scientists the explicit question at hand. 'Before the International Scientific Commission went to China the findings of the Chinese scientists were being disregarded, brushed aside, and ridiculed', he wrote, adding, 'The testimony of the commission is that the Chinese evidence makes sense, and must be taken seriously.'[64] Yet Needham was in the minority, as demonstrated by a debate at the British House of Commons on 24 October 1952 in which the testimony of Chinese researchers came under direct fire. 'The evidence [Needham and the ISC co-authors] have produced is all third-hand', claimed Anthony Nutting, Joint Under-Secretary of State for Foreign Affairs, who went on to say 'It was received through Chinese interpreters from Chinese scientists working on reports from Chinese peasants.' He specifically targeted Needham's endorsement of Chinese researchers, describing the evidence presented in the Commission's report as 'taken at its face value without any regard whatsoever for the inevitable partiality of the people through whom it has passed'.[65] The Foreign Office viewed this declamation as a definitive victory against Needham and his allies.[66]

A full study of the reception of the allegations among non-Chinese publics remains to be written, yet they are still being debated and treated seriously, in multiple languages and countries. In that respect, assertions of Chinese expertise and universal knowledge achieved their goal: serious engagement with the work of Chinese bacteriologists, immunologists, and entomologists on a global scale. I have tried to suggest here that those assertions rested on multiple and shifting approaches to bacteriological knowledge-making. In the years after 1952, the same Chinese researchers who endorsed the accusations of germ warfare led university departments and research institutes that oversaw the development and mass

[62] Henry Dale to Robert Robinson, 29 October 1952, Wellcome Library, Correspondence and papers regarding Dr Joseph Needham, record group WT/A/3/1/7, box 003393/A.
[63] Latour and Woolgar, *Laboratory Life*, pp. 43–104.
[64] Joseph Needham, 'Germ Warfare Report', letter to the editor, *The Times*, 25 October 1952, issue 52452, p. 7.
[65] 'Germ Warfare, Korea', p. 1448. Wellcome Library, Correspondence and papers regarding Dr Joseph Needham, record group WT/A/3/1/7, box 003393/A.
[66] 'The Foreign Office were very pleased with it.' Henry Dale to Robert Robinson, 27 October 1952, Wellcome Library, Correspondence and papers regarding Dr Joseph Needham, record group WT/A/3/1/7, box 003393/A.

production of vaccines and other bacteriological products in service of the Patriotic Hygiene Campaigns. Since their 1952 inception, the campaigns have endured into the twenty-first century as a legacy of Maoist approaches to science and medicine, although memories of bacteriological warfare have faded. So, too, have public memories of Chinese researchers who stood at a unique point in the history of the life sciences – likely facing pressures and incentives to apply scientific expertise to diplomatic ends, yet also finding new means to use that expertise to promote national public health.

15

Soviet Books in Socialist China: Epoch Press and the Making of the Maoist State, 1940–1960

NICOLAI VOLLAND

IN AN UNUSUAL editorial, the journal *The Epoch* (*Shidai*) reminded its readers of its anniversary: 'It has been ten years since the founding of the Epoch Press (Shidai chubanshe) and *The Epoch* on 20 August 1941. Now we are celebrating their tenth anniversary!'[1] Very few Chinese journals had managed to navigate a decade of wartime occupation, civil war, and revolution. To survive three changes in government, political turmoil, economic disruption on a massive scale, and an onslaught of political mass campaigns was a sign of considerable tenacity. The celebratory mood was subdued though. Paradoxically, issue 359, the anniversary issue, was also *Epoch*'s final number: 'While we are celebrating the tenth anniversary of our publishing house, we are transitioning to a new stage, and we harbour even greater confidence and resolution; we are prepared to take on new tasks.'[2] The editors thus announced the end of their journal. With *Epoch*, one of the most potent symbols of Sino-Soviet solidarity from the 1940s disappeared from the newsstands. The new era after 1949 required new solutions – those unspecified 'new tasks' with which the editorial ended.

The Epoch Press and its flagship publications played a crucial if underappreciated role in the Sino-Soviet cultural relationship before and after 1949, and in the making of the Maoist state. In the years leading up to the Communist revolution, the publisher supplied leftist Chinese readers with a steady supply of translated materials from the Soviet Union. After the founding of the People's Republic of China (PRC), the Epoch Press became the leading outlet for Soviet literature and social

[1] 'Shidai shi nian (1941–1951)' [*Epoch* 10 years on], *Shidai* 11.16 (no. 359, 20 Aug. 1951), 9–13, here 9.

[2] 'Shidai shi nian', 13.

Proceedings of the British Academy, **267**, 293–310,

science publications in Chinese translation. In the late 1950s, the press turned into a niche publisher, specialising in the crucial area of Russian-language textbooks and reference works. Despite its prolific output, the Epoch Press is mostly forgotten today. While veteran publishing houses such as the Commercial Press (*Shangwu yinshuguan*) and China Books (*Zhonghua shuju*) have received a significant amount of scholarly attention,[3] Epoch has fallen under a collective amnesia that extends to much of the Sino-Soviet friendship of the 1950s. In the formative years of the Maoist period, however, Epoch played a pivotal role in supplying the models – cultural and otherwise – on which New China's founders drew to build socialism in China. More importantly, Epoch inherited and transformed a cosmopolitan tradition dating back to the late nineteenth and early twentieth centuries. It emerged from the memories of Shanghai's celebration of openness that had survived into the early wartime 'orphan island' (*gudao*, Aug. 1937–Dec. 1941) period. Perpetuating this commitment to cultural worldliness and adapting it to a rapidly evolving environment came to define Epoch. The valourisation of foreign and transnational cultural production were at the heart of the publisher's mission, and its world-minded employees – a virtual who-is-who of Chinese translators of Russian – worked hard to establish a new brand of cosmopolitanism in the socialist nation.

The notion of 'Maoism' has long been associated with Yan'an and the Chinese Communist Party (CCP)'s efforts to adapt Leninism to domestic social and economic realities.[4] As a rich new literature has shown, however, the early PRC was more outward oriented and invested in international politics than this nationalist narrative suggests.[5] The making of Maoism, thus, was very much an event of transnational significance, and it drew on sources of inspiration from beyond China's borders. The Soviet Union in particular offered models that informed the building of New China's political and economic systems, as well as institutions and practices in spheres from education and cultural production to the legal system and the organisation of the military. But how did knowledge about these new models reach China? How did information about things Soviet circulate during the formative years of Maoism? And how did the public image of the Soviet Union as an advanced industrialised nation, and hence a powerful model for New China, take shape? Personal experience played a role. A number of high-ranking CCP leaders had lived and studied in the Soviet Union in the 1920s and 1930s, and more would visit the 'big brother' after 1949. Soviet experts were another crucially important channel for the transfer of technical expertise across all sectors, and especially in

[3] See, for instance, Robert J. Culp, *The Power of Print in Modern China: Intellectuals and Industrial Publishing from the End of Empire to Maoist State Socialism*. New York: Columbia University Press, 2019.

[4] Mark Selden, *China in Revolution: The Yenan Way Revisited*. New York: Routledge, 1995.

[5] Thomas P. Bernstein, Hua-yu Li (eds.), *China Learns from the Soviet Union, 1949-Present*. Lanham: Lexington Books, 2010; Hua-yu Li, *Mao and the Economic Stalinization of China, 1948–1953*. Lanham: Rowman & Littlefield, 2006; Nicolai Volland, *Socialist Cosmopolitanism: The Chinese Literary Universe, 1945–1965*. New York: Columbia University Press, 2017.

industry and technology.[6] The public reach of these personalised ties, however, was limited. By far the greatest impact, with the widest public reach, was generated by the print media: through newspaper and journal articles on the Soviet Union, and especially through translations of Soviet books. A trickle, though an influential one, became a torrent after 1949. An incomplete catalogue of Soviet books in Chinese translation, published in the first five years of the PRC, lists over 2,000 titles.[7] Many of these books appeared in print runs in the tens and sometimes hundreds of thousands, reaching readers all across China. Information from and about the Soviet Union, hence, was ubiquitous in the early Maoist period, and came to shape New China.

A major player – and one with a special status and history – in the effort to make knowledge about things Soviet available in China was the Epoch Press. As I will show in this chapter, Epoch's experience documents the twists and turns of China's transnational engagement during the period Maoism took shape. Founded in 1941 as a foreign-owned enterprise, the publisher was in fact run by the Shanghai bureau of TASS, the Soviet news agency, and served initially as a wartime propaganda outlet for the Soviet Union. After the Japanese surrender in 1945, Epoch became a Sino-Soviet joint venture, a model of transnational cooperation and a potential harbinger of a new age of cooperation. By 1949, the Epoch Press boasted dozens of employees, published a wide range of books and journals, and had become an effective propagandist for the cause of socialist internationalism. Ironically, however, the new government soon had second thoughts about the Soviet role in Epoch and negotiated its transfer to Chinese management, leading to an overhaul of its operations (which included the closing of the journal, *The Epoch*). While the CCP recognised the value of the well-known Epoch imprint, it insisted on having the last word about the shape of New China.

The CCP leadership's distrust of their Soviet brethren and Mao's concern with Chinese sovereignty are well-known. As Lorenz Lüthi, Shen Zhihua, and others have shown, the Sino-Soviet alliance was fragile underneath the surface.[8] The negotiations over the Sino-Soviet friendship treaty (1950) were fraught with disagreements about Chinese sovereignty over railroads and military bases.[9] A decisive

[6] Shen Zhihua, *Sulian zhuanjia zai Zhongguo, 1948–1960* [Soviet experts in China]. Beijing: Xinhua chubanshe, 2009.

[7] *Wo guo fanyi chuban Sulian shuji mulu, 1949.10–1954.6* [Catalog of Chinese translations of Soviet books]. Beijing: Wenhuabu chuban shiye guanliju, 1955.

[8] Lorenz M. Lüthi, *The Sino-Soviet Conflict: Cold War in the Communist World*. Princeton: Princeton University Press, 2008; Zhihua Shen and Danhui Li, *After Leaning to One Side: China and its Allies in the Cold War*. Washington, DC: Woodrow Wilson Center Press, and Stanford: Stanford University Press, 2011. The classic study is Chen Jian, *Mao's China and the Cold War*. Chapel Hill: University of North Carolina Press, 2001. See also Austin Jersild, *The Sino-Soviet Alliance: An International History*. Chapel Hill: University of North Carolina Press, 2014.

[9] See Shen and Li, *After Leaning to One Side*, pp. 3–23; and Shengfa Zhang, 'The Main Causes for the Return of the Chinese Changchun Railway to China and Its Impact on Sino-Soviet Relations', in *China Learns from the Soviet Union, 1949–Present*, eds. Thomas P. Bernstein and Hua-yu Li. Lanham: Lexington Books, 2010, pp. 61–78.

moment in the deteriorating bilateral relationship, in the late 1950s, was a proposal by Khrushchev to set up a joint fleet and long-wave radio stations on Chinese territory.[10] And the presence of thousands of Soviet experts raised unwelcome questions of extraterritoriality.[11] The problem of transnational culture, however, complicates these narratives of deeply distrustful partners, pitting nationalist concerns about sovereignty against the internationalist agenda of the socialist bloc. The Epoch Press, throughout its existence, promoted Soviet cultural production through changing means and channels, demonstrating a deep commitment to the value(s) of transnational engagement and exchange of behalf of mid-level cadres, intellectuals, and readers. The case of Epoch thus not only highlights the importance of these actors in the effort to shape the public image and perception of the alliance and New China's emerging world outlook. It also calls for closer scrutiny of the intermediary zones (in contrast to high-level politics) in which the project of socialist internationalism acquired shape and meaning. Institutional actors like Epoch, while very much subject to the vicissitudes of change in the political line – as the closing of the journal *Epoch* reminds us – showed, at the same time, a remarkable degree of resilience and resolve in negotiating deeply ingrained belief systems with the evolving political and economic environment. Socialist intellectuals with a cosmopolitan outlook had shaped the original identity of Epoch; remarkably, this identity outlasted the departure of most of the early cadre of the publisher's editorial backbone and continued to impact the appropriation of Soviet culture in the early PRC. The case of the Epoch Press, in other words, shows how the project of Sino-Soviet cultural cooperation acquired a life of its own in ways that elude clear-cut categories such as sovereignty and power and that call for a more nuanced understanding of the dynamics of the Maoist state and its relationship with the socialist world.

This chapter traces the institutional history of the Epoch Press, from its founding in 1941 to its disappearance in the early 1960s, scrutinising the publisher's adaptive strategies across two tumultuous decades as well as the evolution of its commitment to bringing Soviet culture to readers in China. In what follows, I will draw on both archival documentation and the Epoch Press's published output of books, newspapers, and journals, which illustrate the publisher's evolving approaches to transnational cultural mediation. Analysing these articulations of Epoch's cosmopolitan commitment, I will show the dynamic interplay of individuals, institutions such as the Epoch Press, and the state in the making of Maoist China's transnational cultural engagement.

[10] See Lüthi, *The Sino-Soviet Split*, pp. 91–5; Shen and Li, *After Leaning to One Side*, pp. 147–66; and You Ji, 'The Soviet Model and the Breakdown of the Military Alliance', in Bernstein and Li, *China Learns from the Soviet Union*, pp. 139–44.

[11] Shen and Li, *After Leaning to One Side*, pp. 117–34; Shen Zhihua, *Sulian zhuanjia zai Zhongguo*, pp. 106–15. Compare also Yang Kuisong, 'The Sino-Soviet Alliance and Nationalism: A Contradiction', *Social Sciences in China* 26.2 (2005): 86–99.

Pan-socialist propaganda in wartime China

The Epoch Press was established in Shanghai in summer 1941 and registered by a Soviet citizen, the expatriate businessman I.M. Zakgeima (dates unknown; in Chinese Zakaimo).[12] As a foreign-owned enterprise, based in the concession area, Epoch thus enjoyed extraterritorial status, which gave it protection from the police in the Japanese-controlled parts of Shanghai surrounding the concessions. Zakgeima seems to have been involved in editing Epoch's Russian-language publications,[13] but the driving force behind the publisher was Vladimir N. Rogov (1906–88), the energetic and multilingual head of the Far East Bureau of TASS, the Soviet news agency.[14] While Chinese accounts try to stress the CCP's agency in the project, there is very little evidence for Chinese initiative during the early years of the press. For all practical purposes, Epoch was set up as a front disseminating Soviet propaganda in China and led by a group of highly cosmopolitan Soviet expatriates. Its propagandistic mission gained importance after the outbreak of the Pacific War a few months later.

The earliest publication of the Epoch Press was *Ėpokha* (The epoch), a Russian-language biweekly that began publication in March 1941 and targeted the sizeable Russian community in wartime Shanghai.[15] This was followed in August 1941 by the *Daily War News*, an English newspaper. The German invasion of the Soviet Union in June that year created great demand for information, which the Epoch Press satisfied by distributing the paper free of charge. The Chinese-language *Epoch* published its first issue a few weeks later, on 20 August 1941. In his memoirs, Jiang Chunfang (1912–87), the head of the cultural branch of the CCP underground

[12] On the Epoch Press's founding see Li Sui'an, 'Shidai chubanshe zai Zhong-Su wenhua jiaoliu shi shang de diwei' [The Epoch Press's place in the history of Sino-Soviet cultural exchange], in Zhongguo Zhong-E guanxi shi yanjiuhui [Chinese historical society on Sino-Russian relations] (ed.), *Zhanhou Zhong-Su guanxi zouxiang, 1945–1960: Zhong-E (Su) guanxi xueshu lunwen xuan* [Trends in post-war Sino-Soviet relations, 1945–1960: Essays on Sino-Soviet (Russian) relations]. Beijing: Zhongguo shehui kexue wenxian chubanshe, 1997, pp. 177–88; and Min Dahong, '"Sushang" Shidai chubanshe yu *Shidai* zhoukan, *Shidai ribao*' [The 'Soviet-owned' Epoch Press, *Epoch* weekly, and *Epoch Daily*], *Xinwen yanjiu ziliao* [Journalism research materials] 36 (1986.2): 134–44. The following account is based on these sources, as well as 'Shidai shi nian (1941–1951)'. The dates for the founding of the Epoch Press are confirmed in SMA B1-1-1882, 5-6.

[13] Very little information is available on Zakgeima, who may have been a Comintern agent. A registration form in the Shanghai Archives dating from November 1945 says that Zakgeima was male, 46 years old, and that he had previously worked as a journalist for Russian newspapers in Shanghai and Harbin. SMA Q6-12-90, 1–7. Chinese sources refer to Zakgeima as a Soviet citizen (*qiaomin*) who merely lent his name to the Soviet venture, but his journalistic credentials imply a bigger role.

[14] Rogov was well-known in the foreign press corps of wartime China. He directed the TASS office in Shanghai from late 1937 to early 1943, but was also working secretly for GRU, the Military Intelligence Directorate of the Soviet Union. For Rogov's role in the founding of the Epoch Press see Jiang Chunfang, 'Wo yu Shidai chubanshe' [I and Epoch Press], *Zhongguo chuban shiliao* [Historical materials on publishing] 16 (Feb. 1989): 9–15, here 10.

[15] For this and the following information see SMA B1-1-1882, 5–6. The journal later switched to a monthly format and operated until 1950.

organisation in Shanghai, recounts that in early July he received instructions from his superiors to contact Rogov with the proposal to set up a Chinese newspaper. The Soviets instead counselled for a more low-key approach, suggesting a weekly paper along the lines of *Ėpokha*. Rogov agreed to underwrite the costs of printing and provided Jiang with the articles – taken from *Ėpokha* and Soviet papers such as *Pravda* – to appear in the journal's inaugural issue. Jiang, a multilingual and cosmopolitan cultural dynamo himself, found two staff members to help him translate and edit the journal's first issue.[16]

The Epoch Press, thus, was a Soviet propaganda organisation, born at the moment when the all-out propaganda battle in East Asia reached a new climax.[17] In this early period, until 1945, initiative rested with the Soviets and the Comintern, who recruited help from the CCP. Epoch represented a rare convergence of interests between the CCP and the Communist Party of the Soviet Union (CPSU) at a time of frequent disagreements. The Soviets had hesitated to get involved in anti-Japanese propaganda in China, but with the German invasion in late June 1941, such reservations disappeared. A Chinese sister publication of *Ėpokha* would extend the reach of Soviet propaganda, and counter the influence of the Germans, who operated a Chinese-language paper of their own. The CCP underground at the time was busy rebuilding their propaganda outlets; the party had lost its veil of protection after an English businessman, under whose name many CCP propaganda outlets had been registered, was bought over by the Japanese and their allies.[18] Their cooperation with the Soviets in Shanghai proved farsighted. Many Chinese and American-registered papers were closed by the Japanese military and the agents of the Nanjing-based Wang Jingwei-regime on 9 December, the day after the Japanese attack on Pearl Harbor. Yet the Japanese respected Soviet neutrality and allowed the Epoch Press to keep operating and even growing its operations.[19] At a time when cultural imports from the Anglophone world dried up, Epoch provided an alternative source of cosmopolitan culture.

The multilingual portfolio of the Epoch Press expanded over the next years. In November 1942, the press inaugurated *Soviet Literature and Art* (Sulian wenyi). *The Epoch* had run frequent reports on cultural affairs, as well as the occasional literary piece, but its focus was current affairs. The new journal was designed to introduce Soviet literature and culture to China.[20] Pressure from the Japanese police

[16] Jiang Chunfang, 'Wo yu Shidai chubanshe', 10–11.

[17] Compare Peter O'Connor, *The English-Language Press Networks of East Asia, 1918–1945*. Folkestone: Global Oriental, 2010, esp. pp. 232–96.

[18] See Lin Ling [Jiang Chunfang], 'Huiyi 'Sushang' Shidai chubanshe de gongzuo' [Recollections of the work of the 'Soviet-owned' Epoch Press], in Zhang Jinglu (ed.), *Zhongguo xiandai chuban shiliao* [Historical material on Chinese modern publishing]. Beijing: Zhonghua shuju, 1954–9, *ding bian*, Vol. 2, pp. 504–9, here p. 505.

[19] Min, ' "Sushang" Shidai chubanshe yu *Shidai* zhoukan, *Shidai ribao*', 136.

[20] See Lin, 'Huiyi 'Sushang' Shidai chubanshe de gongzuo', p. 505; and Min, ' "Sushang" Shidai chubanshe yu *Shidai* zhoukan, *Shidai ribao*', 136. The journal appeared irregularly and had published 37 issues by July 1949, when it ceased publication.

increased in 1943, and Epoch had to submit its publications to censorship from the Wang Jingwei government in Nanjing.[21] In February 1944 the Nanjing government ordered the publisher to halt its Chinese and English operations, on the grounds that foreigners in China were not permitted to publish in languages other than their own.[22] In April 1945, when victory seemed imminent, the press decided to ignore the orders from Nanjing and reopened its Chinese flagship publications, *Epoch* and *Soviet Literature and Art*.[23]

After 1945, with the Japanese gone and the Communist Party regaining operational momentum, the CCP's position within the Epoch Press strengthened considerably. The press consequently moved much closer to becoming a collaborative Sino-Soviet enterprise. The expansion of the CCP's influence was made possible, first and foremost, by the return to Shanghai of numerous CCP cadres, but also by a shift in the press's emphasis from journal to newspaper publishing. In November 1945 the Epoch Press applied for publishing licences for seven periodicals in three languages: a weekly and a biweekly paper in Russian, an English daily, and four Chinese publications.[24] The Chinese papers had the highest projected circulation and were by far the most important: the press proposed to print 5,000 copies of *Soviet Literature and Art*, 7,000 copies of the weekly *Epoch*, and 10,000 copies of a daily newspaper, *The Epoch Daily* (Shidai ribao). Jiang Chunfang became the paper's editor-in-chief, while Zakgeima was listed as publisher, thus affirming the cooperative structure. The paper's editorial staff included Lin Danqiu (1906–81), Chen Bingyi (1916–2008), Ye Shuifu (1920–2002), Xu Leiran (1918–2009), Man Tao (1916–78), and Wang Yuanhua (1920–2008), most of whom were CCP members;[25] the multilingual members of Epoch's office would all play major roles in Sino-Soviet cultural exchanges in the 1950s, shaping the internationalist outlook of Maoism. *The Epoch Daily* featured national and international news, local reports, and a range of columns, as well as letters and advertisements. The paper quickly expanded its editorial section, featuring commentaries written by underground CCP members and veteran journalists. Locally produced content thus featured more prominently in *The Epoch Daily* than in any of the press's earlier publications, an indication of its transition towards a true joint venture. With its new structure, Epoch suggested a pathway of collaboration that signalled a road toward building a new Maoist state.

[21] Min, '"Sushang" Shidai chubanshe yu *Shidai* zhoukan, *Shidai ribao*', 137.

[22] Min, '"Sushang" Shidai chubanshe yu *Shidai* zhoukan, *Shidai ribao*', 138.

[23] Lin, 'Huiyi "Sushang" Shidai chubanshe de gongzuo', p. 508. The publisher's historians note that in the final months before the Japanese surrender, the Epoch Press was run almost singlehandedly by the Soviets, who covered the publisher's operating costs and supplied it with paper imported from the Soviet Union. See ibid., p. 507, and Li, 'Shidai chubanshe zai Zhong-Su wenhua jiaoliu shi shang de diwei', p. 180.

[24] See SMA Q6-12-90, 1–7.

[25] Min, '"Sushang" Shidai chubanshe yu *Shidai* zhoukan, *Shidai ribao*', 139.

The protective umbrella of Soviet ownership, however, remained important for Epoch's survival. With the deterioration of the Kuomintang (KMT)–CCP relationship, the publisher's activities attracted increased scrutiny from the Shanghai Garrison Command, but the friendship treaty between Chiang Kai-shek's government and the Soviet Union, signed in August 1945, restrained the police.[26] Epoch was suspected of secretly printing translations of Soviet pamphlets propagating communism, and of sheltering CCP activists and Party members. The police, of course, were spot on. Epoch worked hand in hand with the CCP underground and complemented the publishing efforts of various CCP-controlled and affiliated publishers. The KMT acted whenever Epoch's communist propaganda became too overt. *The Epoch Daily* was shut down on 16 January 1947, but was allowed to reopen three weeks later.[27] In June 1947 the Shanghai News Bureau called Zakgeima in for a 'chat', complaining about the newspaper's coverage of local student unrest. The government demanded an apology from the paper and threatened its editor, citing the example of the left-leaning *Wenhui Daily* (Wenhuibao), which had been closed down just two weeks earlier.[28] The KMT eventually shut down *The Epoch Daily* for good in June 1948.[29] What is surprising is not *that* the paper was banned, but how long the KMT hesitated to act. And even when it finally did, it was only the paper that was closed, and not the publisher itself. The Epoch Press and its major journals continued to operate until the communist takeover. Extraterritoriality had been abolished in 1943, but the KMT was reluctant to offend a treaty partner – concerns that the CCP had anticipated and exploited.[30] Throughout the Civil War, the Epoch Press thus functioned as an effective propaganda outlet, serving the interests of both the Soviet Union and the CCP. The press had evolved into a model of Sino-Soviet cooperation, laying the foundations for the new Maoist state and its international outlook.

This outlook is most clearly on display in Epoch's publications. From its inception, the journal *Epoch* provided its readers with a mixture of current affairs, articles on Soviet culture, and translations of short stories and reportages, initially focused mostly on the European war and Soviet war efforts. The vast majority of *Epoch*'s content in this early period was translated. Strikingly absent from the pages of *Epoch* – and from its sister publication, *Soviet Literature and Art* – is any mention of China's own war, its resistance against Japanese

[26] The Shanghai police forces had observed Epoch since at least December 1945. Numerous intelligence reports show that the Shanghai authorities were well-informed about the publisher's activities. See SMA Q131-4-3705, Q6-12-90-11 to -13, and in general, Q6-12-90 and Q6-12-89.

[27] Min, ' "Sushang" Shidai chubanshe yu *Shidai* zhoukan, *Shidai ribao*', 139.

[28] See SMA Q6-12-89, 10. For a similar incident, over an article published on 1 September 1947, see ibid., 16.

[29] The last straw had once again been reports on student unrest. See SMA B1-1-1882, 5-6; and Min, ' "Sushang" Shidai chubanshe yu *Shidai* zhoukan, *Shidai ribao*', 141.

[30] The contrast with *Wenhui Daily* is instructive. The latter was not more pro-Communist than *The Epoch Daily*, but, as a Chinese-owned paper, was easier to prosecute.

Figure 15.1 Masthead for the column 'Zhong-Su wenhua jiaoliu' [Sino-Soviet cultural exchange]. Source: *Sulian wenyi* 26 (1947): 161

occupation. Yet the massive anti-Fascist propaganda and the flood of inspirational essays on heroic resistance against foreign aggressors could not but strike a chord with Chinese readers. The journals' editorial strategy helped to promote a sense of solidarity between the Soviet and the Chinese people, both victims of foreign aggression.

Epoch's editorial line shifted after 1945, reflecting the growing Chinese influence in what was now a Sino-Soviet joint venture. Depictions of the Soviet Union, such as in a November 1946 *Epoch* special issue commemorating the October Revolution, moved from wartime heroism to framing the Soviet Union as a socio-economic, political, and cultural success story. Chinese contributions made up a much greater proportion of the journal's content, presenting the voices of Chinese writers such as Guo Moruo (1892–1978) and Mao Dun (1896–1981) praising the Soviet Union as a 'bulwark of peace and democracy' – a model, in other words, that China should follow.[31] An illustration of how Epoch envisioned this new cooperation can be found in Issue 26 of *Soviet Literature and Art*. The issue's last section, called 'Sino-Soviet cultural exchange' (*Zhong-Su wenhua jiaoliu*), visualises the emerging alliance by a handshake in front of a rising sun (see Figure 15.1). The first item in this section is a brief speech with the title 'About Lu Xun' (Tantan Lu Xun), delivered by the prominent writer Konstantin Simonov in his function as vice-head of the Soviet Writers' Union in October 1946 during a ceremony honouring the tenth anniversary of the Chinese writer's death. The other item is a Chinese translation of the preface for Vladimir Rogov's Russian translation of Lu Xun's 'Story of Ah Q', complete with two illustrations – the title page of the Russian collection (published in 1945) and the first page of the Russian 'Ah Q'.[32] While

[31] Guo Moruo, 'Shijie heping de shizhu', *Shidai* 43–4 (1946): 19–21; Mao Dun, 'Heping minzhu de baolei wan sui', ibid., 24.

[32] V. Rogov, trans. Ge Da, '"Ah Q zhengzhuan" E yiben dai xu', *Sulian wenyi* 26 (Jan.-Feb. 1946): 163–9.

readers in China will find little informational value in either essay, the pieces are designed to demonstrate that cultural exchange and literary translation are not a one-way street. They show the journal's Chinese readers that their country's literature is being valued and appreciated in the Soviet Union, that China is entering into a partnership with its socialist neighbour that is based on reciprocity. Just like the handshake in the illustration, the essays emphasise the growing influence of Chinese views in Epoch Press, a Sino-Soviet joint venture with the mission to promote a common socialist culture shared across borders.

In the late 1940s, the Epoch Press also increased its output of books. Apart from literary works, the Press published books on the Soviet Union and on current affairs. Titles such as *The Soviet Constitution* (*Sulian xianfa*, 1948), *The Protection of Mothers and Children in the Soviet Union* (*Sulian dui muxing ji ertong de baohu*, 1947), or *North Korea* (*Bei Chaoxian*, 1948) complemented the coverage in the journal *Epoch*. The publisher's list expanded greatly in 1949; almost two-thirds of 81 titles I have been able to identify were published in the last year before the founding of the PRC.[33] About 40 per cent of Epoch's output were translations of Soviet literary works; social science publications and books on current affairs account for little less than a quarter each, followed by small amounts of works by Lenin and Stalin, and medical literature. Epoch's rapidly growing publishing business made the publisher arguably the most important provider of authoritative information about the Soviet Union, and especially about Soviet culture, on the eve of the founding of the PRC. Its journals and books offered material input for the making of the new Maoist state.

Internationalism and sovereignty: In search of a new balance

The CCP takeover of Shanghai in May 1949, the founding of the PRC four months later, and the Sino-Soviet friendship treaty, negotiated and signed by Mao in Moscow in February 1950, created the parameters for the new Maoist state, and fundamentally altered the political and cultural environment in which the Epoch Press operated. In the weeks after the military occupation of Shanghai, the People's Liberation Army (PLA) forced the closure of foreign media enterprises, ordered numerous resident foreigners to leave the country, and confiscated their companies, all in the name of re-establishing Chinese sovereignty. Shortly after the takeover, the Epoch Press was reorganised and oversight of the publisher transferred to Chinese hands. Jiang Chunfang was appointed the new director of the press; he

[33] Statistics based on books listed by online secondhand bookstores, checked against the online catalogues of the Chinese National Library, the Shanghai Library, and Peking University Library. The editors of *Epoch* provided a somewhat higher number, 130 titles, for the same period. See 'Shidai shi nian', 13.

also became chief editor of its flagship journal, *Epoch*.[34] The Soviet employees of the Epoch Press remained aboard and continued to run its Russian publications.[35]

Under the new regime, the Epoch Press could bring into play its experience in publishing translations from Russian, its access to Soviet sources, and its assets, which had grown significantly in the years leading up to the CCP takeover. Epoch was hence well positioned to benefit from the huge demand for information about the Soviet Union and translations of all kinds of Russian materials. The rapid growth of the Epoch Press is all the more remarkable in light of the financial difficulties faced by other large publishing houses such as the Commercial Press and Zhonghua Books in the immediate takeover years, which struggled and had to be bailed out by the government.[36] Epoch's rise, reflected in the quick expansion of its list, suggests access to Soviet funding. By 1951, the press had branches in Beijing, Nanjing, and Hangzhou, while its main operations were still located in Shanghai. The Shanghai office alone employed 98 people (28 in the main administration, 20 in the editorial department, and 50 in its printing plant), making it one of the biggest Chinese publishers.[37] The press boasted an exceptionally strong team of skilled, multilingual editors and translators, a virtual who-is-who of the leading translators from Russian in the 1940s and 1950s, many of whom would move into key positions in the PRC's cultural diplomacy.

Thanks to its expertise and deep pockets, Epoch quickly became the dominant player in the market for translations from Russian. In 1950 and 1951 alone, the press produced 170 titles, half of them in the category 'literature and art', a third in social sciences, and the rest comprising Russian-language textbooks and scientific and medical literature.[38] In a still chaotic marketplace where one and the same title was often translated by two or three translators for different publishers, it was usually the Epoch translation that stood out, with huge print runs that swamped bookstores. Thanks to its organisational and financial muscle, Epoch occupied a key position in this important segment of the Chinese book market. By the time of its handover to the Chinese government in December 1952, Epoch had published a total of 402 books.[39]

[34] The press announced these decisions in a rare letter to its readers, signed by the 'staff of the Epoch Press' (Shidai chubanshe tongren). 'Guanyu Shidai chubanshe de guoqu he jianglai' [The past and future of the Epoch Press], *Shidai* 19 (1949): 57f. Beginning from issue 20, Jiang Chunfang is named as both chief editor (*bianjiren*) and publisher (*faxingren*) of *Epoch*; hitherto, the journal had listed the Epoch Press (as editor) and Zakgeima (as publisher) in these positions.

[35] *Epokha* published its last issue in 1950. It is unclear what happened to Epoch's Soviet staff after the restructuring of the publisher in the early 1950s.

[36] Nicolai Volland, 'Cultural Entrepreneurship in the Twilight: The Shanghai Book Trade Association, 1945–1957', in *The Business of Culture: Cultural Entrepreneurs in China and Southeast Asia, 1900–1965*, eds. Christopher Rea and Nicolai Volland. Vancouver: UBC Press, 2014, pp. 234–58.

[37] SMA B1-1-1882.

[38] SMA B1-1-1882. A detailed list of titles is attached to this document.

[39] See Min, ' "Sushang" Shidai chubanshe yu *Shidai* zhoukan, *Shidai ribao*', 144.

Epoch's success and its centrality in the book market, paradoxically, made the publisher's reorganisation appear all the more urgent. Its foreign ownership, and the continuing influence of Soviet staff, restricted the Maoist state's ability to shape the image of the Soviet Union and the socialist world on its own terms. Even more so than national key industries, the media and the publishing industry were a highly sensitive field on which the CCP depended to define its image and its legitimacy. Even during the heyday of Sino-Soviet friendship the CCP thus found it difficult to accept market dominance of a foreign company in such an area, and consequently moved to bring Epoch under its control. In order to fully benefit from the opportunities created by the rapid expansion of cultural contacts, the press had to undergo fundamental changes.

First of all, Epoch received a capital injection from the Chinese side that turned it formally into a Sino-Soviet joint venture.[40] In this way, Chinese influence over the company was secured. Second, the CCP engineered the publisher's move from Shanghai to Beijing; in the national capital it would be under the closer control of the Chinese government. As conditions in the capital allowed, the press gradually moved its core staff to Beijing.[41] Next, Epoch had to change its business model. In accordance with the 'division of labour' policy that was implemented across the Chinese publishing sector since 1951, Epoch was forced to let go of its various retail outlets, which were taken over by the new government monopoly, Xinhua Bookstore. Epoch handed its state-of-the-art printing facilities in Beijing over to the Xinhua printing plant.[42] By early 1952, only its editorial department remained, its strongest asset, and Epoch had become a pure publisher.

The CCP then moved to restructure Epoch's editorial offices. Facing an acute shortage of skilled translators, the government transferred the press's best and most experienced personnel to other work-units, effectively bleeding the publisher of its senior personnel backbone. Rather than endorsing Epoch's successful business model, the CCP preferred to exploit its experience by infusing its core staff into the fledgling state-owned publishing houses directly under government control. Epoch's editors moved to work-units such as the newly founded People's Literature Press (Renmin wenxue chubanshe) and the Chinese Writers' Union. In 1952, Jiang Chunfang left Epoch to head the Propaganda Department's office in charge of translating Stalin's complete works.[43]

[40] At the same time, the Soviet government continued to fund Epoch. Jiang Chunfang confirms that the expenses for the new Beijing branch – offices, a printing plant, and employee housing – were covered by the Soviet Council of Ministers. See Jiang, 'Wo yu Shidai chubanshe', 15.

[41] Just like Epoch, all the major publishers – the Commercial Press, China Books, Kaiming, and Sanlian – relocated their headquarters from Shanghai to Beijing in the early 1950s, where they came under tighter Party control.

[42] Li, 'Shidai chubanshe zai Zhong-Su wenhua jiaoliu shi shang de diwei', p. 186; and Min, ' "Sushang" Shidai chubanshe yu *Shidai* zhoukan, *Shidai ribao*', 142.

[43] Jiang, 'Wo yu Shidai chubanshe', 15.

Epoch's periodical publications were dismantled as well. *Soviet Literature and Art* ended publication after the July 1949 issue, without further notice. The Russian language *Epokha* was quietly shut down in 1950. The press's flagship journal *Epoch* was at least allowed to say good-bye to its readers. In the editorial in its final issue, quoted at the beginning of this chapter, the editors recounted the journal's eventful lifetime, reinterpreting its legacy within the new political and cultural framework of the PRC. A cartoon captures the mood remarkably well (Figure 15.2). A huge pen with a razor-sharp tip has already finished off Hitler, Hideki Tojo, and Mussolini, and is now threatening a bomb-wielding Truman, the final incarnation of imperialism. Just like the Axis powers, the latter is literally on his last leg, stumbling in the face of the giant pen that will soon shut his mouth forever. The source of this powerful propaganda is, of course, a giant banner inscribed with *Epoch*'s name – in both Chinese and Russian. It was Sino-Soviet solidarity and, more concretely, the joint propaganda effort symbolised by the Epoch Press, that has led to the downfall of their former enemies. With a hint of defiance the cartoon suggests that *Epoch* is still needed, to fulfil the final task in the anti-imperialist struggle.

In December 1952, TASS formally handed the Epoch Press over to the Chinese government. According to Jiang Chunfang, the understanding was reached as part of a basket agreement that also mandated the transfer of the ports of Dalian to Chinese ownership. After the handover, the Sino-Soviet Friendship Association (Zhong-Su youhao xiehui) took charge of Epoch on behalf of General Press Administration (GPA).[44] The changes at both management and staff levels fundamentally altered the outlook of the Epoch Press, but they also presented new opportunities for Epoch. As the publishing arm of the Sino-Soviet Friendship Association, the Epoch Press was once again at the centre of Sino-Soviet cultural exchange. The changes, however, forced Epoch to alter its list. As I will detail below, the Epoch Press first reduced and then gave up its translations of literature, in order to focus on social science and current affairs publications, and eventually specialised in foreign language textbooks and reference works.

Epoch experienced a second 'golden age' in the mid-1950s, but ultimately lost out in further rounds of reorganisation of the Chinese publishing industry. The Sino-Soviet Friendship Association, its relevance in decline after 1956, was unable to shield the press. In December that year, Epoch was organisationally disbanded, and its remaining employees were transferred to other publishers. Most of them joined the reformed Commercial Press in 1958. Epoch's Shanghai branch became the Shanghai Foreign Literature Press (Shanghai yiwen chubanshe).[45] Despite this dismantling, however, the publisher's name did not disappear immediately. The CCP was acutely aware of the value of the Epoch imprint, its brand recognition and its association with Sino-Soviet friendship. Even after its abolition, several

[44] Information on the period after 1952 is scarce. The most detailed account of Epoch's final years is Wang Shouben, 'Ji Shidai chubanshe' [Remembering the Epoch Press], *Chuban shiliao* 2 (1989): 24–9.
[45] Li, 'Shidai chubanshe zai Zhong-Su wenhua jiaoliu shi shang de diwei', p. 186.

Figure 15.2 '*Shidai* finishes off its enemies' (cartoon). Source: *Shidai* 359 (1951): 15

publishers – especially the Commercial Press – used Epoch's name for their publications, banking on its cultural capital and its reputation for high quality translations of Russian materials.[46] In 1958, for instance, the Commercial Press used the Epoch imprint for the voluminous *Encyclopaedic Dictionary of the Soviet Union* (*Sulian baike cidian*). Epoch's legacy thus outlasted the publisher itself, if only by a few years. Epoch, once a symbol of Sino-Soviet cultural cooperation, had fulfilled its mission, and the Maoist state readied itself for ambitious new experiments.

[46] Wang, 'Ji Shidai chubanshe', 29.

Print and the making of Maoism

So how did Epoch approach its mission to shape the making of the Maoist state? After 1949, books became the most important segment of Epoch's output. After the shuttering of its journal publications, the publisher concentrated its energies on expanding its list, printing new or revised editions of titles published before the founding of the PRC, and embarking on numerous new projects. As Table 15.1 illustrates, Epoch's list changed considerably over the next decade, reflecting the departure of key personnel, the publisher's institutional affiliation, as well as shifts in government policy.[47]

The Epoch Press clearly flourished in the first years after the founding of the PRC. Relying on its team of skilled editors and translators and its financial prowess, Epoch quickly ramped up its output and published 54 titles in 1949, up from just 9 in the year before. In 1950, output rose to 133 titles, its peak in the early-PRC period. Translations of literary works comprised close to half of this figure. Epoch covered all genres, from classics of Russian fiction to socialist realist works, children's literature, popular biographies of Gorky and Leo Tolstoy, and theoretical treatises on Soviet literature. Epoch's access to Soviet sources and its understanding of the latest trends in Soviet letters allowed it to supply readers in New China with the full breadth of Soviet literary works and criticism.

The second largest category in Epoch's list is a broadly defined segment of books classified as social sciences. Titles such as *Consumer Cooperatives in the Soviet Union* (Sulian de xiaofei hezuoshe) or *The Banking System in the Soviet Union* (Sulian yinhang zhidu) were brief introductions of basic organisational and institutional aspects of life in the Soviet Union that acquired high relevance for China as the CCP began to transform the nation. Much of the specialised knowledge in the broader effort of 'learning from the Soviet Union' came from books and pamphlets of this kind that were translated and distributed by the Epoch Press, often in explicit support of emerging policies. Publication of the volume *Love, Marriage, and Family in the Soviet Union* (Sulian de lian'ai, hunyin yu jiating), for example, coincided with the promulgation of the new marriage law and hence attracted much interest, allowing Epoch to sell 40,000 copies. Other books, with titles such as *Hungary* (Xiongyali) and *Czechoslovakia* (Jiekesiluofake) introduced China's new partners in the socialist world. In these efforts, the press built on and expanded fields it had started exploring in the late 1940s.

The departure of Epoch's most experienced staff in 1951 led to a steep drop in its output that was felt across all categories. In 1952, the press managed to publish

[47] Figures quoted in this section have been calculated from *Wo guo fanyi chuban Sulian shuji mulu* for the period from 1949 to June 1954, and for the periods before and after from information listed by online used bookstores (the only available database that allows searches for publishers) and double-checked against the online catalogues of the China National Library, Peking University Library, and the Shanghai Library. The numbers for neither period are entirely complete or accurate, but they provide fairly reliable indicators of the larger trends and shifts in Epoch's output.

Table 15.1 Epoch Press publications, 1945–59

Year	Leaders' Writings [a]	Social Sciences (broadly)	Current Affairs [b]	Science & Technology [c]	Literature & Art	Soviet Encyclopaedia Extracts [c]	Others [d]	Total
1945	0	1	0	-	1	-	0	2
1946	0	0	1	-	2	-	0	3
1947	0	4	4	-	5	-	0	13
1948	0	3	2	-	3	-	1	9
1949	8	**11**	10	-	**21**	-	4	54
Sum	8	19	17	-	32	-	5	81
1949	0	**7**	-	2	**6**	0	0	15
1950	4	**42**	-	15	**62**	0	10	133
1951	0	**18**	-	6	**19**	0	3	46
1952	0	**6**	-	0	**23**	0	3	32
1953	0	**25**	-	0	**14**	0	9	48
1954	3	**69**	-	2	**60**	2	9	145
1955	1	**69**	-	0	32	4	**28**	134
1956	0	**49**	-	3	17	2	**35**	106
1957	1	13	-	1	5	0	**30**	50
1958	1	1	-	1	0	0	**26**	29
1959	0	0	-	0	1	0	2	3
Sum	10	299	-	30	239	8	155	741
Total	18	318	17	30	271	8	160	822

Notes: [a] Works of Marx, Engels, Lenin, Stalin, Mao; [b] only identifiable pre-1949; [c] only identifiable post-1949; [d] pre-1949 mostly medicine, after 1955 chiefly language textbooks and linguistics. Bold print identifies focus areas of Epoch's list.

a mere 32 titles, but soon regained the lost ground as its new owner, the Sino-Soviet Friendship Association, brought in new manpower. Epoch's output mix remained unchanged until the mid-1950s. The press continued to produce bestsellers, especially in the popular category of literature: Elena Il'ina's (1901–64) *The Fourth Degree* (Chetvertaia vysota; chin. Guliya de daolu, trans. Ren Rongrong), a novel for juvenile readers, sold over 700,000 copies in two years.

Epoch experienced a second peak in productivity during the mid-1950s. Between 1954 and 1956, the press produced more than 100 new titles annually, in addition to reprints of older editions. The surge in Epoch's activity, however, coincided with new government directives for the publishing industry that led to stark changes in the composition of the press's list. As the Sino-Soviet Friendship Association was not primarily involved in matters of literature and art, the industry regulator felt that translations of literary texts should be handled by specialised publishers, and the Epoch Press ceased translating new literary titles.[48] From 1955 on, its portfolio narrowed to just two fields, Russian-language textbooks and the social sciences, which emerged as the publisher's main area of activity. The Sino-Soviet Friendship Association sought to promote both popular and specialised knowledge about the Soviet Union and, apart from its own periodicals such as *Sino-Soviet Friendship* (Zhong-Su youhaobao), relied heavily on the Epoch Press for these purposes. Epoch issued books and pamphlets with titles such as *Science and Technology in the Soviet Union* (Sulian de kexue yu jishu) and *The Development of the Soviet Urban Economy* (Lun Sulian chengshi jingji de fazhan). Other volumes provided more general and impressionistic accounts of the contemporary Soviet Union and its people, such as *The Vast and Prosperous Soviet Union* (Di da wu bo de Sulian). But Epoch also published books addressing more specialised audiences, such as the military, with titles like *On the Moral Outlook of Soviet Soldiers* (Sulian junren de daode mianmao). What all these publications have in common is that they promote an image of the Soviet Union as a modern and industrialised society, a model for the new Maoist state. The Sino-Soviet Friendship Association was at the forefront of the campaign to 'learn from the Soviet Union', and the Epoch Press supplied the how-to-do manuals, the blueprints for the Maoist nation-in-the-making.

Conclusion

The Epoch Press, this chapter has argued, played a crucial role at the intersection of pan-Socialist diplomacy and transnational cultural flows, and contributed to the building of the Maoist state. As the main conduit of knowledge about all things Soviet, Epoch helped to construct the image of the contemporary Soviet Union that reached the Chinese public in the 1940s and the 1950s, and to reformulate modern China's international outlook. The very centrality of this question, the young PRC's

[48] On the directions for Epoch see Wang, 'Ji Shidai chubanshe', 29.

proper place within the new world order, however, meant that Epoch operated on contested terrain. Once culture became the battlefield where hearts and minds were won over or lost, once books and journals shaped the worldviews of the new socialist citizens, ultimate control over the institutions and processes of transnational cultural exchange became an issue of great urgency. Paradoxically, it was precisely in the realm of the transnational that national sovereignty was most at stake. Who should define the Soviet Union's image in China? How so, and by what means? The contentious search for answers to these questions resulted in the numerous twists and turns in Epoch's trajectory over the two decades of its existence – the publisher's transformation from a Soviet propaganda outlet to a joint venture, to a Chinese-operated entity within the apparatus of Sino-Soviet friendship, and eventually to a cog and wheel of the state-owned publishing industry. In light of socialist culture's very imbrication with power, however, it is all the more remarkable how much the publisher's institutional identity remained intact. Across multiple shifts in administrative affiliation and policy-mandated redrawings of its field of activity, Epoch exhibited resilience and adaptive agency, continuously redefining its approach without losing sight of its larger mission: to serve the goals of Sino-Soviet cultural understanding, and to help build the culture, institutions, and transnational ties that defined the Maoist state.

As this chapter has shown, Maoism itself is a heterogenous body of knowledge and institutional practices that grew organically if unevenly, in close interaction with a changing environment, and that drew from a range of sources, domestic as well as transnational. The Epoch Press and its complicated organisational history demonstrate the ebb and flow of agency and the contested cooperation of Chinese and Soviet players; it illustrates the shifting modalities of transnationalism. It also highlights the centrality of mid-level players in the making of Maoism: while the political and military leadership defined broad directions, it was up to intermediary institutions such as publishing houses to flesh out the meaning of the Sino-Soviet alliance, and more broadly, the shape of the new state. The case of the Epoch Press, lastly, sheds light on the contested histories and legacies of the early PRC's internationalism. With the Sino-Soviet break of 1960 and, even more so, the rise of a nationalist historiography since the 1990s, the role of transnational agents such as the Epoch Press has been all but written out of history. Whereas Chinese publishers such as the Commercial Press and China Books have enjoyed the limelight of critical and popular attention, players such as the Epoch Press are almost entirely forgotten. Epoch's massive and diverse output, however, is a stark reminder of the centrality of transnational culture and knowledge in the making of Maoism.

16

The Japanese and Korean Roots of Maoist Dance Education, 1951–1952

EMILY WILCOX

EDUCATION WAS AN important part of the communist revolution and the transformation of Chinese society in the twentieth century. It was also an issue of lasting personal interest to Mao Zedong himself, the graduate of a teacher training institute who in his first published essay, Rebecca Karl reminds us, 'wrote passionately on the importance of physical education'.[1] As with many aspects of Mao-era society and culture, scholars have tended to understand the construction of China's immediate post-1949 education system through the lens of learning from the Soviet Union.[2] Nicolai Volland's chapter in this volume demonstrates that the introduction of Soviet ideas and experience was a central component of the intellectual and cultural construction of New China during the 1950s. Nevertheless, it would be overly simplistic to conclude that Soviet knowledge was the sole source of international models and inspiration during this period. In her 1996 study *Radicalism and Education Reform in 20th-Century China: The Search for an Ideal Development Model*, Suzanne Pepper acknowledges this nuance. She writes, 'According to the official chronology, reorganisation was first proposed in mid-1950; the tertiary sector mobilised its forces to resist; thought reform intervened in 1951–1952; and by 1953, the nation's institutions of higher learning had been reconstructed along Soviet lines. *As usual, the actual progression of events was rather more complex than the official chronology suggested*' (emphasis added).[3] One aspect of this complexity that has not been thoroughly explored is the role of knowledge and practices from

[1] Rebecca E. Karl, *Mao Zedong and China in the Twentieth-Century World: A Concise History*. Durham: Duke University Press, 2010, pp. 10–11.

[2] See, for example, Theodore Hsi-en Chen, *Chinese Education since 1949: Academic and Revolutionary Models*. Elmsford: Pergamon Press, 1981; Suzanne Pepper, *Radicalism and Education Reform in 20th-Century China: The Search for an Ideal Development Model*. Cambridge: Cambridge University Press, 1996).

[3] Pepper, *Radicalism and Education Reform in 20th-Century China*, p. 164.

Proceedings of the British Academy, **267**, 311–330, © The British Academy 2025.

other Asian countries – specifically Japan and North Korea – in shaping China's post-1949 education system and, through it, Mao-era Chinese society and culture more broadly. I suggest that continuities in education and culture existed from the Japanese colonial period to the early People's Republic of China (PRC), sometimes filtered through Japanese colonial legacies in North Korea, that have previously been overshadowed by an emphasis on Soviet models. Moreover, I argue that these continuities were often progressive and reformist in nature – at times more so than the Soviet models being introduced at the time – and that even if short-lived or evanescent, they nevertheless contributed meaningfully to the construction of Maoism in the form of revolutionary educational and cultural experiments in Mao-era China.

The case study I examine here is one such experiment in dance education that lasted from March 1951 through the summer or autumn of 1952 and was carried out at China's then newly established national theatre school, the Central Academy of Drama (*Zhongyang xiju xueyuan*, CAD) in Beijing. According to Pepper's 'official chronology', 1951–2 was a time of contestation and thought reform over educational reorganisation before China's higher education was 'reconstructed along Soviet lines' in 1953. This period from 1950 to 1953 has been identified as a lost era needing more attention in examinations of PRC history. In their 2007 book *Dilemmas of Victory: the Early Years of the People's Republic of China*, for example, Jeremy Brown and Paul Pickowicz observe, 'Ever since 1953, when the First Five-Year Plan signalled the close of Mao Zedong's experiment with New Democracy and ushered in the beginning of a transition to socialism, China's early 1950s period has disappeared from the radar screens of successive waves of observers'.[4] In my 2018 book *Revolutionary Bodies: Chinese Dance and the Socialist Legacy*, I respond to Brown and Pickowicz's call and devote an entire chapter to dance developments in 1950–3. In so doing, I reveal a rich period of contestation in China's dance field, as well as the contributions of dance experts of diverse Asian heritages with a wide range of international training, from Trinidad and London to Tokyo, Tashkent, Moscow, and Pyongyang. I show that all of these dance artists laid important foundations for China's dance education prior to the establishment of China's first national dance conservatory, the Beijing Dance School (*Beijing wudao xuexiao*, BDS), under Soviet guidance in 1954.[5] This earlier period, I demonstrate, was one of much more openness and possibility than previously understood. Citing a spring 1950 report on China's first national dance company, established as a student ensemble within CAD in late 1949, I write, '"The new Chinese dance still needs to be created," the report explained, "In the dance troupe, everything is still being explored and tested."'[6] In 1950, what a 'Maoist' system of dance education

[4] Jeremy Brown and Paul Pickowicz (eds), *Dilemmas of Victory: the Early Years of the People's Republic of China*. Cambridge, MA: Harvard University Press, 2007, p. 1.

[5] Emily Wilcox, *Revolutionary Bodies: Chinese Dance and the Socialist Legacy*. Oakland: University of California Press, 2018, chapter 2.

[6] Central Academy of Drama News Group, 'Zhongyang xiju xueyuan san ge wengongtuan', *Guangming ribao*, 23 May 1950, cited in Wilcox, *Revolutionary Bodies*, pp. 50–1.

and cultural construction should look like was still being determined. Soviet models were not the only ones being studied, and what would result ultimately was by no means a foregone conclusion.

It was in this context of exploration, testing, and an infusion of international experience brought by artists of Asian descent trained in places around the world that in 1951 CAD established what is now recognised as the first state-supported effort to systematically train dance professionals in the PRC. It consisted of two programmes: the Dance Movement Cadre Training Class (*Wudao yundong ganbu xunlian ban*) led by a Japan-trained dancer from China Wu Xiaobang (1906–95); and the Choe Seung-hui Dance Research Course (*Cui Chengxi wudao yanjiu ban*) led by a Japan-trained dancer from the Korean peninsula Choe Seung-hui (Choi Seung-hee, 1911–69). The two classes – abbreviated '*Wuyunban*' (Dance Movement Class) and '*Wuyanban*' (Dance Research Class), respectively – ran simultaneously from March 1951 until the summer of 1952, with offshoots that trailed variously through October 1952 (for Choe's group) and June 1953 (for Wu's group). The approximately 155 Chinese students who participated in these two programmes were recruited from performance ensembles and schools across the country, and most returned to their original work-units and emerged as leaders who spread and implemented what they learned at local and national levels. In this way, the course informed, in a fundamental manner, how dance work was formulated and carried out through the rest of the Mao era and beyond.[7]

Several features of the 1951–2 CAD courses make them demonstrably different from the approach taken to dance education established later at BDS under Soviet guidance in the mid-1950s. First, both CAD courses were led by dancers whose artistic lineages traced to Japanese modern dance – or, as it was known at the time 'New Dance' – rather than to Russian and Soviet ballet. Wu Xiaobang received his early dance training in Japan from 1929, first with Takada Seiko (formerly Hara Seiko, 1895–1977) and later with Eguchi Takaya (1900–77). Choe Seung-hui received her early dance training in Japan from 1926 with Ishii Baku (1886–1962). Moreover, these artistic lineages were reflected in the curricula of the two CAD programmes, which both included courses in modern dance technique. Thus, although modern dance was taught explicitly in the two CAD courses, it was excluded from the Beijing Dance School curriculum established in the mid-1950s on ideological grounds, a practice that continued until the 1980s.[8] Second, the CAD courses differed from later BDS programmes in their integration of dance theory, creative methods, research, and social engagement as core components of

[7] For a detailed overview of the programme, see Tian Jing and Li Baicheng (eds), *Xin Zhongguo wudao yishujia de yaolan*. Beijing: Zhongguo wenlian chubanshe, 2005.

[8] Emily Wilcox, 'When Place Matters: Provincializing the "Global"', in Larraine Nicholas and Geraldine Morris (eds), *Rethinking Dance History: Issues and Methodologies*, 2nd ed. Abingdon and New York: Routledge, 2018, pp. 160–72. For a thorough investigation of modern dance in China, including its suppression from the mid-1950s through the early 1980s, see Nan Ma, *When Words are Inadequate: Modern Dance and Transnationalism in China*. New York: Oxford University Press, 2023.

the teaching content. The CAD courses presented dance as a multi-faceted artistic endeavour and fostered students' abilities in a variety of skills, of which movement technique was one of many. In line with this approach, despite the short length of the CAD programme, students were exposed to a large range of different dance forms and approaches to making dance that continued to inform their later careers.

In *Dilemmas of Victory*, a major focus in exploring the early 1950s was issues of governance – how the Chinese Communist Party (CCP) established and maintained its authority, how new policies were disseminated, implemented, and/or resisted, and how groups potentially threatening to the new regime were integrated, neutralised, or eliminated. The question of arts education, as I pursue it here, shifts our attention away from governance and toward the creation of new resources, specifically, human resources. Education is fundamentally a project of *making* – it consists of moulding and shaping human beings into people with particular sets of skills, ideas, knowledge, and aspirations. In the case of dance education, the entire field of Maoist dance production would not have been possible without these trained individuals, who were necessary to create and perform the many dance productions that became such a fundamental component of the cultural landscape of China during and after the Mao era. What were the types of people and skillsets the CCP deemed important for the new society and culture it was seeking to create? How did educational institutions go about cultivating such people and through what kinds of methods? In the case of the CAD programme, it seems clear that two goals were central: dancers in New China should be able to combine theory and practice; and they should learn through a combination of classroom training and social engagement, whether service work in factories or staged public performances of new choreography. Beyond the impressive range of dance styles included in the CAD curriculum, what is notable about Wu's and Choe's programmes from the perspective of pedagogy is the ways in which they challenged dancers to engage in creative reflection and put their knowledge directly into practice. In this early programme, dancers were conceived of both as technicians and thinkers, performers as well as active members of society. These lessons were key to Maoist educational theory and prepared the first generation of dancers in the PRC to do the immediate work for which they were needed – to build the dance field of Mao's China.

The CAD dance training courses of 1951–2 offer a useful case study for exploring the early development of Maoist arts education. Unlike other fields in the performing arts such as music or theatre, dance was not well developed as an independent art form in China before the late 1940s, and most of the dance schools that did exist were run by foreigners.[9] Thus, when cultural planners in the early PRC decided to make dance an important part of China's new culture, there were few local precedents they could rely on for the critical task of training skilled practitioners for this new field. The courses that emerged resulted from deliberate

[9] Wilcox, *Revolutionary Bodies*.

planning and choice. Consequently, they reflect the vision and ideals of the early-PRC cultural leadership. More specifically, these early educational experiments implemented at the start of the 1950s before the mass arrival of Soviet experts demonstrate the concrete ways in which PRC leaders initially translated Mao's theoretical guidelines on the development of new arts and culture into daily practice and institutional structures at the national level.

What appears in the CAD dance programme, I contend, is a regime that placed a high value on what we today might consider humanistic or liberal arts values in higher education. These include the development of the whole person; exposure to and familiarity with international art developments; knowledge of and respect for China's traditional cultural practices and communities; and an emphasis on creative ability, the application of knowledge in practice, and making art relevant to wider society. In fact, these general principles are all evident in Maoist cultural doctrine as early as 1942, when Mao Zedong gave his influential 'Talks at the Yan'an Form of Literature and the Arts', thereby setting the agenda for cultural policy under his leadership for decades to come.[10] In these 1942 'Talks', Mao emphasised issues such as the importance of personal transformation through education, applying theoretical knowledge in practice, and deriving theoretical knowledge from practice, combining international knowledge with a respect for and commitment to traditional culture, and encouraging creative activities that are socially engaged through direct interactions with diverse communities.[11] In this sense, the dance education model implemented at CAD was in line with Mao's early writings on arts and culture, and it provides an alternative understanding of what aspects were valued in a 'Maoist' system of dance education beyond the traditional focus on the technical vocational middle school programme initially instituted at BDS. Exploring the CAD dance training courses provides insight into how arts educators in the early PRC embodied these ideas in practice to train a new generation of artists and cultural workers in a range of skills, technical, theoretical, social, and creative.

Theoretical foundations: Bringing 'New Dance' from Japan and Korea to China

Wu Xiaobang and Choe Seung-hui were strikingly different as individuals and in their approach to dance performance, teaching, and choreography. Wu was a male Chinese intellectual who made his name writing and lecturing on dance theory and creating dances on contemporary political themes, most famously depicting

[10] For a detailed analysis of how Mao's educational vision and policies instituted in Yan'an informed educational reform in China after 1949, see Pepper, *Radicalism and Education Reform in 20th-Century China.*

[11] Mao Zedong, 'Talks at the Yan'an Forum on Literature and Art', in *Selected Works of Mao Zedong*, Vol. 3, May 1942, accessed 28 January 2017, www.marxists.org.

soldiers during the Chinese Civil War (1946–9).[12] He was well connected in the People's Liberation Army (PLA) cultural work troupe system and had spent some time in Yan'an and the CCP-occupied northeast before emerging as a key leader in the PRC dance field after 1949.[13] Choe Seung-hui, on the other hand, was a female Korean stage star who became famous in Asia and abroad for her beauty and her modernised, glamorous renditions of traditional Asian imagery, such as Buddhist statues and classical heroines. By the late 1940s, Choe had settled in North Korea, where she became the leading architect of North Korean dance and was familiar to both PRC and Democratic People's Republic of Korea (DPRK) state leaders through her role in high-level diplomatic events.[14] However, Choe's main artistic network in China was among traditional theatre actors such as Mei Lanfang, whom she had befriended and worked with during the Japanese occupation era.[15]

Despite their differences, Wu and Choe shared some qualities in common that left a mark on the CAD dance training programmes. First, both were highly respected as artists and leaders in their communities, and both had significant experience outside China. Both Wu and Choe had studied and performed in Japan, and Choe had also toured in Korea, as well as across Europe, and North and South America.[16] Second, both distinguished themselves as innovative and inspiring teachers. Wu spent years lecturing and giving workshops at theatre schools and performing arts ensembles in China, while Choe had set up her own dance research institutes and schools in China and Korea. Finally, both Wu and Choe were known for their highly analytical approaches to dance technique and composition, as well as their development of deliberate systems of dance pedagogy grounded in rigorous theoretical and practical foundations. Significantly, both Wu and Choe treated dance as research, and they excelled at conveying passion for dance while also building its connections to other cultural, social, and political endeavours. The selection of Wu and Choe to lead the CAD dance training programmes, as well as the design and content of their curricula, thus ensured that students received a dance education that was at once internationalised, comprehensive, and intellectually substantive. To better understand their teaching methods, it is helpful to delve further into Wu and Choe's dance philosophies and their connection to dance in Japan in the pre-1949 period.

[12] Feng Shuangbai, *Xin Zhongguo wudao shi 1949–2000*. Changsha: Hunan meishu chubanshe, 2002.
[13] Emily Wilcox, *Revolutionary Bodies*.
[14] Suzy Kim, 'Choe Seung-hui Between Ballet and Folk: Aesthetics of National Form and Socialist Content in North Korea' in Katherine Mezur and Emily Wilcox (eds), *Corporeal Politics: Dancing East Asia*. Ann Arbor: University of Michigan Press, 2020, pp. 203–22.
[15] For a detailed discussion of Choe's career in China during the 1940s and how it informed her later work in China, see Emily Wilcox, 'Crossing Over: Choe Seung-hui's Pan-Asianism in Revolutionary Time', *The Journal of Society for Dance Documentation and History*, 51 (December 2018), 65–97.
[16] Li Aishun, 'Cui Chengxi yu Zhongguo wudao', *Beijing wudao xueyuan xuebao* 4 (2005), 16–22 at 17; Tian and Li, *Xin Zhongguo wudao*.

Wu Xiaobang was born in Jiangsu Province and received his secondary education in China before he ventured to Japan in 1929 to study European classical music. Shortly after arriving in Tokyo, Wu adopted the given name 'Xiaobang' after his favourite composer, Chopin. However, after seeing a stirring modern dance performance called *A Group of Ghosts* at Waseda University, Wu was inspired to begin studying dance. Between 1929 and 1936, Wu studied on and off with female Japanese dancer Takada Seiko, where Wu was exposed to a style of ballet influenced by the early American modern dance of Isadora Duncan. In the summer of 1936, Wu attended a three-week course taught by male Japanese dancer Eguchi Takaya, a former student of Takada who had recently returned from studying with Mary Wigman in Germany.[17] Nan Ma describes the training Wu received from Eguchi as follows:

> [T]he founders of the German School strove to theorize and systemize the 'natural law' of bodily movement and the general method of composing dance works from scratch ... The 'natural law' and the methodology of dance composition were the foci of the three-week course ... Eguchi's training emphasized the movements of every [body] part. Wu was not used to it at first, and felt pain all over his body, especially in his chest and abdomen, which were seldom exercised in ballet training.
> A unique feature of Eguchi's training methodology was an independent session called 'the theory and techniques for the practice of composition,' in which students were required to create their own dance movements according to the 'natural law' ... Unlike other dance training systems, which usually force students to practice the existing formal vocabulary, codes, and repertoire for many years before they are permitted to choreograph their own dances, Eguchi's method, inherited from that of Wigman, allowed each student, at the very beginning, to be an 'inventor' who creates new movements for each dance work based on personal life experience and inner inspiration ... .[18]

After Wu returned permanently from Japan in late 1936, he worked tirelessly to bring the 'New Dance' he had learned in Japan to China.[19] Wu initially used the Chinese translation '*xinxing wuyong*' for New Dance, maintaining the Japanese neologism '*buyō*' (pronounced in Chinese as *wuyong*), rather than the more common Chinese term for dance, '*wudao*'. In an interview introducing New Dance in 1937, Wu traced the origins of New Dance to Isadora Duncan, Rudolf von Laban, and Mary Wigman and described it as a dance form that 'places importance on expression, technique, and consciousness and is close to the lives of the masses'.[20] By 1939, in the midst of the Second Sino-Japanese War, Wu was

[17] Wu Xiaobang, *Wo de wudao yishu shengya*. Beijing: Zhongguo xuju chubanshe, 1982; Nan Ma, 'Transmediating Kinesthesia: Wu Xiaobang and Modern Dance in China, 1929–1939', *Modern Chinese Literature and Culture* 28, 1 (2016), 129–73.

[18] Ma, 'Transmediating Kinesthesia', pp. 138–9.

[19] This term 'New Dance' was most likely based on the German '*Neuer Tanz*'.

[20] 'Wu Xiaobang tan xinxing wuyong', *Diangying zhoukan* (Shanghai) 29 (1937), 1264. All translations from Chinese are mine unless otherwise noted.

advocating New Dance as a tool for energising Chinese society and promoting nation-building.[21] He had also developed a curriculum based on the 'natural principle' (translated into Chinese as *ziran faze*) that he was using to train Chinese theatre actors.[22] Over the next 10 years, Wu travelled across war-torn China promoting New Dance in CCP base areas, among progressive youth and artists, and in the military.[23]

In 1949, just two years before the opening of his CAD course, Wu published his first book on New Dance theory, *Introductory Course in New Dance Art*. Here, the earlier Japanese-derived term *xinxing wuyong* is replaced with the more localised *xin wudao yishu* (New Dance art). Yet, the content remains closely connected to the German-inspired 'natural principle' approaches Wu had learned in Japan. The book opens with a preamble about how life, especially the life of labour, is the ultimate source of New Dance movements and that New Dance must express the life of the masses, a clear nod to Maoist art theory of the time.[24] Wu also emphasises that New Dance is based on 'scientific principles' (*kexue faze*) and requires the activation of body and mind together. Proposing an analogy of using both hands, Wu writes, 'it is only by being able to use the mind (have thought) that one will be able to make both hands perform to their fullest ability—and make technique continue to develop'.[25] Wu then walks the reader through his movement theory, which he describes as being grounded in 'the natural principle of human body movement, or what is also called the scientific principle'.[26] Wu argues that dance movement should be understood in terms of what he calls 'the three major joints' (shoulders/chest, waist, and knees), that each action begins with a counteraction, that all movement originates in the alternation of relaxation and straightening of the joints, that there is a sequential relationship between movements in the 'primary joints' and the 'secondary joints', that New Dance advances a particular approach to beauty that is grounded in the life of the workers and peasants, principles for the development, rhythm, and unity of movement, and so on.[27] Wu then goes on to detail a series of dance exercises to practice these principles, starting with movements to relax and straighten each joint, followed by sequences of steps, jumps, and turns, and etc. This section is heavily illustrated with diagrams showing a barefoot male dancer wearing only shorts executing the different exercises, with dotted lines and arrows to indicate pathways and directions of movement.[28] Finally, Wu outlines rules for composition. These range from specific instructions such as 'always move on the diagonal' to strategies for finding inspiration through guided reflection on personal

[21] Wu Xiaobang, 'Wo weishenme dao Yiyoushe', *Yiyou* 2 (1939), 4–5 and 18.
[22] Wu Xiaobang, 'Wuyong mantan', *Xiju zazhi* 2, 4 (1939), 152–4.
[23] Wu Xiaobang, *Wo de wudao.*
[24] Wu Xiaobang, *Xin wudao yishu chubu jiaocheng*. Wuhan: Huazhong Xinhua shudian, 1949, pp. 1–4.
[25] Wu Xiaobang, *Xin wudao yishu,* p. 4.
[26] Wu Xiaobang, *Xin wudao yishu,* p. 4.
[27] Wu Xiaobang, *Xin wudao yishu,* pp. 4–15.
[28] Wu Xiaobang, *Xin wudao yishu,* pp. 16–47.

memories, music, images, etc.[29] The book ends with a short discussion on how to choreograph for different size groups, from duets and trios to mass dances.[30]

Choe Seung-hui was born to a declining *yangban* family in Japanese-occupied Seoul and moved to Japan in 1926 after seeing a performance by Japanese modern dancer Ishii Baku. Choe's early career embodied the nexus of two key historical developments: one, the engagement with Euro-American stage dance in Japan and the space it opened up for new, 'modern' ways of presenting women's bodies; and, two, the emergence of Japanese imperialism and its politics of ethnic and national representation, including Pan-Asianism.[31] Reflecting these two related but divergent phenomena, Choe's performances ranged from choreographies closely aligned with early US and German modern dance aesthetics to those that explicitly repurposed materials from existing Asian performance practices. These latter works became the basis for contemporary Korean and Chinese folk and classical dance forms that Choe herself helped devise in both North Korea and China.[32]

Ishii Baku, Choe's first dance teacher, was part of the inaugural cohort of Japanese dancers trained in ballet by Italian teacher Giovanni Vittorio Rossi at the Imperial Theater in Tokyo beginning in 1911. Wu Xiaobang's first teacher, Takada Seiko, had also been a member of this cohort. A few years into his study, Ishii rejected ballet and pursued what he called 'creative dance' or 'lyric dance', inspired by collaborations with US- and Europe-trained Japanese artists working in music and theatre. Like Wu's teacher Eguchi, Ishii also spent time in Europe and was influenced by German dancer Mary Wigman, as well as Swedish music educator Émile Jaques-Dalcroze, who developed the Dalcroze eurhythmics method for expressing music through movement. Choe began studying with Ishii in 1926, just two years after he returned from his travels abroad and established his own studio, the Ishii Baku Dance Research Center on the outskirts of Tokyo.[33]

[29] Wu Xiaobang, *Xin wudao yishu,* pp. 47–68.

[30] Wu Xiaobang, *Xin wudao yishu,* pp. 68–74.

[31] There is extensive scholarship on Choe's career in the Japanese colonial period and the implications of colonial ideology on her dance choreography and career. See, for example, Judy Van Zile, *Perspectives on Korean Dance*. Middletown: Wesleyan University Press, 2001; Sang Mi Park, 'The Making of a Cultural Icon for the Japanese Empire: Choe Seung-hui's U.S. Dance Tours and "New Asian Culture" in the 1930s and 1940s', *positions: asia critique* 14, 3 (2004), 597–632; E. Taylor Atkins, *Primitive Selves: Koreana in the Japanese Colonial Gaze, 1910–1945*. Berkeley: University of California Press, 2006; Young-Hoon Kim, 'Border Crossing: Choe Seung-hui's Life and the Modern Experience', *Korea Journal* (Spring 2006), 170–97; Judy Van Zile, 'Performing Modernity in Korea: The Dance of Ch'oe Sung-hui', *Korean Studies* 37 (2013), 124–49; Faye Yuan Kleeman, *In Transit: The Formation of the Colonial East Asian Cultural Sphere*. Honolulu: University of Hawai'i Press, 2014; Wilcox, 'Crossing Over'.

[32] Takashima Yusaburo and Chong Pyong-ho, *Seiki no bijin buyoka Sai Shoki*. Japan: MT Publishing Company, 1994; Won Jongsun (dir), *Choe Seung-hui: The Story of a Dancer*. Seoul: Arirang TV, 2008); Emily Wilcox, *Revolutionary Bodies*; Emily Wilcox, 'Locating Performance'; Choe Seung-hui, 'East Asian Modernisms, and the Case for Area Knowledge in Dance Studies', in Susan Manning, Janice Ross, and Rebecca Schneider (eds), *The Futures of Dance Studies*. Madison: University of Wisconsin Press, 2019, pp. 505–22; Wilcox, 'Crossing Over'; Kim, 'Choe Seung-hui Between Ballet and Folk'.

[33] Kleeman, *In Transit*, pp. 191–5.

Unlike Wu, who studied and performed a few times in Japan but developed his professional performance career mainly in China, Choe had a very successful career as a professional dancer and even became a celebrity and household name in Japan and Korea before she arrived in China. Choe's first performances in China, initially as part of Japanese state-sponsored 'comfort tours' for imperial army soldiers, occurred in the early 1940s. Recurring themes in Chinese accounts about Choe include the intellectual sophistication of Choe's performances, her effectiveness as a teacher, and her role as a visionary for the future development of dance in China. In a 1943 review, for example, Chinese critic Liu Junsheng wrote that what distinguished Choe's dances from those of other performers in China at the time was that they provided not only sensorial pleasure but also 'intellectual inspiration' (*zhishi shang de qifa*).[34] A work that embodied this for Liu was Choe's *Samantabhadra Bodhisattva*, which 'makes people feel a sublimely beautiful sculptural feeling, presenting the beauty of the meanings of Buddhist art … realizing the central consciousness of dance art'.[35] Two others were *Song of Wang Zhaojun* (Mingfei qu), which explored the complex emotions of this legendary Chinese woman at the moment she was forced to leave her homeland to marry a foreign king, and *Scattered Tunes* (San diao), a dance in Choson costume and set to Choson classical music that explores a young woman's sexual awakening. Liu noted the skilful use of rhythm in Choe's dances as well as the high level of her students, three of whom also performed in the show. According to Liu, the students 'show long training and demonstrate that Choe's dance has already achieved the status of a sect/school (*zongpai*)'.[36] Liu ends by commending Choe's plans for developing dance in China, which Choe shared at a 1943 symposium in Shanghai. 'During the symposium, Choe expressed herself, and we know that she not only has deep knowledge of dance technique but also has specialised research on theory. In her dances, she advocates "getting rid of all unnecessary elements, making it as simple as possible, and only putting forth the most important materials to make it complete." This is the highest principle of modern art'.[37]

This minimalist drive was just one part of Choe's vision for what she called the new Chinese dance art. Equally important, according to Choe, was the use of revised traditional materials as the foundation for this new form. In the published Chinese report on the 1943 symposium, the following passages offer some insight into Choe's motivations and plan:

> When I was in Beiping, I saw performances by a famous Kunqu performer Han Shichang. I felt that many of the movements already were quite close to Western dance art. If we get rid of the excess and keep the essential, deleting some unnecessary scenes and excitement, certainly we can create from Chinese theatre a kind of new Chinese dance art. … So, we can use new techniques to make ancient heritage

[34] Liu Junsheng, 'Lun Cui Chengxi de wudao', *Zazhi* 12, 2 (1943), 26–30 at 27.
[35] Liu, 'Lun Cui Chengxi', 28.
[36] Liu, 'Lun Cui Chengxi', 29.
[37] Liu, 'Lun Cui Chengxi', 30.

> climb the stage, make it advance and become a modern thing. Over the past 10 years, I've been putting great effort into this, taking what is good in the old heritage of Japan and Korea and struggling to add new creation, making it become a new form of dance. China has youth who are committed to Chinese dance art. If they can work hard at this, why not open a new path, founding a new Chinese dance art? … In the past the East was in the conservative period of art. Masters transmitted art to their disciples, and disciples didn't dare use their own brains to create and make it improve and become greater. However, now is the time to construct. As an artist, one should discover Eastern dance art heritage and create it into something new.[38]

Choe took a major step toward implementing her plan when she opened her first dance school in China just over a year later – the Oriental Dance Research Institute (*Dongfang wuyong yanjiusuo*) in central Beiping. As explained in reports and interviews published in China at the time, Choe opened this institute with the express purpose of teaching Chinese students and developing New Dance styles by applying her dance research and compositional methods to traditional Chinese theatre and other East Asian materials.[39] During this time, Choe studied with leading Peking opera actors including Mei Lanfang, Shang Xiaoyun, Zhang Junqiu, Ye Shenglan, and Ma Lianliang, and she learned to perform some Peking opera scenes as part of her dance research.[40] This was similar to what she had done studying with Korean folk performers such as Han Sŏng-jun in Japan and Korea to create her Korean-style dance choreographies during the early 1930s.[41] The Oriental Dance Research Institute lasted for only about a year because the Japanese surrender changed Choe's political status in China, and she eventually returned to Korea, first travelling to her hometown of Seoul and then settling in Pyongyang in the spring of 1946. Choe's move to Pyongyang likely had several motivations, including the criticism she received in Seoul because of her collaboration with the Japanese government, the fact that her husband An Mak was a leftist writer who had already settled in Pyongyang, and Kim Il-sung's welcoming attitude at the time toward artists and intellectuals.[42] Throughout the late 1940s and 1950s, Choe's dance research received full support of the North Korean government. Her writings, choreography, and teaching materials during this period reflected the ideals of socialist realism, including its emphasis on representing everyday life and folk culture. She also continued to tour internationally and showed her work across the socialist bloc.[43]

Choe travelled again to China in December 1949, now as a representative of the DPRK. She attended the Asia Women's Conference in Beijing and gave performances with her dance company, which were attended by top PRC leaders,

[38] 'Cui Chengxi wudao zuotanhui', Zazhi 12, 2 (1943), 33–8.
[39] Luo Chuan, 'Cui Chengxi er ci lai Hu ji', *Zazhi* 15, 2 (1945), 84–8.
[40] Wilcox, 'Crossing Over'.
[41] Van Zile, *Perspectives*; Van Zile, 'Performing'.
[42] Won, *Choe Seung-hui.*
[43] Kim, 'Choe Seung-hui between Ballet and Folk'. See also Suzy Kim, *Among Women across Worlds: North Korea in the Global Cold War*. Ithaca: Cornell University Press, 2023, chapter 5.

including members of the Ministry of Culture.[44] After the eruption of the Korean War, an agreement between the PRC Foreign Ministry and the DPRK government allowed Choe to relocate to Beijing in late 1950 with her Korean students. This created the conditions for Choe's special dance training course at CAD the following year.[45]

Cultivating dancers for New China: Study and life in the *Wuyunban* and *Wuyanban*

Wuyunban and *Wuyanban* were both launched at CAD in mid-March 1951, and both were designed to train a cohort of dance leaders to develop the dance field of the new PRC. Nevertheless, the two classes, each led by a different teacher, were separate programmes with distinct institutional designs and their own courses, extracurricular activities, staff, and students. *Wuyunban*, led by Wu, had a total of 68 students, including 39 women and 29 men, all of whom were from China. *Wuyanban*, led by Choe, had 120 students, of whom 84 were from China and 36 from North Korea.[46] Of the 84 Chinese students in *Wuyanban*, 27 were students of Korean nationality from Yanji, a city in Jilin Province near the border with North Korea. The rules of admission were the same for both classes: prospective students were to be between the ages of 17 and 26 and to have either worked for at least one year as employed dance cadres (for the Chinese students, this typically meant being a member of a national, provincial, local, or military cultural work troupe or other performance ensemble) or be students with a high school education level. Prospective students also had to pass an entrance exam. Recruitment notices for both *Wuyunban* and *Wuyanban* were sent out on 15 January 1951, to 25 cultural work troupes across the country encouraging them to send qualified representatives to Beijing and Tianjin for entrance exams in late February. After completion of their respective CAD course, students who had been dance cadres before were to return to their original work-units to contribute to the further development of dance in their home institutions. Students who were recruited from high schools and universities were to be appointed to dance-related work upon completion of the course. While participating in the CAD courses, students received stipends the equivalent of 135 jin (62.5kg) of millet per month for working cadres and 115 jin (57.5kg) of millet per month for high school and university students. Classes met in Mianhua

[44] Chen Ji, 'Chaoxian funü daibiaotuan fangwen ji', *Renmin ribao* (17 December 1949); 'Wenhua bu, quanguo wenlian jüxing wanhui "Cui Chengxi biaoyan Chaoxian wuyong"', *Guangming ribao* (13 December 1949).

[45] Tian and Li, *Xin Zhongguo wudao.*

[46] The precise number of students in each course varies slightly in different accounts. These numbers are based on those reported in the Li and Tian (2005) volume, which was compiled by course alumni based on records held in CAD archives and other historical sources. The volume also includes the name of each student who participated.

Hutong in central Beijing, where the CAD campus is still located today and just a short walk from where Choe had established her Oriental Dance Research Institute six years earlier.[47]

Opening ceremonies for *Wuyunban* were held on 12 March 1951, and for *Wuyanban* on 17 March 1951. As reported in *Guangming Daily*, the *Wuyanban* opening ceremony featured comments by CAD president Ouyang Yuqian, PRC Ministry of Culture Director Zhou Yang, Choe Seung-hui, Mei Lanfang, a representative from the North Korean Embassy, and the Chinese and Korean class representatives. After their comments, several of the Korean teachers and Chinese and Korean students performed basic movements of the four major styles of dance that would be taught in the course: Chinese dance, Korean dance, Southern dance (*nanfang wu*, a dance style Choe created based on South and Southeast Asian materials), and New Dance (*xinxing wu*, Choe's version of modern dance).[48] Apart from these four main styles, the article explained, *Wuyanban* students would receive a comprehensive education as follows:

> The Research Course curriculum, in addition to theory courses in politics, literature, music, composition methods, and introduction to dance history, will include dance training classes mainly focusing on the basic movements of Chinese dance and Korean dance, so that when students graduated they will have a mastery of the introductory foundations of national dance (*minzu wudao*), which will serve as the basis for further research into national dance. Additionally, there will be classes in southern dance, which is filled with the national styles of the Orient, the progressive country USSR dance, ballet with its elegant body and posture expressions, new dance that exercises boldness and control of the body's center of gravity, as well as dance rhythm training. These classes will make students gain foundational training in various dance basics and will serve as guidance for national dance choreography.[49]

The *Wuyunban* opening ceremony was not reported in the major newspapers. However, *Dance News*, a bulletin for the dance community that was edited by Wu Xiaobang, included a very brief story about the course in its inaugural issue released on 7 July 1951. The full story reads:

> In March of this year, the Central Academy of Drama set up the Dance Department Dance Movement Course. The establishment of this course will help greatly with the launching of the nation-wide dance movement, because every region is currently in urgent need of dance cadres. Dance movement work must be connected to real circumstances and elevate a step starting from the foundation of popularization. Therefore, most of the students in the Dance Movement Course have been sent by cultural work troupes in different locations across the country. These cadres transferred for training possess practical work experience and have been cultivated from within the movement. Another smaller group are educated youth who have tested in from Beijing

[47] Tian and Li, *Xin Zhongguo wudao*, pp. 9–11, 283–5.

[48] Fang Ming, 'Peiyang Zhong Chao wudao gongzuo ganbu Cui Chengxi wudao yanjiuban chengli', *Guangming ribao* (20 March 1951).

[49] Fang, 'Peiyang'.

> and Tianjin. The entire class has 57 students[50], and the classes mainly focus on theory and composition, taught by Department Chair Wu Xiaobang. Technique classes include three types: basic training for dance, Chinese traditional opera (*xiqu*) dance, and Soviet ballet. Additionally, there are classes in dance notation, music, literature, fine art, literature and art thought, and political theory. The study period is one year. Within two months of starting, students begin to engage in composition and practical training. This causes the technique training to be digested into creative practice and be united with reality. The principle is to connect with the needs of the movement. Under the leadership of Chairman Mao's direction for literature and the arts, after one year, on the wide path of constructing the new dance movement, they should become a group of powerful cadres.[51]

The teaching curricula and respective foci of the two courses reflected in these contemporary accounts align with the descriptions provided in retrospective materials, including a 641-page collection published by alumni of the two courses in 2005.[52] In all of these materials, it is clear that both *Wuyunban* and *Wuyanban* placed emphasis on breadth in their curricula, both in terms of balancing 'technique' and 'theory' and in teaching a wide variety of dance styles. Additionally, both courses incorporated significant opportunities for students to quickly put their newly acquired knowledge into practice. In the case of Choe's *Wuyanban*, students engaged in extensive public performances and touring with Choe and their other teachers during their time in the course, and this gave them the opportunity to learn critical skills of stage performance, as well as other technical and artistic considerations when bringing dance from the classroom to the stage. Reflecting on this experience of learning through performance, alumni recalled:

> We were fortunate to watch close up as Madame Choe performed her representative works *Braving Wind and Waves* and *Mother*. Madame Choe's performances and her choreographies became a key that further opened the door of the dance world in our hearts. This is very similar to how traditional opera artists in China often learned their crafts—they would hide themselves in the wings and concentrate their attention watching their teacher's performance, treating it as a valuable learning experience, or so-called 'stealing the drama'. We were really lucky and would often loose ourselves in watching Madame Choe attain perfection in her performances. We became so intoxicated in the beautiful scenes that we nearly forgot to change costumes ourselves and missed our cues.[53]

In the case of Wu's *Wuyunban*, students gained their applied experience by carrying out field research and promoting dance activities with workers. To prepare for their final projects in the spring of 1952, the students broke into eight groups that were stationed in different factories or mines on the outskirts of Beijing. Groups 1 and 2 went to Shijingshan steel factory; groups 3 and 8 went to Mentougou coal

[50] According to Li and Tian (2005), students continued to enroll while the course was running, so the remaining 11 students most likely joined after this report.
[51] 'Quanguo wudao yundong ganbu xunlian ban chengli', *Wudao tongxun* 1 (1951), 4.
[52] Tian and Li, *Xin Zhongguo wudao*.
[53] Tian and Li, *Xin Zhongguo wudao*, p. 300.

mine; group 4 went to Shulihe cement factory; group 5 went to Qinghe woollen cloth factory; group 6 went to Beijing agricultural machinery factory; and group 7 went to Renli carpet factory. The students spent 15 days in their field sites, personally experiencing different kinds of manual labour involving smelting furnaces, mine pits, and looms. They also got to know the workers and through trial and error devised collective dance activities that accorded with the workers' busy schedules, interests, and abilities. For example, sometimes they offered five or six group dance classes in one day to accommodate the workers' three-shifts system. Finally, these experiences became the basis for the students' own dance creations, some of which directly represented the factory and mine communities they had worked with on stage. For example, one of the class's graduation pieces, titled *Red Flag Competition*, reflected how steel workers participating in a competition devised clever improvements to their manufacturing tools. Another, titled *Woolen Cloth Weavers' Dance*, used the different characteristics of fine spinning and machine weaving to reflect the lives of weavers. At the graduation performance, the students presented these pieces for large and diverse audiences, including 150 workers from three of the factories the students had visited.[54]

The significant impact that the *Wuyunban* and *Wuyanban* had on the development of the dance field in Maoist China can be seen in the career trajectories of the courses' graduates:

> Of the approximately 150 Chinese alumni of the two courses, five became vice chairs of the Chinese Dancers Association, over twenty became chairs or vice chairs of provincial or municipal dancers associations, twenty-four became directors or vice directors of local provincial, municipal, or military song and dance ensembles, over 60 have high level rank as professors, researchers, choreographers, performers, teachers, and art directors; many also continue arduous work at the grassroots level in remote areas, with outstanding results.[55]

For anyone familiar with the history of dance in the PRC, the list of alumni of the two CAD courses is quite extraordinary. Wu Yun (*Wuyunban*), Siqintariha (*Wuyanban*), and Baoyin Batu (*Wuyanban*) were leaders in the development of dance in Inner Mongolia and among the first dancers to represent the PRC in international dance competitions in Eastern Europe; Sun Ying (*Wuyunban*), Gao Dakun (*Wuyunban*), Li Zhengyi (*Wuyanban*), and Gao Jinrong (*Wuyanban*) founded the three major styles of Chinese classical dance – Sun Ying created the Han-Tang style, Gao Dakun and Li Zhengyi created the xiqu-based style (along with Tang Mancheng, the only major figure in the establishment of Chinese classical dance styles who was not a member of either course), and Gao Jinrong the Dunhuang style. Luo Xiongyan (*Wuyunban*) established the academic discipline of Chinese folk dance cultural studies and became a leading dance theorist. Shu Qiao (*Wuyanban*), Jiang Zuhui (*Wuyanban*), Zhang Minxin (*Wuyanban*), and

[54] Tian and Li, *Xin Zhongguo wudao*, pp. 18–24.
[55] Tian and Li, *Xin Zhongguo wudao*, p. 6.

Wang Shiqi (*Wuyanban*) all became famous dance drama choreographers known for their contributions to canonical Mao-era productions such as *Dagger Society* (1959), *Red Detachment of Women* (1964), *East is Red* (1964), and *Fish Beauty* (1959). Several members of *Wuyanban* also led the development of the teaching curriculum, theoretical writings, and performance repertoires of Korean dance in China, while still others made foundational contributions to the advancement of military dance and mass dance programmes among children and the middle aged and elderly.

As was intended, the CAD courses fulfilled the Ministry of Culture's goal of preparing a cohort of dedicated cadres who were equipped technically, theoretically, and in terms of practical experience to establish dance as an important cultural field in the early PRC. During my years of fieldwork with professional dancers in China, I have had the pleasure of talking with alumni of the CAD classes about their experiences and how the courses impacted their lives. Sun Ying, for example, recalled with humour how Wu took a special liking to Sun even though Sun did not practice hard, because during composition sessions Sun's choreography was creative and did not follow the common trend.[56] Zhang Minxin recalled enjoying studying modern dance, to which she had some exposure from Sophia Delza, an American teacher in Shanghai, before she joined *Wuyanban*. Zhang also was appreciative that despite not being a coordinated performer, the training she received from Choe led her eventually to a satisfying career as a choreographer.[57] Siqintariha recalled how she was impressed by the way Choe's Korean dance curriculum had clearly systematised a national dance form so that it could be passed on to later generations, and she said this experience laid the foundation for her own work to systematise Mongol dance decades later.[58] Shu Qiao spoke of Choe's particular method of breaking down complex movements from Chinese opera, Indian dance, Korean dance, and other Asian dance forms to create new combinations of these movements set to regular eight-count rhythms and oriented in different directions in space. She argued that this had set the pathway for how Shu's classmates would go on to construct Chinese classical dance in later years and that because Shu herself had seen and experienced how these combinations were originally created, she also saw their flaws and could be more critical.[59] In talking with graduates of the CAD courses, I have been struck by their level of self-reflexivity and confidence, the deeply analytical and theoretical approaches they bring to their dance work, as well as their enduring commitment to making dance have meaning for society. I believe the CAD courses played a role in developing these particular qualities and values, which proved vital to their success in contributing to the development of dance in Maoist China.

[56] Sun Ying, Interview with the author (Beijing, 26 September 2008).
[57] Zhang Minxin, Interview with the author (Beijing, 1 October 2008).
[58] Siqintariha, Interview with the author (Hohhot, 24 June 2014).
[59] Shu Qiao, Interview with the author (Shanghai, 8 July 2014).

Conclusion

Given the age of students recruited to the CAD courses and the content of their curricula and teaching methods, these courses laid the foundation more for university-level dance pedagogy, which exploded and took shape in China after the Mao era, than for the early age technical training that dominated the early Beijing Dance School and other institutions modelled after it during the Mao years. It is no surprise, then, that when the first graduate programme in dance was established in China in the early 1980s, Wu was selected to head this programme. The CAD programmes remind scholars of the need to look beyond the dominant narratives and institutions to fully understand the complexity of the Mao era. In fact, apart from the CAD programmes, there was much going on in dance education in China from the 1950s to the 1970s that does not fit the technique-heavy training model for young students implemented at BDS. When conducting field research, I learned that a majority of professional dancers who grew up during the Mao years were trained not in formal schools but instead in apprenticeship programmes housed in professional performance troupes – what were known as 'ensemble-led class' (*tuan dai ban*) arrangements.[60] Even BDS offered several short-term study advanced courses in specialised topics such as dance drama choreography, and many dancers also learned their skills in an ad hoc manner during so-called 'performance meetings' – when many ensembles gathered in a single location to exchange works – or during encounters with foreign ensembles and teachers during international tours or visits.[61] The ubiquitous 'discussion meetings' (*zuotanhui*), 'creation groups' (*chuangzuo zu*), and even political campaigns with their many meetings, study sessions, and mass activities all contributed to the theoretical, political, and social skillset of dancers and other cultural workers in China during the Mao era. These activities are similar to the pedagogical approaches offered by Choe and Wu in the CAD courses in that they emphasised hands-on learning and integration between theory and practice, as well as exposure to a wide variety of dance styles. To fully understand dance education in Maoist China, then, scholars should look at this wide range of educational programmes and activities. This lesson can also be extrapolated to other fields.

Another lesson one can take from the CAD dance training courses is the importance of inter-Asia exchange before and during the Mao years in the history and cultural development of the PRC. In this specific case, it is clear how Wu's and

[60] Emily Wilcox, 'The Dialectics of Virtuosity: Dance in the People's Republic of China, 1949–2009' (Ph.D. thesis, University of California, Berkeley, 2011).

[61] Emily Wilcox, 'Performing Bandung: China's Dance Diplomacy with India, Indonesia, and Burma, 1953–1962', *Inter-Asia Cultural Studies* 18, 4 (2017), 518–39; Emily Wilcox, 'Aesthetic Politics at Home and Abroad: *Dagger Society* and the Development of Maoist Revolutionary Dance Drama' in Xiaomei Chen, Siyuan Liu, and Tarryn Chun (eds), *Rethinking Chinese Socialist Theaters of Reform: Performance Practice and Debate in the Mao Era*. Ann Arbor: University of Michigan Press, 2021, pp. 162–86.

Choe's experiences in 1920s and 1930s Tokyo remained salient to their later professional activities and artistic pursuits, which persisted during the Mao years despite significant changes in artistic, political, and social contexts of their work, as well as in their own personal activities and commitments. At the same time, in both cases, but especially for Choe, the Japanese empire and its expansion into China brought unexpected consequences, including genuine cultural exchanges that took place across and between colonised groups – in this case between Choe as a colonised Korean and Chinese opera actors such as Mei Lanfang – even though the factor that brought them together were paradoxically affordances of imperial violence and an oppressive occupation. It is important to consider how the legacies of Japanese imperialism were filtered through North Korea and reintroduced to the PRC through the inter-Asia socialist pathway of Korean War-era Sino-Korean interaction and mutual aid. Choe's story and the story of the CAD dance training courses open our eyes to these possibilities. In turn, they suggest other circuitous routes and transnational or transtemporal genealogies that fed into the making of Maoism – complicating existing pictures of socialist transnationalism and imperial historical inheritances, and helping to better map out the many sources of inspiration that circulated their way in and out of Maoist culture. From this perspective, it is perhaps not surprising that the first ballet version of *The White-Haired Girl* – predecessor to the renowned Chinese revolutionary ballet of the Cultural Revolution – was staged by a Japanese ballet company in Tokyo in 1955 and then first brought to China in 1958, where it was enthusiastically welcomed by Chinese audiences and dance and theatre experts, seven years before the premiere of the first Chinese ballet version.[62] The old visions of Maoist China as culturally closed off, internally monolithic, or a simple clone of the Soviet Union are increasingly tendentious. What new narratives can be built in their place to tell a new story of socialist China and even to redefine that social and cultural landscape that has come to be called 'Maoism'?

A number of surprises emerge from this story of the CAD training courses: modern dance emerges as a core ingredient in the broader picture of dance in Maoist China; North Korea comes to precede the Soviet Union as a source of professional dance expertise, teaching methods, and choreographic inspiration; it is a theatre school, rather than a dance school, that becomes the so-called 'birthplace' of specialised dance training; a formative moment for institutionalising dance theory is also a site of extreme emphasis on applied knowledge and 'practical training'; and the beginnings of a dance genre, Chinese classical dance, that embodies highly nationalist ideas about culture and aesthetics emerges as an outgrowth of a Pan-Asianist project, and so on. The names of these two courses – Dance Movement Class and Dance Research Class – reveal much about the history of the Mao era and the methods through which it was created. 'Movement' here refers to social movement and 'research' refers to the process of accumulating knowledge through

[62] Emily Wilcox, 'Sino-Japanese Cultural Diplomacy in the 1950s: The Making and Reception of Matsuyama Ballet's *The White-Haired Girl*', *Twentieth-Century China* 48, 2 (2023), 130–58.

experiment and theorisation. The unification of these two programmes under one roof represents the heart of Maoist method: it is a process of accumulating knowledge through engaged practice using the best resources at hand at any given time. Better than conceptual abstractions such as 'socialist realism', 'historical materialism', or 'Sino-Soviet alliance', this practical model aids understandings of what Maoism was and how it was made.

Coda

17

Point Counterpoint: Temporal Interplay in the Soviet and Chinese Revolutions

JULIANE FÜRST AND JOCHEN HELLBECK

IN THE INTRODUCTION to this volume devoted to pathways of socialist construction in Communist China the editors cite Mao Zedong's 1958 use of a Hunan proverb to characterise the Chinese revolution: 'Straw sandals have no pattern—they shape themselves in the making'.[1] Mao's words also helpfully frame our intervention as Soviet historians reflecting on the building of socialism in Communist China. When set in the context of its time and place, Mao's statement reads as an assertion of a distinctly Chinese approach toward a socialist future. In the Chairman's rendering, the Hunan proverb asserted a Maoist revolutionary temporality, steeped in the idiom of the Chinese peasantry, and challenging Soviet claims to historical and philosophical precedence. As Mao's words suggest, the relationship between Communist China and the Soviet Union evolved in a complex temporal dynamic, in which the sometimes synchronised, sometimes syncopated tempi of socialist construction imparted a characteristic rhythm to the larger history of global socialism.[2] The Soviet Union was the first socialist society, but when the Chinese Communist Party (CCP)'s revolution was beginning to remake China, the Union was no longer in its revolutionary phase. Many features of Maoist society remind acutely of the early years of the Soviet project, while the Soviet Union under Khrushchev, its actual contemporary, functioned according to different laws and could look back

[1] The first scholar to quote Mao's use of the proverb was John Bryan Starr, *Continuing the Revolution: The Political Thought of Mao*. Princeton: Princeton University Press, 1979, ix.

[2] The history of global socialism in terms of parallel and conflicting temporalities remains to be written. For reflections on socialism and time, Stephen E. Hanson, *Time and Revolution: Marxism and the Design of Soviet Institutions*. Chapel Hill: University of North Carolina Press, 1997; Martin Malia, *The Soviet Tragedy: A History of Socialism in Russia*. New York: Simon & Schuster, 2008; Stephen Kotkin and Catherine Evtuhov (eds.), *The Cultural Gradient: The Transmission of Ideas in Europe, 1789–1991*. Lanham: Rowman and Littlefield, 2003; Nikolai Ssorin-Chaikov, *Two Lenins: A Brief Anthropology of Time*. Chicago: University of Chicago Press, 2017.

on three generations of Sovietness. This inherently makes direct comparison, based on a simple historical chronology, a challenge. When reading the chapters in this collection with Soviet eyes, we hence noticed two possible approaches to thinking about socialism in the People's Republic of China (PRC) and the Soviet Union: one in step with the temporal chronology as two communist countries co-existing in a post-war world, and another that is measuring time in revolutionary life cycles, comparing revolutionary society in relation to their respective age.

Soviet influences loom large in this volume, even if they are not always made fully explicit. The reason for this is historiographical: the trope of Chinese 'learning from the Soviet Union' is well studied. Several chapters point out that the Soviet vector was far from the only transnational force shaping Communist China.[3] Emily Wilcox seeks to part with 'old visions' of Maoist China as merely a Soviet clone, by highlighting Korean instructors and Japanese cultural forms as significant actors in Maoist dance education. In their chapters on China's geological profession and the Knowledge Series, Shellen Xiao Wu and Robert Culp similarly explore hitherto overlooked inter-Asia lineages of transnational ideological transmission. These discoveries do not invalidate the enormous Soviet hold on culture and science in Maoist China: founded in 1961, the Chinese *Knowledge Series* echoed activities of the vastly influential Soviet Knowledge Society, which had formed in 1947; Soviet conceptions of science such as agroscientist Trofim Lysenko's teachings on vernacularisation, fell on extremely fertile soil in China; and the Soviet ballet tradition was foundational for the Beijing Academy of Dance, established in 1954. Nicolai Volland's chapter identifies the 1940-founded Chinese Epoch Press as one of the earliest transmitters of Soviet values and cultural forms. The Soviet imprint on Chinese literary publication lasted well beyond Epoch's closing in 1960, with which the chapter concludes.[4]

Any revolution arguably unfolds in a materialising process, rather than on the strength of revolutionary declarations alone. This said, the Soviet revolution revealed a distinct verbal, or more specifically, graphic core, as the body of works by Marx, Engels, Lenin, and Stalin came to be valourised as the canonical essence of the revolution, and as revolutionary men of action conceived of much of their action in terms of the voracious reading and writing of texts that would fuse their

[3] For a superb volume on Soviet influences, see: Thomas P. Bernstein, Hua-yu Li (eds.), *China Learns from the Soviet Union, 1949-Present*. Lanham: Lexington Books, 2010.

[4] On the hitherto little-studied Soviet knowledge society, see Vera Rich, 'Russians Get Their Fun from Learning about Science', *Nature*, Vol. 279, 14 June 1979: 569; on Chinese and Soviet agroscience, see Gregory Rohlf, 'The Soviet Model and China's State Farms', in: *China Learns from the Soviet Union*, 197–230. Volland's own recently published monograph on Socialist cosmopolitanism makes clear that Soviet literature kept being published in Chinese translation throughout the 1960s and that much of China's revolutionary literature modelled itself on Soviet patterns. Nicolai Volland, *Socialist Cosmopolitanism: The Chinese Literary Universe, 1945–1965*. New York: Columbia University Press, 2017. See also on the Chinese reception of an influential Stalin-era novel: Donghui He, 'Coming of Age in the Brave New World: The Changing Reception of the Soviet Novel, How the Steel Was Tempered, in the People's Republic of China', in: *China Learns from the Soviet Union*, 393–420.

personal lives with the trajectory of the revolution.[5] Shortly after coming to power, Communist Party officials launched a massive project to document the history of the Communist Party by stenographing curated group discussions among rank-and-file party members in the Soviet capitals and other cities. It was through such shared storytelling that the 'October Revolution' congealed into an event for participating actors, and one that was fastened to their personal lives.[6] The endeavour to fuse subjective horizons to the revolution continued into the Stalin period, with the large-scale collection of testimonies from former fighters in the Russian Civil War, builders of the factories of the first Five-Year Plan, and Soviet soldiers and civilians who defended the Soviet lands against the Nazi invaders.[7] Many of these projects were overseen by professional writers whom Stalin had famously tasked as engineers of human souls.[8] Their equal parts literary and documentary projects were to distil the ideal subjectivity that would propel forward in time a revolution hemmed in by 'objective' constraints and therefore considered premature and doomed by 'bourgeois' critics. 'Time, Forward!', the title of Valentin Kataev's 1932 industrial production novel, encapsulated the spirit of voluntarist time compression that was at the heart of the Soviet revolution, in both its Leninist and Stalinist guises. Provoked by fascism, Soviet ideologues in the 1930s built up this literary bedrock to cast the Soviet Union as a guardian of world literature – from Shakespeare to Goethe and Pushkin – and defender of humanism against 'medieval' and 'barbaric' book-burning fascists. Within the Soviet Union between the 1920s and early 1950s, literature was extolled as the highest of all Soviet art forms.[9]

The Soviet focus on texts as histories of revolution and catalysts of revolutionary action travelled East in the guise of Chinese Communists who had been forged in Moscow's Comintern offices in the 1920s and 1930s. Mao, who vied for leadership over the CCP at the time, without ever having visited the Soviet Union,

[5] Regis Debray locates the 'life-cycle of socialism, that great fallen oak of political endeavour, within the last 150 years of the graphosphere'. Regis Debray, 'Socialism - A Life-Cycle', *New Left Review* 46 (July/August 2007): 5–28. The Soviet House of Government, built up across the Moscow Kremlin as a futurist apartment complex for the new Soviet elite, became filled with Old Bolsheviks and their literary dreamworlds about communism. Yuri Slezkine, *The House of Government: A Saga of the Russian Revolution* (2017).

[6] Frederick Corney, *Telling October: Memory and the Making of the Bolshevik Revolution*. Ithaca: Cornell University Press, 2004.

[7] Elaine MacKinnon, 'Writing History for Stalin: Isaak Izrailevich Mints and the *Istoriia grazhdanskoi voiny*', *Kritika: Explorations in Russian and Eurasian History* 6.1 (2005): 5–54; Katerina Clark, 'The History of the Factories as a Factory of History', in *Autobiographical Practices in Russia*, eds. Jochen Hellbeck and Klaus Heller. Göttingen: Vandenhoeck & Ruprecht, 2004, 251–78; Jochen Hellbeck, *Stalingrad: The City that Defeated the Third Reich*. New York: Public Affairs, 2015.

[8] Elizabeth A. Papazian, *Manufacturing Truth: The Documentary Moment in Early Soviet Culture*. DeKalb: Northern Illinois University Press, 2009; Josette Bouvard, 'L'injonction autobiographique dans les années 1930: G.A. Medynskij et l'histoire du métro de Moscou', *Cahiers du monde russe* 50.1 (2009): 69–92.

[9] Katerina Clark, *Moscow, the Fourth Rome: Stalinism, Cosmopolitanism, and the Evolution of Soviet Culture, 1931–1941*. Cambridge, MA: Harvard University Press, 2011.

experienced the Soviet textual mould as a challenge to his own legitimacy as aspiring party leader. His 1930 essay, 'Opposing Bookism', railed strongly, if implicitly, against domestic rivals who preached Soviet textual dogmata: 'Books cannot walk and you can open and close a book at will; this is the easiest thing in the world to do, a great deal easier than it is for the cook to prepare a meal, and much easier than it is for him to slaughter a pig.'[10] Mao did not discount all Soviet texts. He avowedly prescribed the Chinese translation of *The Short Course of The History of the All-Union Communist Party (Bolshevik)*, first published in Russian in 1938, as a vade mecum of 'revolutionary practice' and must-read for CCP cadres in Yan'an. With its emphasis on Lenin's innovation in adapting Marxism to Russia, the *Short Course* did much useful work to legitimate Mao's own unorthodox pathway to building socialism in China, grounded in the 'concrete study of China's present conditions'.[11]

For Mao, China's 'concrete conditions' dictated a wager on the peasantry – rather than the party's urban followers (numerically insignificant, to begin with, and thoroughly repressed during the purge of 1927) – as an insurrectionary force. They dictated adopting low-brow methods of propaganda that appealed to illiterate and semiliterate audiences and were referred to as 'agitation', rather than moulding the minds of a more developed party through print media and other texts, an activity that Communists referred to as 'propaganda'. Seen through this lens, China did not simply lag behind Soviet developments by 30 years, in the sense that Chinese revolutionaries in the late 1940s took their cues from Russian revolutionaries in 1917 and the Civil War. The popular slogan, 'The Soviet Union of today is our tomorrow', masked a more complex temporal relationship. Compared to their Soviet precursors, Mao and his comrades placed much more weight on practice in order to challenge orthodox laws of historical development that even Soviet revolutionaries who advocated for time compression in their own country believed to be unavoidable in China and elsewhere on the globe.[12] China's trajectory of building socialism differed significantly from the Soviet one in terms of its wager on the peasantry

[10] Cited in Julia Lovell, *Maoism: A Global History*. New York: Random House, 2019.

[11] Thomas P. Bernstein, 'Introduction: The Complexities of Learning from the Soviet Union', and Xiaojia Hou, '"Get Organized": The Impact of the Soviet Model on the CCP's Rural Economic Strategy, 1949–1953', in: *China Learns from the Soviet Union*, 5 and 170.

[12] Hence Stalin's dismissive view of the Chinese Reds as 'not real Communists' but 'margarine Communists'. Bevin Alexander, *The Strange Connection: U.S. Intervention in China, 1944–1972,* 4–6. Stalin occasionally cited folk wisdom, but never lost sight of his overriding commitment toward an urban industrial modernity. In a speech given on the 12th anniversary of the October Revolution, the Soviet leader vowed to 'leave behind the age-old "Russian" backwardness' by 'putting the U.S.S.R. on an automobile, and the muzhik on a tractor'. Once his country of metal would become a reality, Stalin dared the 'capitalists who boast so much of their "civilisation"': 'try to overtake us! We shall yet see which countries may then be "classified" as backward and which as advanced.' J.V. Stalin, 'A Year of Great Change: On the Occasion of the Twelfth Anniversary of the October Revolution', in: idem, *Works*, Vol. 12, April 1929–June 1930, (Moscow, 1954), 124–41.

rather than the working class, and the idea that a revolution carried out by less than fully literate followers could nonetheless thrust China into the Communist future.

This temporal mandate frames Christine Ho's chapter on mass muralism and mass creativity during the Great Leap Forward (GLF). Inconceivable in the Soviet setting, where communist contempt for peasantry prevailed even past transformation of muzhik into collectivised kolkhoz worker, Chinese officials sought to mobilise hundreds of thousands of villagers as artistic agents of the revolutionary countryside. The thousands of mural 'art galleries' that sprung from this campaign of communist agitation in turn became agents of further agitation, on the part of non-expert peasant artists propelled by the power of 'raw revolutionary subjectivity' (Ho). Significantly, Ho notes, this campaign of mass artistic expression was not born in the heat of the GLF, but had roots reaching back to wartime Yan'an. Yan'an storytelling, standing for the practice of collective study and self-criticism which the CCP leadership first prescribed for its literate followers in 1942, was to forge the revolutionary consciousness that would release itself in mass mural projects.[13]

How did this rural-centred revolutionary culture with its strong oral and pictorial focus inflect on urban intellectuals? This is a question that weaves itself through Aaron Moore's chapter on the theme of work and study in personal diaries during the founding years of the Chinese Communist order. Moore designates diary-writing as a technology of the self that outsourced Yan'an's initial injunction to make it fit the much vaster requirements of a national revolution. All six diarists that his chapter examines used their journals to inculcate the revolutionary discourse. A key part of this endeavour appears to have been the intellectual processing ('learning something until you're convinced') of group study sessions and other aspects of the public work sphere. As intellectuals, the six diarists sought proof of attainment of a socialised and universalised self in the demonstration of a clear break with the past, as one of them, Jiang Jianmin from Nanjing, succinctly pointed out: 'At the level of consciousness, we will start anew and draw a clear line [with the past].'

Establishing a temporal break with the pre-revolutionary past mattered to the authors of these texts, especially as all of them had come of age in former Kuomintang (KMT) strongholds: they entered Communist China with potentially polluted souls. These cases bring to mind the personal writings of the Russian intelligentsia with an educational background in Late Imperial Russia, which staged veritable killings of the unreformed 'old intelligentsia' self in order for the new and pure revolutionary consciousness to assert itself.[14] Strikingly, the urgent and insistent tone of self-accusation that one encounters in these diaries, their moral panic

[13] David E. Apter and Tony Saich, *Revolutionary Discourse in Mao's Republic*. Cambridge, MA: Harvard University Press, 1994, 282–4.

[14] Jochen Hellbeck, *Revolution on My Mind: Writing a Diary under Stalin*. Cambridge, MA: Harvard University Press, 2006; Igal Halfin, *Terror in My Soul: Communist Autobiographies on Trial*. Cambridge, MA: Harvard University Press, 2003.

about hidden political enemies lurking in the dark, including in the recesses of the author's own soul, appears largely absent from this selection of Chinese diaries, which could indicate a much lower presence of political surveillance in Maoist China, compared to the Soviet Union. Then again, some of the diaries *were* confiscated during the Cultural Revolution, and the fate of their authors is not known. What is clear, though, is that the Chinese revolution shared with its Soviet counterpart a strong vector of individual self-realisation. To gauge this theme more fully, it would be interesting to compare the personal diaries kept by members of the first generation of Chinese citizens who were fully socialised under the communist regime with the writings of the first socialist generation in the Soviet Union.[15]

While Mao was raising the first revolutionary Chinese generation, however, those who had been in their shoes in the Soviet Union were approaching middle or even old age. They were certainly not young radicals anymore – not necessarily because they had lost enthusiasm, but more because they had witnessed and survived purges and war. The innocence of utopianism and the belief that revolution alone will make a new man had been shattered. While Mao was invoking Lenin and Stalin, Mao's counterpart in Moscow, Nikita Khrushchev, was busy navigating a reform course designed to mobilise but also to heal people from Stalinism. The Soviet Union that existed for most of the 1950s was not a country in the process of revolutionary construction, but one trying to redefine its political system through careful reassessment of revolutionary strategy and experimentation with consumerism.

China's relationship with the Soviet Union under Mao's reign was thus suspended between two poles: the experience of Soviet Union making and implementing revolution into the societal fabric, and the tension that arose from the fact that the Soviet Union was grasping its way forward to a 'developed' form of socialism. Soviet leadership was unsure where this path led (resulting, at times, in ambivalent policies), but post-Stalin leaders were quite clear what they wanted to get away from: Stalinism, and by extension, the kind of radical and ruthless policies Mao embodied. From the standpoint of chronological contemporality, Mao's China and the post-Stalin Soviet Union were moving in different directions and pursuing opposite goals. Thaw-era and Maoist societies were in different parts of what Amir Weiner has called 'the revolutionary cycle'.[16] The Soviet revolution was on

[15] To be sure, Stalin era personal writings could appear in large numbers only as part of the massive delegitimisation of Soviet history that set in during the 1980s and helped to bring about the Soviet collapse. The wave of documentary publications was such that findbooks appeared in the 1990s, listing recently published personal records from the Soviet era (see e.g.: *Otkrytyi arkhiv. Spravochnik opublikovannykh dokumentov po istorii Rossii XX veka iz gosudarstvennykh i semeinykh arkhivov. 1985–1996*, ed, I.A. Kondakova. Moscow: Rosspen, 1999. Absent such a temporal break, Maoist era personal diaries are unlikely to appear in large numbers.

[16] Amir Weiner, 'Robust Revolution to Retiring Revolution: The Life Cycle of the Soviet Revolution, 1945–1968', *The Slavonic and East European Review* 86.2 (2008): 208–31. See also Michael David-Fox, 'Toward a Life Cycle Analysis of the Russian Revolution', *Kritika: Explorations in Russian and Eurasian History* 18.4 (2017): 741–83.

its second, if not third, generation. The maintenance of power was as important to Soviet leaders, indeed probably more important, as the mobilisation of revolutionary and iconoclastic energy.[17] In Mao's China everything was still up for grabs, and every aspect of life awaited the impact of revolution. In the Soviet Union the revolution *was* the establishment.

Yet as Soviet and Chinese research demonstrates – not least the articles in this volume – this conclusion skips over many nuances and more complex trajectories that connected rather than divided the two communist societies even when comparing an older Soviet Union and a young PRC. First of all, Soviets and Chinese existed in the same twentieth-century globality that entangled not only the fates of their countries, but turned the whole world into an interconnected platform, on which crucial questions of modernity played out. On a very simple but fundamental level both were post-war societies. This fact is most pronounced in Toby Lincoln's article on the reconstruction of Changsha, which exemplified the Maoist commitment to put the past into the service of the present, and the blurry boundaries between successive waves of trauma and destruction and between destroyers and constructors. Changsha became a showcase of communist construction built on the foundations of KMT efforts to do the same. Changsha stands here for many locales where the new ideology was cemented into bricks and mortars but failed to eradicate the underlying layers. The post-war Soviet Union also used the urban voids created by the destruction of World War II (WWII) to reconstruct cities in the Stalinist image and at the same time emphasise both heroism and victimhood of their residents. Kyiv, Sevastopol, and Minsk cityscapes are testament to this process of remodelling and rewriting history.[18] The complete reconstruction of Moscow and Yerevan into modern twentieth-century cities largely devoid of medieval features, which were still present during the 1930s and 1940s, demonstrates that the desire to rebuild was not limited to places where war and natural catastrophes had created opportunity, but consciously and deliberately included the destruction of the old and undesired.[19] At the same time certain imperial structures were appropriated and preserved to suggest continuity of power, greatness and domination, with the two processes often overlapping as demonstrated in the Arbat neighbourhood in Moscow and

[17] Juliane Fürst and Stephen V. Bittner, 'The Aging Pioneer: Late Soviet Socialist Society, Its Challenges and Challengers', in *The Cambridge History of Communism*. Cambridge: Cambridge University Press, 2017, 281–306.

[18] Karls D. Qualls, *From Ruins to Reconstruction: Urban Identity in Soviet Sevastopol after World War II*. Ithaca: Cornell University Press, 2009; Martin J. Blackwell, *Kyiv as Regime City: The Return of Soviet Power after Nazi Occupation*. Rochester: University of Rochester Press, 2016; Franziska Exeler, *Ghosts of War: Nazi Occupation and Its Aftermath in Soviet Belarus*. Ithaca: Cornell University Press, 2022.

[19] Timothy J. Colton, *Moscow: Governing the Socialist Metropolis*. Cambridge, MA: Harvard University Press, 1995; Karen Azatyan, Madlena Igitkhanyan, and Anush Ohanyan, 'Challenges to Residential Quarter Reconstruction: The Case of the Center of Yerevan City', *Journal of Architectural and Engineering Research* 3 (2022): 10–31.

elsewhere.[20] Interestingly, preservationism became one of the loci around which independent civil society during Perestroika developed, most famously gathering force in the fight for the Hotel Angleterre in Leningrad, which was the site of the suicide of the Soviet poet Sergei Esenin.

The needs – or indeed desires – of the post-war period are also apparent in Sarah Mellors Rodriguez's article on the production of sexual knowledge in 1950s China. State-sponsored sex education is directly linked to policies of procreation and definitions of behavioural normativity. That was also true in early revolutionary Russia where hygiene guides expressed fears of contamination through bourgeois or foreign elements. Yet it was also, and again, true in the late-Stalinist Soviet Union, when concerns over the loss of millions of Soviet citizens during WWII resulted in policies and propaganda that were both re-affirming to the patriarchy through strict marriage and divorce regulations while also subtly encouraging procreation outside the marital bond by making alimony claims against non-spouses impossible.[21] The traumas of a post-war society and the latter's inherent desire for stability are at the heart of Mellors Rodriguez's observations about what Maoist sex guides were telling people beyond rather mechanical instructions of how to fulfil the sexual act – instructions that in their sexual bluntness were unthinkable in the Soviet Union. Mellors Rodriguez points to the continuities with the Republican period, when family policy also sponsored patriarchal structures, hence putting into question their Maoist particularity and indeed their revolutionary nature. This is somehow reminiscent of what has been called 'the great retreat' by Nicholas Timasheff, initiating a vivid debate about the nature of Stalinism that is still ongoing. Timasheff's observation of a return of 'imperial and conservative values' questioned not only Stalin's communist credentials but also the novelty of the Soviet project overall. Just as Mellors Rodriguez describes it for Mao's China, the abolition of the right to abortion and the tightening of divorce laws in the 1930s formed, together with the veneration of Tsarist heroes and the resurrection of Russian nationalist thought, a package of policies that seemed keen to undo some of the ruptures caused by revolution. The term and its implication have been fiercely debated with many historians pointing out that despite pandering to certain 'conservative' ideals, the context and execution of Stalinist policies were nonetheless of very different quality, which evidenced the formation of Stalinist 'civilisation' in its own right rather than a simple imitation.[22] Interestingly, Khrushchev relegalised abortion in 1955, but carefully avoided most of the other revolutionary experiments of the early years that aimed

[20] Stephen V. Bittner, *The Many Lives of Khrushchev's Thaw: Experience and Memory in Moscow's Arbat*. Ithaca: Cornell University Press, 2008; Katherine Zubovich, *Moscow Monumental: Soviet Skyscrapers and Urban Life in Stalin's Capital*. Princeton: Princeton University Press 2020; Steven Maddox, *Saving Stalin's Imperial City: Historic Preservation in Leningrad, 1930–1950*. Bloomington: Indiana University Press, 2015

[21] Mie Nakachi, *Replacing the Dead: The Politics of Reproduction in the Postwar Soviet Union*. New York: Oxford University Press, 2021.

[22] Nicholas S. Timasheff, *The Great Retreat: The Growth and Decline of Communism in Russia*. New York: E.P. Dutton & Co., Inc., 1946. On more current discussions, see Matthew E. Lenoe, 'In Defense

to redefine love, partnership, and sexuality.[23] The back and forth of policies points to the fact that large sways of everyday Soviet life was not codified in eternal norms and values but rather open to negotiation, which at times could include elements of previous normative systems, just as Mao too drew on a large number of different texts not shying away of drawing on traditional Chinese thought to construct his own brand of civilisation.

In other respects, Khrushchev pursued a policy of making the private public. The Khrushchev era has often been shown as more meddling in the personal affairs of Soviet citizens than the repressive Stalin years. Public shaming, comrade courts, citizen patrols, and extensive propaganda for normative emotional behaviour all flourished in the Thaw era, a time period that is often mistaken as sponsoring liberalism and individuality but was in fact a rerun of Bolshevik mobilisation without the most egregious violence.[24] This care for – or intrusion into – personal life also included Khrushchev's massive housing programme that saw millions of people moving out of communal apartments into private flats, which came in standardised sizes and were to be furnished with new standardised and mass-produced interior design elements. As Jennifer Altehenger's article on Maoist furniture demonstrates, in this respect the Soviet Union was far ahead of the PRC, whose ambition was an equalised, accessible way of life, but whose Maoist reality was more of a 'make-do' nature. But the sign of the times firmly pointed to a vision of a China whose lifestyle, rather than mode of production, reflected socialist affluence.

Similarly, Fabio Lanza's piece on the urban communes of the Mao movement seems to point to the communal experiments of young people in the years immediately following the Russian Revolution. Yet in terms of popular mobilisation the Khrushchev era was not without similar short-lived successes. Khrushchev oversaw not only the very popular Virgin Lands Campaign, in which young volunteers were tasked to turn the Kazakh step into fertile agricultural ground. His era also witnessed a number of grassroots self-governing initiatives such as the communard movement (self-governing pioneer groups) and the Komsomol patrols (youth-led vigilante groups), both of which empowered youthful agency in the name of perfecting socialism. Even Brezhnev, whose era quickly became known as the period of *zastoi* – stagnation – dreamed of mobilising youth and simultaneously socialising them into the spirit of collectivity, such as when he initiated the construction of the Baikal-Amur Railway – a gigantic engineering task and mass participation

of Timasheff's Great Retreat', and Jeffrey Brooks, 'Declassifying a 'Classic', *Kritika* 5.4 (2004): 709–19 and 720–30; Mikhail Shmatov, ' "The Great Retreat" or "the Great Maneuver": N. Timasheff's Concept and Ideological Changes in the USSR of the 1930s', *Ideas and Ideals* 2.3 (2018): 169–84.

[23] Deborah A. Field, *Private Life and Communist Morality in Khrushchev's Russia*. New York: Peter Lang, 2007.

[24] Oleg Kharkhordin, *The Collective and the Individual in Russia: A Study of Practices*. Berkeley: University of California Press, 1999.

drive at the same time.[25] These mobilisation campaigns were nothing compared to the Red Guards movement and their rustication into the countryside. Nonetheless, they were fuelled by the same sentiment of keeping the revolution alive and securing future generations to the cause while keeping them in check. Youth in both systems always represented a duality: promise and threat.

This ambiguity vis-a-vis youth ran through a number of Soviet policies, which oscillated between embracing the wider world and shunning it in favour of a shared us-versus-them paradigm. Childhood and youth were worshipped in the Soviet Union, because from the very moment of revolutionary success these two demographics contained within themselves the promise of achieving true socialism and becoming the fabled new Soviet person. Yet youth in particular also had the potential to squander what their parents, and soon grandparents, had achieved. The big corruptor of youth was the capitalist bourgeoisie, embodied initially by reactionary anti-Bolshevik forces, but increasingly represented by what was termed 'the West', which meant first and foremost the US and its Cold War allies in Europe. Indeed, the Soviet post-war years were characterised by an intense and largely one-sided conversation with Western culture, which was at once worshipped, translated, and sponsored as part of world civilisation, and bedevilled, forbidden, and persecuted as a foreign agent.[26] The tension that resulted from this dichotomy ran the entire societal gamut and affected every area of knowledge production and dissemination. The chapters by Brazelton, Culp, and Wu demonstrate that China, too, struggled with the question to what extent knowledge and cultural production should be comprehensive and diverse or to what extent ideological purity was called for to shore up the revolution. The Soviet Union never found a good answer to this problem, oscillating between opening and closure, between trust and suspicion. Ultimately the state lost the initiative over the process since large parts of society started to consume Western culture in private, mostly in the form of popular music and its fashions, but also as literature, radio, material items and more.[27] As the hardliners always suspected, in the wake of seemingly apolitical items came Western political ideas, not least because the Soviet state was so busy making every cultural item political.

[25] Allen Kassof, *The Soviet Youth Program: Regimentation and Rebellion*. Cambridge, MA: Harvard University Press, 1965; Simon Huxtable, *News from Moscow: Soviet Journalism and the Limits of Postwar Reform*. Oxford: Oxford University Press, 2022, 123–56; Christopher J. Ward, *Brezhnev's Folly: The Building of BAM and Late Soviet Socialism*. Pittsburgh: University of Pittsburgh Press, 2009.
[26] Abbott Gleason, *Robert D. English. Russia and the Idea of the West: Gorbachev, Intellectuals and the End of the Cold War*. New York: Columbia University Press, 2000; Eleonor Gilburd, *To See Paris and Die: The Soviet Lives of Western Culture*. Cambridge, MA: Harvard University Press, 2018.
[27] Alexey Golubev, *The Things of Life: Materiality in Late Soviet Russia*. Ithaca: Cornell University Press, 2020; Stephen Lovell, *Russia in the Microphone Age: A History of Soviet Radio, 1919–1970*. Oxford: Oxford University Press, 2015; Sergei I. Zhuk, *Rock and Roll in the Rocket City: The West, Identity, and Ideology in Soviet Dniepropetrovsk, 1960–1985*. Baltimore: Johns Hopkins University Press, 2017; Juliane Fürst, *Flowers through Concrete: Explorations in Soviet Hippieland*. Oxford: Oxford University Press, 2021; Artemyi Troitskii, *Back in the USSR: The True Story of Rock in Russia*. London: Faber & Faber, 1987; Leslie Woodhead, *How the Beatles Rocked the Kremlin: The Untold Story of a Noisy Revolution*. London: Bloomsbury, 2013.

One of the most interesting aspects of Soviet post-war ambiguity was the status of religion in these staunchly atheist systems. A closer look at what seems at first glance an uncompromising policy reveals a number of internal contradictions. Wang shows in his chapter not only how in the early 1960s a kind of religious revival occurred in Wenling County, but argues that indeed the very conditions created by communist policies facilitated such a revival. The state tolerated it as a valve to let off the steam that built up during tough periods such as the GLF. It is hard not to think of Stalin's elevation of the Orthodox Church during the Great Fatherland War, which semi-legalised religiosity again, despite continued persecution. Victoria Smolkin has convincingly shown that even later the state condoned and tolerated a number of other spiritual valves, even turning its own atheist propaganda into a type of spiritual discussion in order to satisfy a societal thirst for identity beyond the Soviet.[28] Since the collapse of the Soviet Union, Russian state leaders have been propping up the Russian Orthodox Church into a powerful resource of Russia's new national identity. This identity blends Orthodox Christianity with militarism, patriotism, and an imperial mission to recover the Russian speaking areas from former Soviet times that lie outside of Russia's present state borders.[29] Recent years have also witnessed increasing political and ideological bonds between Russia and China, in line with Russia's economic pivot to China.

The new alliance initially sat on heterogeneous ground as it at once extolled communist-era heroic norms, claimed victim status in a language that appeared to derive from the remembrance of past genocides, and adopted a pronounced anti-Western tone. (Xi was the only major foreign guest at the 2015 Victory Day parade to commemorate WWII. Western leaders had declined the invitation to express their protest against Russia's annexation of Crimea the year before.) Increasingly, though, the axis between Moscow and Beijing bespeaks a unified temporality: decolonisation is its tune, and the theme is not entirely novel. In the 1950s and 1960s, Maoism succeeded in establishing itself on the world stage, in good part because Mao's anti-imperial furore coincided with the wave of decolonisation sweeping Africa, Asia, and the Middle East. Mao used these liberation wars to proclaim himself a vanguard leader of the decolonising world, and to wrest China from ideological dependence on the post-Stalinist Soviet state.[30] Today, it is especially President Vladimir Putin who seeks to portray Russia as a vanguard nation in a global struggle for decolonisation, fought against a US-led global empire. The new alliance shows a reversed power relationship: Moscow, formerly the Elder Soviet Brother, is now relegated to the junior partner position, sorely dependent on

[28] Victoria Smolkin, *A Sacred Space Is Never Empty: A History of Soviet Atheism*. Princeton: Princeton University Press, 2018.

[29] Shaun Walker, *The Long Hangover: Putin's New Russia and the Ghosts of the Past*. New York: Oxford University Press, 2018; idem: 'Angels and Artillery: A Cathedral to Russia's New National Identity', *The Guardian*, 20 October 2020, accessed on 15 March 2023, www.theguardian.com/world/2020/oct/20/orthodox-cathedral-of-the-armed-force-russian-national-identity-military-disneyland.

[30] Lovell, *Maoism*.

the Chinese Big Brother. It is now Russia that is looking towards China, keen to learn not only from the way China reformed its economy while leaving the political monopoly of the Communist Party intact, but also how to harness modern technology for the state while not making it a tool of power in the hands of its subjects. Yet in essence both Russia and China have clung to a rhetoric of national and cultural superiority, which does not sit well with collaboration with or subservience to another great empire. Despite much praise for the 'new alliance' from both sides, the future might bring an updated version of the Sino-Soviet splits, confirming that similar ideologies and ambitions have proven remarkably unstable glue for the two big (post-)communist nations of the long twentieth century.

Regardless of the fate of this new alignment, if history provides any measure, there is reason to expect that Chinese and (post-)Soviet historians will continue to have a lot to talk about.

Glossaries by chapters

Chapter 1

danwei 单位 work-unit
fanshen 翻身 liberate or rise up
Ha Xiongwen 哈雄文
jianzheng 建政 construct a government
Wan Li 万里

Chapter 2

aiguo juewu 爱国觉悟 nationality consciousness
bianjiang weiji 边疆危机 frontier crisis
bla brang 拉卜楞寺 Labrang Monastery
dahanzuzhuyi 大汉族主义 Great Han Chauvinism
daminzuzhuyi 大民族主义 Great Nationality Chauvinism
difang zhuyi 地方主义 'local nationalism'
Dorjé rdo rje 多日吉
Gengya T. rgan gya; C. 甘加
Gyelwo T. rgyal bo; C. 加务
Jinyintan 金银滩 Gold and Silver Grasslands
Liao Hansheng 廖汉生
Lin Yi 林艺
Ling Zifeng 凌子風
Liu Geping 刘格平
minzu tongzhi jieji 民族统治阶级 nationality ruling class
minzu tuanjie 民族团结 nationality unity
Neimenggu chunguang 内蒙古春光 Inner Mongolian Spring
Neimenggu renmin de shengli 内蒙古人民的胜利 Victory of the People of Inner Mongolia
rong bon ang suo 隆务昂锁 Rongwo nangso
sanda fabao 三大法宝 'three magic weapons'
sha bo tshe ring 夏吾才朗 Shawo Tsering
shaoshu minzu ticai yingpian 少数民族题材影片 'minority nationalities film'
tiaojie jiufen 调节纠纷 'mediation of disputes'
tongyi zhanxian 统一战线 United Front
Trashi Namgyel bkra shis rnam rgyal 扎西昂嘉
Wang Feng 汪锋

Wang Zhenzhi 王震之
xia'ai minzu zhuyi 狭隘民族主义 'narrow nationalism'
Zhang Zhongliang 張忠良
Zhao Shoushan 赵寿山
Zhou Renshan 周仁山

Chapter 3

chengshi renmin gongshe 城市人民公社 urban people's commune
danwei 单位 work-unit
fuli 福利 welfare
gongjizhi 供给制 supply system
Guan Feng 关锋
jiti suoyouzhi 集体所有制 collective ownership
jumin 居民 (urban) residents
pingjunzhuyi 平均主义 egalitarianism
quanmin suoyouzhi 全民所有制 ownership by the whole people
Zhang Chunqiao 张春桥

Chapter 4

chushen 出身 (family) background
daosuan 倒算 illegal reseizing
fan xingwei 反行为 counteraction
fankang 反抗 resistance
fuye 副业 side-line production
kouliang 口粮 ration
pohuai 破坏 sabotage
taogou 套购 fraudulent purchase
tonggou tongxiao 统购统销 unified purchase and sale
touji 投机 speculation

Chapter 5

gongzuo yu xuexi 工作与学习 work and study
jianguo / jianshe 建国、建设 nation-building / development
piping / minzhu jiancha 批评、民主检查 critical / democratic inspection
qunzhong 群众 (the) masses
sixiang gaizao 思想改造 thought work / reform
tanwu fenzi 贪污分子 corrupt official

Chapter 6

Qiongsi 琼斯 Jones (surname)
tifa 提法 'watchwords' or official rules about word usage
xinao 洗脑 brainwashing
xixin gemian 洗心革面 to turn over a new leaf
xixin 洗心 heartwashing
xizao 洗澡 bathing

Chapter 7

beidong 被动 sexually passive
duanzheng sixiang 端正思想 correct thinking
funü mo 妇女膜 hymen
qingbai 清白 pure and clean
qi 气 the essence of matter
seqingkuang 色情狂 erotomania
yin 淫 obscene
yin 阴 and *yang*阳 feminine and masculine aspects of qi
yuanqi 元气 vitality

Chapter 8

biaozhunhua 标准化 standardisation
gaizao 改造 transformation
gong 工 one workday
jiaju xishu 家具系数 furniture coefficient
Quanguo jiaju jishu zu 全国家具技术组 National Furniture Technical Group
tougong jianliao 偷工减料 theft of labour and raw materials
xiao mianji zhuzhai sheji 小面积住宅设计 small-space residential housing design
xincun 新村 new villages
zonghe liyong 综合利用 integrated use

Chapter 9

bihua yundong 壁画运动 mural movement
gongnongbing meishu 工农兵美术 worker-peasant-soldier art
Gu Yuan 古元
nongminhua 农民画 peasant painting
Pixian 郫县

puluo yishu 普罗艺术 proletarian art
qunzhong yishu 群众艺术 mass art
Wang Zhaowen 王朝闻
Wu Zuoren 吴作人
yeyu yishu 业余艺术 amateur art

Chapter 10

fangshan 方山 Mount Fangshan
yunxiao si 雲霄寺 Cloud Heaven Temple
yuchan gong 玉蟾宮 Jade Toad Palace
yangjiao dong 羊角洞 Goat Horn Cave
fangyankou 放焰口 (the rite of) feeding flaming mouths
shenwei 神位 spirit tablet
yangfu dashen 楊府大神 Great Lord Yang
Kefang 可方 Abbot Kefang
Liu Peihua 刘佩华 Abbess Liu Peihua
wupo 巫婆 female spiritual medium

Chapter 11

Bai Meichu 白眉初
dida wubo 地大物博 natural resources
difangzhi 地方誌 local gazetteers
diguang renxi 地廣人稀 broad territories sparse populations
diguang renxi 地廣人稀
dixue 地學 geosciences
fanshu 藩屬 outer dependencies
Guo Weiping 郭維屏
Huang Guozhang 黄國璋
kaihuang zhongdi 開荒種地 Reclaim Wastelands and Plant Crops Campaign
kaiken 開墾 land reclamation
Li Anzhai 李安宅
Li Chengsan 李承三
Lin Chao 林超
shokumin 殖民 colonialism
shubu 屬部 dependent regions
shuguo 屬國 dependency
Tang Qiyu 唐 啟宇
tunken 屯墾 land reclamation
Zhang Xiangwen 張相文

Zhu Jiahua 朱家驊
Zhu Kezhen 竺可楨

Chapter 12

changjiang weiyuanhui 长江委员会 Yangzi River Commission
Li Rui 李锐
Lin Yishan 林一山
Yichang 宜昌
Zeng Siyu 曾思玉
Zhang Guangdou 张光斗
Zhang Tixue 张体学

Chapter 13

Bao Zhijing 包之静
cankao 参考 reference
changshi 常识 ordinary knowledge
Chen Hanbo 陳翰伯
Hu Yuzhi 胡愈之
Jin Canran 金灿然
Lu Dingyi 陸定一
Qi Yanming 齊燕銘
Shijie zhish congshu 世界知识丛书 World Knowledge Series
Wang Li 王力
Wang Ziye 王子野
xuexi 学习 study
Zhishi congshu 知识丛书 Knowledge Series
Zhou Yang 周扬

Chapter 14

buliao liaozhi 不了了之
Fang Gang 方刚
He Cheng 贺诚
He Qi 何琦
Li Dequan 李德全
Tang Feifan 湯飛凡
Wei Xi 魏曦
Wu Zaidong 吳在东

Wu Zhili 吴之理
Xie Shaowen 謝少文
Zheng Lingcai 鄭玲才
Zhong Huilan 钟惠澜

Chapter 15

Chen Bingyi 陳冰夷
gudao 孤島 'orphan island'
Jiang Chunfang 姜椿芳
Lin Danqiu 林淡秋
Man Tao 滿濤
qiaomin 僑民 Soviet citizen
Shidai 時代 Epoch
Shidai chubanshe 時代出版社 Epoch Press
Shidai ribao 時代日報 Epoch Daily
Sulian wenyi 蘇聯文藝 Soviet Literature and Art
Wang Yuanhua 王元化
Xu Leiran 許磊然
Ye Shuifu 葉水夫
Zakaimo 匝開莫 I.M. Zakgeima
Zhong-Su youhao xiehui 中蘇友好協會 Sino-Soviet Friendship Association

Chapter 16

Choe Seung-hui 崔承喜
Cui Chengxi wudao yanjiu ban 崔承喜舞蹈研究班 Choe Seung-hui Dance Research Course
Dongfang wuyong yanjiusuo 东方舞踊研究所 Oriental Dance Research Institute
Eguchi Takaya 江口隆哉
Ishii Baku 石井漠
Wu Xiaobang 吴晓邦
Wudao yundong ganbu xunlian ban 舞蹈运动干部训练班 Dance Movement Cadre Training Class
xin wudao yishu 新舞蹈艺术 new dance art
xinxing wuyong 新兴舞踊 new dance
ziran faze 自然法则 natural principle

Index